AF581728

Mathematics in Software Reliability and Quality Assurance

Mathematics in Software Reliability and Quality Assurance

Editors

Tadashi Dohi

Shaoying Liu

MDPI • Basel • Beijing • Wuhan • Barcelona • Belgrade • Manchester • Tokyo • Cluj • Tianjin

Editors
Tadashi Dohi
Hiroshima University
Japan

Shaoying Liu
Hiroshima University
Japan

Editorial Office
MDPI
St. Alban-Anlage 66
4052 Basel, Switzerland

This is a reprint of articles from the Special Issue published online in the open access journal *Mathematics* (ISSN 2227-7390) (available at: https://www.mdpi.com/journal/mathematics/special_issues/mathematics_software_reliability_quality_assurance).

For citation purposes, cite each article independently as indicated on the article page online and as indicated below:

LastName, A.A.; LastName, B.B.; LastName, C.C. Article Title. *Journal Name* **Year**, *Volume Number*, Page Range.

ISBN 978-3-0365-3799-3 (Hbk)
ISBN 978-3-0365-3800-6 (PDF)

Contents

About the Editors

Tadashi Dohi received his B.S.E., M.S. and Dr. of Engineering degrees from Hiroshima University, Japan, in 1989, 1991 and 1995, respectively. Since 2002, he has been Full Professor at Hiroshima University. In 1992 and 2000, he was a Visiting Researcher in the Faculty of Commerce and Business Administration, University of British Columbia, Canada, and Hudson School of Engineering, Duke University, USA, respectively, on a leave of absence from Hiroshima University. He was appointed as the Dean of School of Informatics and Data Science and the associated Dean of Graduate School of Advanced Science and Engineering, Hiroshima University, in 2022. His research areas include software engineering, reliability engineering and dependable computing. He is a regular member of ORSJ, IEICE, IPSJ, REAJ, and IEEE. He also serves as a member of Editorial Board of IEEE Transactions on Reliability, among others.

Shaoying Liu is Professor of Software Engineering at Hiroshima University, Japan, IEEE Fellow, and BCS Fellow. He received his Ph.D. in Computer Science from the University of Manchester, UK, in 1992. His research interests include formal engineering methods, specification verification and validation, specification-based program inspection, automatic specification-based testing, testing-based formal verification, and intelligent software engineering environments. He has published a book entitled Formal Engineering for Industrial Software Development with Springer Verlag, edited 12 conference proceedings, and published over 250 academic papers in refereed journals and international conferences. He proposed use of the terminology "formal engineering methods" in 1997 and has established formal engineering methods as a research area based on his extensive research on the structured object oriented formal language (SOFL) method since 1989, and the development of ICFEM conference series since 1997. In recent years, he has served as a General Co-Chair of QRS 2020 and ICECCS 2022 and a PC member for numerous international conferences. He is Associate Editor of IEEE Transactions on Reliability and Innovations in Systems and Software Engineering, respectively. He is a member of IPSJ and IEICE, Japan.

Preface to "Mathematics in Software Reliability and Quality Assurance"

Despite the well-known fact that every complex system is controlled by software, how to develop fault-free software in industry has been a challenging issue because it is almost impossible to guarantee or prove in advance that the software systems under consideration are fault-free. In this sense, software reliability and quality assurance play a central role in modern software engineering and science.

Over the last 5 decades, much effort has been expended to develop the mathematical aspects of software reliability and quality assurance. In the formal verification and validation, mathematical logic is frequently used. Discrete event systems, automata theory and model checking are also significant tools to perform static analysis of software. In software testing, many kinds of optimization techniques and meta-heuristics are applied to derive feasible solutions. Software reliability, availability and safety assessment is based on stochastic processes and their statistical inference.

In preparing this Special Issue of the journal Mathematics (MDPI), we sent out a call for the latest research results on software reliability and quality assurance, including formal methods and design, automatic program generation, automatic software testing, software verification and validation, program analysis and language theory, coalgebra theory, automata theory, hybrid system, software reliability modeling and assessment, software safety and security, software fault tolerance and dependability. Finally, we received several high-quality submissions and 11 high-quality papers were finally accepted for publication. This monograph, consisting of these articles, aims to promote the latest research results in software reliability and quality assurance.

The article "Fuzzy Automata as Coalgebras" by Ai Liu et al. focuses on the coalgebraic method, which is of great significance to research in process algebra, modal logic, object-oriented design and component-based software engineering. The authors propose different types of fuzzy automata as coalgebras with a monad structure capturing fuzzy behavior and define a notion of fuzzy language to consider several versions of bi-simulation for fuzzy automata.

Rong Wang et al. propose "Mutated Specification-Based Test Data Generation with a Genetic Algorithm" as a specification-based testing method that does not depend on having knowledge of the program structure. The authors provide a new method that combines formal specifications with a genetic algorithm (GA) to effectively generate test data. More specifically, formal specifications are reformed by GA such that they can be used to generate input values that kill as many mutants of the target program as possible. The results, through two examples, show that the proposed method can assist in effectively generating test cases to kill program mutants, which contributes to further maintenance of the software.

The article "A Divide and Conquer Approach to Eventual Model Checking" by Moe Nandi Aung et al. proposes a new technique to mitigate the state of explosion in eventual model checking, where the technique is dedicated to eventual properties and divides an original eventual model checking problem into multiple smaller model checking problems and tackles each of these. The authors prove the theorem that the multiple smaller model checking problems are equivalent to the original eventual model checking problem and conduct a case study that demonstrates the power of the proposed technique.

Question answering (QA) enables the machine to understand and answer questions posed in natural language, which has emerged as a powerful tool in various domains. The article "A Metamorphic Testing Approach for Assessing Question Answering Systems" by Kaiyi Tu et al.

proposes to apply the technique of metamorphic testing (MT) to evaluate QA systems from the users' perspectives toward helping them better understand the capabilities of these systems and to then select appropriate QA systems for their specific needs. To demonstrate the approach, the authors study two typical categories of QA systems, identify a total number of 17 metamorphic relations (MRs), and apply MT to four QA systems by using all the MRs. The experiment results demonstrate the capabilities of the four subject QA systems from various aspects, revealing their strengths and weaknesses.

Junjun Zheng et al. performs "Availability Analysis of Software Systems with Rejuvenation and Checkpointing" by means of a composite stochastic Petri reward net and its associated non-Markovian availability model to capture the dynamic behavior of an operational software system in which time-based software rejuvenation and checkpointing are both aperiodically conducted. They focus on human-error factors during checkpointing and solve the stationary solution of the non-Markovian availability model, which is derived on the basis of the reachability graph of stochastic Petri reward nets and is actually not one of the trivial stochastic models, such as the semi-Markov process nor the Markov regenerative process, and the phase-expansion approach is considered.

Dongping Xiang et al. develop "DICER 2.0: A New Model Checker for Data-Flow Errors of Concurrent Software Systems". While Petri nets are widely used to model concurrent software systems, there are different kinds of Petri net tools that can analyze system properties such as deadlocks, reachability and liveness. The authors take on the challenge of modeling the control flows and data flows of concurrent software systems to resolve the state–space explosion problem and pseudo-states. Through some case studies and experiments, they demonstrate the effectiveness and advantage of DICER 2.0.

The article "Application of EM Algorithm to NHPP-Based Software Reliability Assessment with Generalized Failure Count Data" by Hiroyuki Okamura et al. summarizes expectation maximization (EM) algorithms for non-homogeneous Poisson process (NHPP)-based software reliability models and provide proof of the global convergence properties. The authors derive the EM-step formulas for 12 basic NHPP-based SRMs and conduct numerical experiments to present the convergence property of the EM algorithms. These results are useful in implementing the software reliability model as an automatic reliability assessment tool because the general-purpose optimization algorithms for the maximum likelihood estimation of NHPP-based software reliability models strongly depend on the initial guess and cannot guarantee convergence.

In the article "An Enhanced Evolutionary Software Defect Prediction Method Using Island Moth Flame Optimization", Ruba Abu Khurma et al. are concerned with the use of software defect prediction (SDP) to locate defects and defect-prone software modules and deal with a feature selection (FS) problem with polynomial time complexity. The authors apply the moth flame optimization (MFO) algorithm as an interesting swarm intelligence algorithm and propose the island BMFO (IsBMFO) model by dividing the solutions in the population into a set of sub-populations named islands. Twenty-one public software datasets are analyzed for evaluating the proposed method. The results of the experiments show that improved classification results are obtained when using IsBMFO to solve FS.

In the article "Performance of Enhanced Multiple-Searching Genetic Algorithm for Test Case Generation in Software Testing" by Wanida Khamprapai et al., the multiple-searching genetic algorithm is applied to improve test case generation. The enhanced multiple-searching genetic algorithm (EMSGA), which involves a few additional processes for selecting the best chromosomes in the GA process, is evaluated in terms of the performance through comparison with seven different search-based techniques, including random search. The experimental results show that EMSGA increased the efficiency of testing compared with conventional algorithms and could detect more software faults.

Kyawt Kyawt San at al. consider "Deep Cross-Project Software Reliability Growth Model Using Project Similarity-Based Clustering" to predict the potential number of software bugs from the beginning of a development project. The authors propose a new software reliability modeling method called a deep cross-project software reliability growth model (DC-SRGM), which is a cross-project prediction method that uses features of previous projects' data through project similarity. Specifically, it applies cluster-based project selection for the training data source and modeling using a deep learning method. Experiments involving 15 real datasets from a company and 11 open source software datasets show that DC-SRGM can more precisely describe the reliability of ongoing development projects than existing traditional SRGMs and the LSTM model.

The Nervos CKB (common knowledge base) is a public permissionless blockchain designed for the Nervos ecosystem, and its consensus protocol is the key protocol to improving the limit of the consensus performance for Bitcoin. The article "Modeling and Verifying the CKB Blockchain Consensus Protocol" by Meng Sun et al. develops the formal model of the CKB consensus protocol using timed automata. Based on the model, the authors formally verify various important properties of the Nervos CKB to provide a sufficient trustworthiness assurance. In particular, the security of the Nervos CKB against selfish mining attacks of the protocol is investigated.

The editors are proud to be able to edit such a high-quality monograph based on the above 11 articles and believe that it will be of use in considering software reliability and quality assurance problems in practice. Finally, we thank all the authors and reviewers for their contributions to the publication of this monograph. Our special thanks go to the Managing Editors of MDPI.

Tadashi Dohi and Shaoying Liu
Editors

mathematics

MDPI

Article

Fuzzy Automata as Coalgebras

Ai Liu [1], Shun Wang [2], Luis Soares Barbosa [3] and Meng Sun [2,*]

[1] Graduate School of Advanced Science and Engineering, Hiroshima University, Hiroshima 739-8511, Japan; liuai@hiroshima-u.ac.jp
[2] School of Mathematical Sciences, Peking University, Beijing 100871, China; wshun94@pku.edu.cn
[3] INL (International Iberian Nanotechnology Laboratory) & INESC TEC, Universidade do Minho, 4704-553 Braga, Portugal; lsb@di.uminho.pt
* Correspondence: sunm@pku.edu.cn

Abstract: The coalgebraic method is of great significance to research in process algebra, modal logic, object-oriented design and component-based software engineering. In recent years, fuzzy control has been widely used in many fields, such as handwriting recognition and the control of robots or air conditioners. It is then an interesting topic to analyze the behavior of fuzzy automata from a coalgebraic point of view. This paper models different types of fuzzy automata as coalgebras with a monad structure capturing fuzzy behavior. Based on the coalgebraic models, we can define a notion of fuzzy language and consider several versions of bisimulation for fuzzy automata. A group of combinators is defined to compose fuzzy automata of two branches: state transition and output function. A case study illustrates the coalgebraic models proposed and their composition.

Keywords: fuzzy automata; coalgebra; fuzzy language; bisimulation; composition

Citation: Liu, A.; Wang, S.; Barbosa, L.S.; Sun, M. Fuzzy Automata as Coalgebras. *Mathematics* **2021**, *9*, 272. https://doi.org/10.3390/math9030272

Academic Editor: Tadashi Dohi

Received: 17 December 2020
Accepted: 25 January 2021
Published: 29 January 2021

Publisher's Note: MDPI stays neutral with regard to jurisdictional clai-ms in published maps and institutio-nal affiliations.

1. Introduction

Control logic plays an important role in component-based programming in deciding a run-time mechanisms and rules of composition. Precise control needs meticulous implementation so that many applications may be expensive and inefficient. To tackle this problem, there is an increasing interest in using fuzzy logic in many new areas. As a very efficient method for handling imprecise properties, fuzzy logic then provides a systematic approach to incorporating approximate reasoning into such systems so that fuzzy implementations are not only cheaper and faster than precise ones, but also more understandable for users [1,2]. Therefore, some devices that profit from the use of vagueness in their overall operation have emerged and the related theory is described in [3]. For instance, the fuzzy principal component analysis method, based on the variance contribution rate of the principal component combined with the fuzzy theory to obtain a reasonable correction weight, is used to refine quantitative and qualitative index data of innovation service capability [4]. Moreover, this approach makes sense not only at the control level, but also at the test level [5].

Fuzzy control systems incorporate a number of components driven by fuzzy logic [6]. Most of them are rule-based systems that exchange information through interfaces. Technically, the modeling approach of fuzzy control systems contains three aspects: an input stage, a processing stage and an output stage, whose details are as follows.

- The input stage transforms an input into a value. The method is to abstract the relation of an input and its corresponding vague value into a point in a coordinate system, where the horizontal axis stands for the input domain and the vertical axis stands for the vagueness domain.
- The processing stage involves inference rules and generates a result for each input, and then combines the results of the rule. In this stage, logical inference rules are used to describe the connection between cause and effect. The rules are of the form

$$\texttt{If}\ \langle \textit{condition} \rangle\ \texttt{then}\ \langle \textit{conclusion} \rangle.$$

Such rules provide information for the decision of control variables.

- The output stage processes the combined results from the processing stage and converts them to a specific control value. For instance, common techniques for conversion process includes max-min inference, max-membership principle and mean-max membership.

Automata theory has a long history in modeling systems and applications which can be realized as a set of states and transitions between them depending on some inputs. Fuzzy finite-state automata (FFA) incorporate fuzziness into the internal state representation and output of these computational systems [7]. Depending on the non-fuzzy output labels associated with (final) states or transitions, there are different classes of FFA: FFA with final states, FFA without final states, Fuzzy Moore FA and Fuzzy Mealy FA [8]. There are also works considering fuzzy output maps, such as fuzzy Mealy machines and fuzzy Moore machines [9,10]. Fuzzy automata have been studied from different aspects. In order to study behavior control, a novel method to compute the membership values of the next states of a fuzzy automaton with an averaging function between the membership value of the input, and the membership value of the current state is proposed in [11]; the behaviors of lattice-valued nondeterministic fuzzy automata are compared through two language equivalence relations which have different discriminating power in [12]. Categories of deterministic fuzzy automata and fuzzy languages based on a complete residuated lattice with zero divisors are introduced in [13], a common framework for fuzzy type automata is developed in relationships with morphisms of monads in [14], and the concept of fuzzy regular language accepted by fuzzy finite automata is purposed in [15]. Describing systems that behave in the same way in the sense that one system simulates the other and vice versa, several notions of (approximate) bismulation relations are investigated in [16–18].

Along the past two decades, coalgebra has emerged as a well established general framework for the study of the behavior of various kinds of automata [19–21]. There is in particular a generalized determinization construction from automata to coalgebras, including partial Mealy machines, (structured) Moore automata, Rabin probabilistic automata, and pushdown automata [22]. A survey and hierarchy of probabilistic systems as coalgebras is discussed in [23]. It connects probabilistic verification with coalgebraic modeling and compares expressiveness of system types by natural transformations between functors. Hybrid automata specifying both discrete and continuous behavior can also be modeled as coalgebras [24]. A coalgebraic perspective supporting a generic theory of hybrid automata with a rich palette of definitions and results is studied in [25]. In addition, a coalgebraic semantics framework for quantum systems is developed in [26]. One obvious advantage of the coalgebraic view is that it induces a simple and intuitive notion of *bisimulation* between coalgebras, a notion originally stemming from the world of labeled transition systems and process algebra [27–29]. Witnessed by the notion of *coalgebra homomorphism*, bisimulation on coalgebras can be defined by commutative diagrams and shown to be formally dual to congruence on algebra [30,31]. Moreover, there is a general framework for the study of components as concrete coalgebras and the development of the corresponding calculi [32].

A recent thesis [33] proposes a coalgebraic approach to fuzzy automata, which obtains the following results: (a) a coalgebraic definition of the fuzzy language recognized by a fuzzy automaton, (b) the definition of a functor describing the determinization process of a fuzzy automata via a generalization of the powerset construction, (c) a coalgebraic definition of bisimulation on fuzzy automata allowing the construction of a quotient fuzzy automaton. However, it only considers the output as the current membership value for the current state. Moreover, a coalgebraic theory of fuzzy transition systems and their concrete fuzzy bisimulation is studied in [34]. The authors resort to relational lifting that is one of the most used methods in bisimulation research, leading to an algorithm for testing bisimulation in [35], and group-by-group fuzzy bisimulation and its corresponding modal logic in [36]. Nevertheless, the output stage is omitted. To consider different types of fuzzy automata, our main contributions are as follows:

- Explore the fuzzy-set monad to serve as the basis to a coalgebraic approach;
- Provide a coalgebraic framework for different types of fuzzy automata, where the notions of fuzzy language and bisimulation can be addressed;
- Define appropriate combinators for composing fuzzy automata from two branches:state transition and output function.

Thus, we not only consider fuzzy language respecting the controlling behavior and bisimulation relations for fuzzy automata, but also study the composition mechanism in our coalgebraic framework.

This paper is structured as follows. Section 2 introduces different types of fuzzy automata. Section 3 recalls the definition of the fuzzy-set monad and studies its properties. Section 4 defines the coalgebraic models for fuzzy automata, the notion of fuzzy language and considers several versions of bisimulation. Section 5 develops a series of combinators for composing fuzzy automata. Section 6 discusses a case study. Section 7 concludes and raises some topics for future work.

2. Fuzzy Automata

In a complex controlled system driven by fuzzy logic, a fuzzy automaton is the basic unit which contains fuzzy processors and input/output interfaces. Considering fuzzy output maps, we focus on three types of fuzzy automata: Fuzzy Moore Automata (FMrA), Fuzzy Mealy Automata (FMlA) and Fuzzy Unified Automata (FUA). FMrA and FMlA are obtained by modifying the definitions of fuzzy Moore machine and fuzzy Mealy machine in [8]. Unlike the definition of fuzzy Mealy machine in [8] requiring two functions, one to describe the next state and the other to describe the output, a fuzzy Mealy machine is equipped with one fuzzy function to characterize completely the next state and the output produced in [9]. For distinction between them, we name the latter one as FUA. For simplicity, initial and final states are ignored for the moment.

Definition 1 (Fuzzy Moore Automata (FMrA)). *A fuzzy Moore automaton is a 5-tuple $p = (X, I, O, \alpha, e)$ where*

- *X is a set of states.*
- *I is a set of input symbols.*
- *O is a set of output symbols.*
- *$\alpha : X \times I \to [0,1]^X$ is a fuzzy transition function.*
- *$e : X \to [0,1]^O$ is a fuzzy output function.*

Note that each non-fuzzy output map $e' : X \to O$ corresponds to a function $e : X \to [0,1]^O$ such that $e(x) = \delta_{e'(x)}$, where $\delta_k(t) = \delta(t-k)$ and δ is the Dirac function.

Definition 2 (Fuzzy Mealy Automata (FMlA)). *A fuzzy Mealy automaton is a 5-tuple $p = (X, I, O, \alpha, e)$ where*

- *$X, I, O, \alpha : X \times I \to [0,1]^X$ are defined as in FMrA.*
- *$e : X \times I \to [0,1]^O$ is a fuzzy input-output function.*

Note that each non-fuzzy output map $e' : X \times I \to O$ corresponds to a fuzzy input-output function $e : X \times I \to [0,1]^O$ where $e(x,i) = \delta_{e'(x,i)}$.

Given an FMrA (X, I, O, α, e), it is easy to construct an FMlA (X, I, O, α, e') where $e'(x,i) = e(x)$ without loss of information, so we regard it as a subcase of FMlA and concentrate on the study of FMlA as coalgebras.

Definition 3 (Fuzzy Unified Automata (FUA)). *A fuzzy unified automaton is a 4-tuple $p = (X, I, O, \beta)$ where*

- *X, I, O are defined as in FMrA.*
- *$\beta : X \times I \to [0,1]^{X \times O}$ is a fuzzy input-transition output function.*

In classical methods, two operations $F_1 : [0,1] \times [0,1] \to [0,1]$ and $F_2 : [0,1]^* \to [0,1]$ should be defined to to define the language accepted by an automaton [7]. Instead, we intend to define the notion of fuzzy language with the aid of the fuzzy-set monad.

3. Fuzzy-Set Monad

3.1. Fuzzy Set

The fuzzy set theory [37] was developed by Lotfi A. Zadeh in 1965. The main purpose of using fuzzy sets is to deal with vague data under some given properties. For example, consider a finite set of real numbers $S \subseteq \mathbb{R}$ and the property "close to 0". This property seems ambiguous because there is not an explicit criterion to judge whether objects are closed to 0. We want to ask within what distance we can say "one real number is close to 0". To make it precise, the one should figure out a function which fits the property. For example,

$$\psi_S(x) = max\{0, 1 - \frac{1}{m}|x|\}, \quad x \in [-m, m]$$

where $m = max_{s \in S}|s|$. This function is called *the membership function* and indicates that the closer the data $s \in S$ is to 0, the closer the membership value $\psi_S(s)$ is to 0. Obviouly, data from which the distance to 0 are equal have the same membership value, i.e., $\psi_S(s) = \psi_S(-s)$. However, the selection of membership function is not unique and usually depends on the goal of application.

Definition 4 (Residuated Lattice [33])**.** *A residuated lattice is an algebra $\mathcal{K} = (K, \wedge, \vee, \otimes, \to, 0, 1)$ with four binary and two nullary operations satisfying:*

1 *$(K, \wedge, \vee, 0, 1)$ is a lattice with the partial order $\leq$ which is defined by "$x \leq y$ if and only if $x \vee y = y$". The greatest (least) element is 1(0) that for all $x \in K$, $x \leq 1 (x \geq 0)$;*
2 *$(K, \otimes, 1)$ is a commutative monoid with the unit element 1;*
3 *For $x, y, z \in K$, $x \leq y \to z$ if and only if $x \otimes y \leq z$.*

Especially, if $(K, \wedge, \vee, 0, 1)$ is a complete lattice, then $\mathcal{K}$ is called a complete residuated lattice.

Residuated lattices are the algebraic structure that characterizes fuzzy components.

Example 1. *The Boolean algebra $(\mathbf{2}, \wedge, \vee, \neg)$ is a residuated lattice $(\mathbf{2}, \wedge', \vee', \otimes', \to', 0, 1)$. In this expression, $\mathbf{2} = \{0, 1\}$ is the set of elements. $\wedge', \vee'$ correspond to $\wedge$ and $\vee$ operations in Boolean algebra, respectively. Multiplication $\otimes'$ is defined as $\wedge$. The residuate operation $\to'$ comes as $x \to' y := \neg x \vee y$.*

Definition 5 (Fuzzy Subset [33])**.** *Given a set X, a fuzzy subset over $\mathcal{K}$ of X is a function $\phi : X \to K$ that assigns to each object $x \in X$ a membership value. The set of all fuzzy subsets of X is denoted by $\mathcal{Z}_{\mathcal{K}}(X)$ and obviously $\mathcal{Z}_{\mathcal{K}}(X) = K^X$. In the sequel, we use the shorthand notation $\mathcal{Z}(X)$ to represent $\mathcal{Z}_{\mathcal{K}}(X)$.*

Note that $\mathcal{Z}$ can be interpreted as an endofunctor on Set where

$$\begin{aligned} \mathcal{Z}(f) :& K^X \to K^Y \\ & \kappa \mapsto \lambda y. \bigvee_{x \in f^{-1}(y)} \kappa(x) \end{aligned}$$

for any $f : X \to Y$. Note that

$$\bigvee_{x \in X} \kappa(x) = \vee\{\kappa(x) | x \in X\}.$$

3.2. Properties of Fuzzy-Set Monad

The fuzzy-set monad on Set is defined in [33]. In this section, we will firstly recall the definition and then prove this monad is strong and commutative. Although every monad in Set is strong, we include the explicit contribution to build up intuitions.

Definition 6 (The fuzzy-set Monad [33]). *Fuzzy-set monad $\mathbb{Z} = (\mathcal{Z}, \eta, \mu)$ over $\mathcal{K} = (K, \wedge, \vee, \otimes, \rightarrow, 0, 1)$ satisfies for a set X*

- *$\eta : Id \Rightarrow \mathcal{Z}$ satisfies that*

$$\eta_X(x)(y) = \begin{cases} 1 & x = y \\ 0 & otherwise, \end{cases} \qquad x, y \in X,$$

- *$\mu : \mathcal{Z}^2 \Rightarrow \mathcal{Z}$ satisfies that*

$$\mu_X(\Phi) = \bigcup_{\psi \in \mathcal{Z}(X)} \Phi(\psi) \circledast \psi, \qquad \Phi \in \mathcal{Z}^2(X).$$

where

$$(\bigcup_{i \in I} \phi_i)(x) = \bigvee_{i \in I} \phi_i(x) \quad x \in X, \phi_i \in \mathcal{Z}(X)$$

and

$$(a \circledast \phi)(x) = a \otimes \phi(x) \quad a \in K, x \in X, \phi \in \mathcal{Z}(X)$$

Definition 7 (Strong monad [21]). *A strong monad is a monad $\mathbb{T} = (T, \eta, \mu)$ equipped with a left tensorial strength $\sigma_{X,Y} : T(X) \times Y \to T(X \times Y)$ that commutes with the unit and multiplication of the monad:*

$$\begin{array}{ccc} X \times Y & \xleftrightarrow{\mathsf{id}} & X \times Y \\ {\scriptstyle \eta_X \times \mathsf{id}} \downarrow & & \downarrow {\scriptstyle \eta_{X \times Y}} \\ T(X) \times Y & \xrightarrow[\sigma_{X,Y}]{} & T(X \times Y) \end{array} \qquad \begin{array}{ccccc} T^2(X) \times Y & \xrightarrow{\sigma_{T(X),Y}} & T(T(X) \times Y) & \xrightarrow{T(\sigma_{X,Y})} & T^2(X \times Y) \\ {\scriptstyle \mu_X \times \mathsf{id}} \downarrow & & & & \downarrow {\scriptstyle \mu_{X \times Y}} \\ T(X) \times Y & & \xrightarrow[\sigma_{X,Y}]{} & & T(X \times Y) \end{array}$$

Theorem 1. *The triple $\mathbb{Z} = (\mathcal{Z}, \eta, \mu)$ is a strong monad.*

Proof. Firstly, define a left tensorial strength with components $\sigma_{X,Y} : \mathcal{Z}(X) \times Y \to \mathcal{Z}(X \times Y)$ as

$$\sigma_{X,Y}(\psi, y) = \lambda x, \lambda y'.(\psi(x) \otimes \eta_Y(y)(y'))$$

that commute appropriately with trivial projection and associativity isomorphisms for $f : X \to Z$ and $g : Y \to W$:

$$\begin{array}{ccc} \mathcal{Z}(X) \times 1 & \xrightarrow{\sigma_{X,1}} & \mathcal{Z}(X \times 1) \\ & {\scriptstyle \pi_1} \searrow & \downarrow {\scriptstyle \mathcal{Z}\pi_1} \\ & & \mathcal{Z}(X) \end{array} \qquad \begin{array}{ccc} \mathcal{Z}(X) \times Y & \xrightarrow{\sigma_{X,Y}} & \mathcal{Z}(X \times Y) \\ {\scriptstyle \mathcal{Z}(f) \times g} \downarrow & & \downarrow {\scriptstyle \mathcal{Z}(f \times g)} \\ \mathcal{Z}(Z) \times W & \xrightarrow[\sigma_{Z,W}]{} & \mathcal{Z}(Z \times W) \end{array}$$

$$\begin{array}{ccccc}
(\mathcal{Z}(X) \times Y) \times Z & \xrightarrow{\sigma_{X,Y} \times \mathsf{id}} & \mathcal{Z}(X \times Y) \times Z & \xrightarrow{\sigma_{X \times Y, Z}} & \mathcal{Z}((X \times Y) \times Z) \\
\downarrow \cong & & & & \downarrow \cong \\
\mathcal{Z}(X) \times (Y \times Z) & & \xrightarrow[\sigma_{X, Y \times Z}]{} & & \mathcal{Z}(X \times (Y \times Z))
\end{array}$$

For the unit,

$$\begin{array}{ll}
& \sigma_{X,Y} \cdot (\eta_X \times \mathsf{id})(x,y) \\
= & \{ \text{ Definition of } \times \ \} \\
& \sigma_{X,Y}(\eta_X(x), y) \\
= & \{ \text{ Definition of } \sigma \ \} \\
& \otimes \cdot (\eta_X(x) \times \eta_Y(y)) \\
= & \{ \text{ Definition of } \eta \ \} \\
& \eta_{X \times Y}(x,y).
\end{array}$$

For the multiplication, we have to show that $\mu_{X \times Y} \cdot \mathcal{Z}(\sigma_{X,Y}) \cdot \sigma_{\mathcal{Z}_{\mathcal{K}}(X),Y} = \sigma_{X,Y} \cdot (\mu_X \times \mathsf{id})$. For a pair $(\Phi, y) \in \mathcal{Z}^2(X) \times Y$,

$$\begin{array}{l}
\mu_{X \times Y} \cdot \mathcal{Z}(\sigma_{X,Y}) \cdot \sigma_{\mathcal{Z}(X),Y}(\Phi, y) \\
= \{ \text{ Definition of } \sigma \ \} \\
\quad \mu_{X \times Y} \cdot \mathcal{Z}(\sigma_{X,Y})(\otimes \cdot (\Phi \times \eta_Y(y))) \\
= \{ \text{ Definition of } \mathcal{Z}, \sigma \ \} \\
\quad \mu_{X \times Y}(\bigcup_{(\psi, y') \in \mathcal{Z}(X) \times Y} \otimes \cdot (\Phi \times \eta_Y(y))(\psi, y') \circledast \eta_{\mathcal{Z}(X \times Y)}(\sigma_{X,Y}(\psi, y'))) \\
= \{ \otimes \cdot (f \times g)(x,y) = f(x) \otimes g(y) \ \} \\
\quad \mu_{X \times Y}(\bigcup_{(\psi, y') \in \mathcal{Z}(X) \times Y} (\Phi(\psi) \otimes \eta_Y(y)(y')) \circledast \eta_{\mathcal{Z}(X \times Y)}(\sigma_{X,Y}(\psi, y'))) \\
= \{ \text{ Definition of } \eta \ \} \\
\quad \mu_{X \times Y}(\bigcup_{\psi \in \mathcal{Z}(X)} \Phi(\psi) \circledast \eta_{\mathcal{Z}(X \times Y)}(\sigma_{X,Y}(\psi, y))) \\
= \{ \text{ Definition of } \sigma \ \} \\
\quad \mu_{X \times Y}(\bigcup_{\psi \in \mathcal{Z}(X)} \Phi(\psi) \circledast \eta_{\mathcal{Z}(X \times Y)}(\otimes \cdot (\psi \times \eta_Y(y)))) \\
= \{ \text{ Definition of } \mu \ \} \\
\quad \bigcup_{\psi' \in \mathcal{Z}(X \times Y)} (\bigcup_{\psi \in \mathcal{Z}(X)} \Phi(\psi) \circledast \eta_{\mathcal{Z}(X \times Y)}(\otimes \cdot (\psi \times \eta_Y(y)))(\psi')) \circledast \psi' \\
= \{ \text{ Definition of } \eta \ \} \\
\quad \bigcup_{\psi \in \mathcal{Z}(X)} \Phi(\psi) \circledast (\otimes \cdot (\psi \times \eta_Y(y))
\end{array}$$

For the right side of the equation,

$$
\begin{aligned}
&\sigma_{X,Y} \cdot (\mu_X \times \mathrm{id})(\Phi, y) \\
=&\{ \text{ Definition of } \mu \text{ } \} \\
&\sigma_{X,Y}(\bigcup_{\psi \in \mathcal{Z}(X)} \Phi(\psi) \circledast \psi, y) \\
=&\{ \text{ Definition of } \sigma \text{ } \} \\
&\otimes \cdot ((\bigcup_{\psi \in \mathcal{Z}(X)} \Phi(\psi) \circledast \psi) \times \eta_Y(y)) \\
=&\{ \text{ Distributive law: } \otimes \cdot (\cup_i f_i \times g) = \cup_i (\otimes \cdot (f_i \times g)) \text{ } \} \\
&\bigcup_{\psi \in \mathcal{Z}(X)} \otimes \cdot (\Phi(\psi) \circledast \psi \times \eta_Y(y)) \\
=&\{ \text{ Constant } \Phi(\psi) \text{ } \} \\
&\bigcup_{\psi \in \mathcal{Z}(X)} \Phi(\psi) \circledast (\otimes \cdot (\psi \times \eta_Y(y)).
\end{aligned}
$$

□

In the proof of Theorem 1, we defined a left tensorial strength σ with components $\sigma_{X,Y} : \mathcal{Z}(X) \times Y \to \mathcal{Z}(X \times Y)$ as

$$\sigma_{X,Y}(\psi, y) = \otimes \cdot (\psi, \eta_Y(y)) = \lambda x, \lambda y'.(\psi(x) \otimes \eta_Y(y)(y')).$$

Of course, a "swapped" tensorial strength σ' with components $\sigma'_{X,Y} : X \times \mathcal{Z}(Y) \to \mathcal{Z}(X \times Y)$ can be obtained by applying swapping operation from the left tensorial strength:

$$(X \times \mathcal{Z}(Y) \xrightarrow[\cong]{\mathsf{s}} \mathcal{Z}(Y) \times X \xrightarrow{\sigma_{Y,X}} \mathcal{Z}(Y \times X) \xrightarrow[\cong]{\mathcal{Z}(\mathsf{s})} \mathcal{Z}(X \times Y)).$$

where $\mathsf{s} \mathrel{\widehat{=}} \langle \pi_2, \pi_1 \rangle$ is product communicating. Formally,

$$\sigma'_{X,Y} = \otimes \cdot (\eta_X(x) \times \phi) = \lambda x'.\lambda y.(\eta_X(x)(x') \otimes \phi(y)).$$

With both $\sigma_{X,Y}$ and $\sigma'_{X,Y}$, there are two ways to obtain $\mathcal{Z}(X) \times \mathcal{Z}(Y) \to \mathcal{Z}(X \times Y)$, as depicted in the following diagram. If the diagram commutes, then $\mathcal{Z}$ is commutative with left and right strength natural transformations $\sigma_{X,Y}, \sigma'_{X,Y}$. We use $\gamma : \mathcal{Z}(X) \times \mathcal{Z}(Y) \to \mathcal{Z}(X \times Y)$ to denote the composed arrow.

$$
\begin{array}{ccccc}
 & \mathcal{Z}(X \times \mathcal{Z}(Y)) & \xrightarrow{\mathcal{Z}(\sigma'_{X,Y})} & \mathcal{Z}^2(X \times Y) & \\
{}^{\sigma_{X,\mathcal{Z}(Y)}}\nearrow & & & & \searrow{}^{\mu_{X \times Y}} \\
\mathcal{Z}(X) \times \mathcal{Z}(Y) & & & & \mathcal{Z}(X \times Y) \\
{}_{\sigma'_{\mathcal{Z}(X),Y}}\searrow & & & & \nearrow{}_{\mu_{X \times Y}} \\
 & \mathcal{Z}(\mathcal{Z}(X) \times Y) & \xrightarrow[\mathcal{Z}(\sigma_{X,Y})]{} & \mathcal{Z}^2(X \times Y) &
\end{array}
$$

Theorem 2. *The triple $(\mathcal{Z}, \eta, \mu)$ is a commutative monad.*

Proof. To show the diagram is commutative, select a pair of membership functions $(\psi_1, \psi_2) \in \mathcal{Z}(X) \times \mathcal{Z}(Y)$, then

$$\begin{aligned}
& \mu_{X\times Y} \cdot \mathcal{Z}(\sigma'_{X,Y}) \cdot \sigma_{X,\mathcal{Z}(Y)}(\psi_1, \psi_2) \\
=& \{ \text{ Definition of } \sigma \ \} \\
& \mu_{X\times Y}(\mathcal{Z}(\sigma'_{X,Y})(\otimes \cdot (\psi_1 \times \eta_{\mathcal{Z}(Y)}(\psi_2)))) \\
=& \{ \text{ Definition of } \mathcal{Z} \ \} \\
& \mu_{X\times Y}(\bigcup_{(x,\psi)\in X\times\mathcal{Z}(Y)} \otimes \cdot (\psi_1 \times \eta_{\mathcal{Z}(Y)}(\psi_2))(x,\psi) \circledast \eta_{\mathcal{Z}(X\times Y)}(\sigma'_{X,Y}(x,\psi))) \\
=& \{ \otimes \cdot (f \times g)(x,y) = f(x) \otimes g(y) \ \} \\
& \mu_{X\times Y}(\bigcup_{(x,\psi)\in X\times\mathcal{Z}(Y)} (\psi_1(x) \otimes \eta_{\mathcal{Z}(Y)}(\psi_2)(\psi)) \circledast \eta_{\mathcal{Z}(X\times Y)}(\sigma'_{X,Y}(x,\psi))) \\
=& \{ \text{ Definition of } \eta \ \} \\
& \mu_{X\times Y}(\bigcup_{x\in X} \psi_1(x) \circledast \eta_{\mathcal{Z}(X\times Y)}(\sigma'_{X,Y}(x,\psi_2))) \\
=& \{ \text{ Definition of } \sigma' \ \} \\
& \mu_{X\times Y}(\bigcup_{x\in X} \psi_1(x) \circledast \eta_{\mathcal{Z}(X\times Y)}(\otimes \cdot (\eta_X(x) \times \psi_2))) \\
=& \{ \text{ Definition of } \mu \ \} \\
& \bigcup_{\psi'\in\mathcal{Z}(X\times Y)} (\bigcup_{x\in X} \psi_1(x) \circledast \eta_{\mathcal{Z}(X\times Y)}(\otimes \cdot (\eta_X(x) \times \psi_2))(\psi') \circledast \psi') \\
=& \{ \text{ Definition of } \eta \ \} \\
& \bigcup_{x\in X} (\psi_1(x) \circledast (\otimes \cdot (\eta_X(x) \times \psi_2))) \\
\widehat{=}& \{ \text{ Denotation } \} \\
& f_1
\end{aligned}$$

For the right side of the equation,

$$\begin{aligned}
& \mu_{X\times Y} \cdot \mathcal{Z}(\sigma_{X,Y}) \cdot \sigma'_{\mathcal{Z}(X),Y}(\psi_1, \psi_2) \\
=& \{ \text{ Definition of } \sigma \ \} \\
& \mu_{X\times Y}(\mathcal{Z}(\sigma_{X,Y})(\otimes \cdot (\eta_{\mathcal{Z}(X)}(\psi_1) \times \psi_2))) \\
=& \{ \text{ Definition of } \mathcal{Z} \ \} \\
& \mu_{X\times Y}(\bigcup_{(\psi,y)\in\mathcal{Z}(X)\times Y} \otimes \cdot (\eta_{\mathcal{Z}(X)}(\psi_1) \times \psi_2)(\psi,y) \circledast \eta_{\mathcal{Z}(X\times Y)}(\sigma_{X,Y}(\psi,y))) \\
=& \{ \otimes \cdot (f \times g)(x,y) = f(x) \otimes g(y) \ \} \\
& \mu_{X\times Y}(\bigcup_{(\psi,y)\in\mathcal{Z}(X)\times Y} (\eta_{\mathcal{Z}(X)}(\psi_1)(\psi) \otimes \psi_2(y)) \circledast \eta_{\mathcal{Z}(X\times Y)}(\sigma_{X,Y}(\psi,y))) \\
=& \{ \text{ Definition of } \eta \ \} \\
& \mu_{X\times Y}(\bigcup_{y\in Y} \psi_2(y) \otimes \eta_{\mathcal{Z}(X\times Y)}(\sigma_{X,Y}(\psi_1,y))) \\
=& \{ \text{ Definition of } \sigma \ \}
\end{aligned}$$

$$\mu_{X\times Y}(\bigcup_{y\in Y}\psi_2(y)\circledast\eta_{\mathcal{Z}(X\times Y)}(\otimes\cdot(\psi_1\times\eta_Y(y))))$$
$$=\{\text{ Definition of }\mu\ \}$$
$$\bigcup_{\psi'\in\mathcal{Z}(X\times Y)}(\bigcup_{y\in Y}\psi_2(y)\circledast\eta_{\mathcal{Z}(X\times Y)}(\otimes\cdot(\psi_1\times\eta_Y(y)))(\psi')\circledast\psi')$$
$$=\{\text{ Definition of }\eta\ \}$$
$$\bigcup_{y\in Y}(\psi_2(y)\circledast(\otimes\cdot(\psi_1\times\eta_Y(y))))$$
$$\widehat{=}\{\text{ Denotation }\}$$
$$f_2$$

Note that $f_1(x,y) = \psi_1(x)\otimes\psi_2(y) = \otimes\cdot(\psi_1\times\psi_2)(x,y) = f_2(x,y)$. Hence the diagram commutes. □

4. Going Coalgebraic

4.1. Coalgebraic Models

Since the automata introduced in Section 2 are defined over the interval $[0,1]$, we assume the fuzzy-set monad $\mathbb{Z} = (\mathcal{Z},\eta,\mu)$ is also defined over some complete residuated lattice $([0,1], min, max, \otimes, \rightarrow, 0, 1)$. The corresponding coalgebraic models are based on the fuzzy-set monad.

Example 2 ([33]). *Note that $([0,1], min, max, 0, 1)$ is a complete lattice. Then there are several ways to construct complete residuated lattices $([0,1], min, max, \otimes, \rightarrow, 0, 1)$; namely*

- *Define*
$$x\otimes y = max(x+y-1,0)$$
$$x\rightarrow y = min(1-x+y,1)$$
for $x,y\in[0,1]$. Then, $([0,1], min, max, \otimes, \rightarrow, 0, 1)$ is a complete residuated lattice corresponding to the standard Lukasiewicz algebra.
- *Define*
$$x\otimes y = min(x+y-1,0)$$
$$x\rightarrow y\begin{cases}1 & \text{if } x\le y\\ y & \text{if } y<x\end{cases}$$
for $x,y\in[0,1]$. Then, $([0,1], min, max, \otimes, \rightarrow, 0, 1)$ is a complete residuated lattice corresponding to the standard Gödel algebra.
- *Define*
$$x\otimes y = x\cdot y$$
$$x\rightarrow y\begin{cases}1 & \text{if } x\le y\\ \frac{y}{x} & \text{if } y<x\end{cases}$$
for $x,y\in[0,1]$. Then, $([0,1], min, max, \otimes, \rightarrow, 0, 1)$ is a complete residuated lattice corresponding to the standard product algebra.

Consider the two functors $F_{I,O} = \mathcal{Z}(-\times O)^I$ and $T_{I,O} = \mathcal{Z}(-)^I\times\mathcal{Z}(O)^I$. Given a FMIA (X,I,O,α,e), the corresponding $T_{I,O}$-coalgebra is $(X,\langle\overline{\alpha},\overline{e}\rangle : X\rightarrow\mathcal{Z}(X)^I\times\mathcal{Z}(O)^I)$ where $\overline{f}$ is the curried version of f. Given a FUA (X,I,O,β), the corresponding $F_{I,O}$-coalgebra is $(X,\overline{\beta}: X\rightarrow\mathcal{Z}(X\times O)^I)$. Obviously, there is a natural transformation θ from $T_{I,O}$ to $F_{I,O}$:
$$\theta(\langle f,g\rangle)(i) = \gamma(\langle f(i),g(i)\rangle)$$

for $f \in \mathcal{Z}(X)^I, g \in \mathcal{Z}(O)^I$ and $i \in I$. In the sequel, $F_{I,O}$-coalgebras provide a universal framework for defining fuzzy language and bisimulation for different fuzzy automata while $T_{I,O}$-coalgebras serve as a basis for composition calculi of fuzzy Mealy automata.

4.2. Fuzzy Language

In [33], fuzzy automata with initial fuzzy subsets and final fuzzy subsets are equipped with the notion of fuzzy language over a set of input symbols. Due to the type of their initial/final fuzzy subsets, that notion can not be naturally extended to the case involving output. Here we consider the notion of fuzzy language over a set of input symbols and a set of output symbols based on $F_{I,O}$-coalgebras.

Definition 8. *Let (X, f) be an $F_{I,O}$-coalgebra. Define $f^* : X \to \mathcal{Z}(X \times O^*)^{I^*}$ as follows:*

$$
\begin{aligned}
f^*(x)(i)(y,o) &= f(x)(i)(y,o)\\
f^*(x)(\emptyset)(y,\emptyset) &= \begin{cases} 1 & \text{if } x = y \\ 0 & \text{if } x \neq y \end{cases}\\
f^*(x)(i)(y,\emptyset) &= 0\\
f^*(x)(\emptyset)(y,o) &= 0\\
f^*(x)(wi)(y,vo) &= \bigvee_{z \in X} f^*(x)(w)(z,v) \otimes f^*(z)(i)(y,o)
\end{aligned}
$$

for $\forall x, y \in X, i \in I, o \in O, w \in I^*, v \in O^*$. Note that $\emptyset$ represents the empty input/output.

Lemma 1. *Given an $F_{I,O}$-coalgebra (X, f), $\forall x, y \in X, w \in I^*, v \in O^*$, if $|w| \neq |v|$ then*

$$f^*(x)(w)(y,v) = 0.$$

Proof. First, we prove the result for $|w| > |v|$ by induction on $|w| = n$. Let $x, y \in X$, $w \in I^*$, $v \in O^*$. If $n = 0$, there exists no v such that $|v| < 0$ and hence the result holds. If $n = 1$, then $v = \emptyset$ and the result holds by the Definition 8. Assume that the result is true for all $|w| \in I^*$ such that $|w| = n-1, n > 1$. Now there are two cases: $|v| = \emptyset$ and $|v| \neq \emptyset$. For the case $|v| = \emptyset$, let $|w| = w'i$, where $|w| = n, i \in I$, and then

$$f^*(x)(w'i)(y,\emptyset) = \bigvee_{z \in X} f^*(x)(w')(z,\emptyset) \otimes f^*(z)(i)(y,\emptyset).$$

By the induction hypothesis, $f^*(x)(w')(z,\emptyset) = f^*(z)(i)(y,\emptyset) = 0$ and thus the result holds. For the case $|v| \neq \emptyset$, let $w = w'i, v = v'o$ where $|w| = n > |y|, i \in I, o \in O$ and then

$$f^*(x)(w'i)(y,v'o) = \bigvee_{z \in X} f^*(x)(w')(z,v') \otimes f^*(z)(i)(y,o).$$

By the induction hypothesis, $f^*(x)(w')(z,v') = 0$ and hence $\forall z \in X, f^*(x)(w')(z,v') \otimes f^*(z)(i)(y,o) = 0$. Therefore, the result holds.

Second, by a similar proof, we can prove the result holds for $|w| < |v|$ by induction on $|y| = n$. □

Lemma 2. *Given an $F_{I,O}$-coalgebra (X, f), $\forall x, y \in X$, $w_1, w_2 \in I^*$, $v_1, v_2 \in O^*$, if $|w_1| = |v_1|$ and $|w_2| = |v_2|$, then*

$$f^*(x)(w_1w_2)(y,v_1v_2) = \bigvee_{z \in X} f^*(x)(w_1)(z,v_1) \otimes f^*(z)(w_2)(y,v_2)$$

Proof. The results can be proved by induction on $|w_2| = n$. If $n = 0$, then $w_2 = v_2 = \emptyset$ and $w_1 w_2 = w_1, v_1 v_2 = v_1$. Since $f^*(x)(\emptyset)(y,\emptyset)$ is 1 when $x = y$ and $f^*(x)(\emptyset)(y,\emptyset)$ is 0 otherwise,

$$f^*(x)(w_1)(y,v_1) = \bigvee_{z\in X} f^*(x)(w_1)(z,v_1) \otimes f^*(z)(\emptyset)(y,\emptyset)$$

holds, which completes the proof of the base case. Now assume that the result is true for all $|w_2| = n-1, n > 0$. Let $w_2 = w'i$ and $v_2 = v'o$, where $|w'| = |v'| = n-1, i \in I$, $o \in O$. Then

$$\begin{aligned}
& f^*(x)(w_1w_2)(y,v_1v_2) \\
=& f^*(x)(w_1w'i)(y,v_1v'o) \\
=& \bigvee_{z\in X} f^*(x)(w_1w')(z,v_1v') \otimes f^*(z)(i)(y,o) \\
=& \bigvee_{z\in X} (\bigvee_{r\in X} f^*(x)(w_1)(r,v_1) \otimes f^*(r)(w')(y,v')) \otimes f^*(z)(i)(y,o) \\
=& \bigvee_{z\in X} (\bigvee_{r\in X} f^*(x)(w_1)(r,v_1) \otimes f^*(r)(w')(y,v') \otimes f^*(z)(i)(y,o)) \\
=& \bigvee_{r\in X} (\bigvee_{z\in X} f^*(x)(w_1)(r,v_1) \otimes f^*(r)(w')(y,v') \otimes f^*(z)(i)(y,o)) \\
=& \bigvee_{r\in X} (f^*(x)(w_1)(r,v_1) \otimes \bigvee_{z\in X} f^*(r)(w')(y,v') \otimes f^*(z)(i)(y,o)) \\
=& \bigvee_{r\in X} f^*(x)(w_1)(r,v_1) \otimes f^*(r)(w'i)(y,v'o) \\
=& \bigvee_{r\in X} f^*(x)(w_1)(r,v_1) \otimes f^*(r)(w_2)(y,v_2)
\end{aligned}$$

□

Now we consider a generic fuzzy language for $F_{I,O}$-coalgebras and naturally obtain the definition for the fuzzy language accepted by a fuzzy automaton.

Definition 9 (Fuzzy language)**.** *A fuzzy language over an input set I and an output set O (with membership values over $\mathcal{K}$), is a fuzzy subset of $(IO)^*$, that is a function $\phi : (IO)^* \to [0,1]$.*

Example 3. *For instance, let $I = \{i_1, i_2\}, O = \{o_1, o_2\}$. A fuzzy language ϕ can be defined as $\phi(i_1o_1) = 0.6, \phi(i_1o_2) = 0.8, \phi(i_2o_1) = 0.5, \phi(i_2o_2) = 1$ and $\phi(s) = 0, \forall s \in (IO)^*, |s| \neq 2$.*

Definition 10. *Consider an $F_{I,O}$-coalgebra $(X, f : X \to \mathcal{Z}(X \times O)^I)$. For $w = i_1o_1i_2o_2\cdots \in (IO)^*$, define $w_i = i_1i_2\cdots$ and $w_o = o_1o_2\cdots$. Given an initial fuzzy state $\epsilon \in \mathcal{Z}(X)$ and a final fuzzy state $\tau \in \mathcal{Z}(X)$, the fuzzy language L_f recognized by (X, f) is defined by*

$$L_f(w) = \bigvee_{x,y\in X} \epsilon(x) \otimes f^*(x)(w_i)(y,w_o) \otimes \tau(y), \ \ w \in (IO)^*.$$

Naturally, the fuzzy language recognized by a FUA (X, I, O, β) is the one recognized by its corresponding $F_{I,O}$-coalgebra $(X, \overline{\beta})$.

When considering the language recognized by an FMIA, the membership values of the next state and the output must be integrated, which can be captured by the natural transformation θ.

Definition 11. *The fuzzy language recognized by a FMIA (X, I, O, α, e) is the one recognized by the corresponding $F_{I,O}$-coalgebra $(X, \theta(\langle \overline{\alpha}, \overline{e} \rangle))$*

4.3. Bisimulation

Let us now discuss the notion of bisimulation for fuzzy automata. In fact, coalgebra theory provides a generic notion of bisimulation on H-coalgebras for any functor H [20].

Definition 12 (H-bisimulation). *Given two H-coalgebras $(X, f : X \to H(X))$ and $(Y, g : Y \to H(Y))$, an H-bisimulation between them is a relation $R \subseteq X \times Y$ such that there exists an H-coalgebra $(R, h : R \to H(R))$ making the following diagram to commute.*

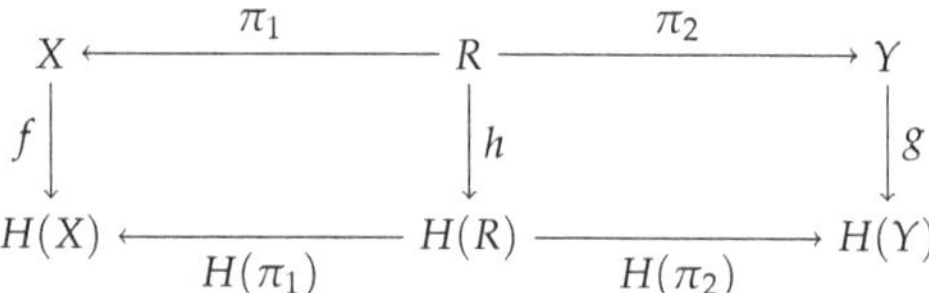

Theorem 3. *Given two $T_{I,O}$-coalgebras (X, f) and (Y, g), if $R \subseteq X \times Y$ is a $T_{I,O}$-bisimulation, then R is an $F_{I,O}$-bisimulation between $(X, \theta \circ f)$ and $(X, \theta \circ g)$.*

Proof. The proof of the result is immediate from the definition. □

We now consider concrete bisimulations for different types of fuzzy automata. Since FMrA can be easily transformed to FMlA, we only focus on bisimulation for FMlA and FUA. Given a FMlA (X, I, O, α, e), denote a transition $x \xrightarrow[o,v_2]{i,v_1} x'$ if $\alpha(x,i)(x') = v_1, e(x,i)(o) = v_2$. Given a FUA (X, I, O, β), denote a transition $x \xrightarrow{i|v|o} x'$ if $\beta(x,i)(x',o) = v$.

Definition 13 (Bisimulation for FMlA). *Given two FMlA (X, I, O, α, e) and (Y, I, O, α', e'), $R \subseteq X \times Y$ is a concrete bisimulation if it satisfies the following properties.*

- *For $(x,y) \in R$, if $x \xrightarrow[o,v_2]{i,v_1} x'$, there exists $y' \in Y$, such that $y \xrightarrow[o,v_2]{i,v_1} y'$ and $(x', y') \in R$.*
- *For $(x,y) \in R$, if $y \xrightarrow[o,v_2]{i,v_1} y'$, there exists $x' \in X$, such that $x \xrightarrow[o,v_2]{i,v_1} x'$ and $(x', y') \in R$.*

Definition 14 (Bisimulation for FUA). *Given two FUA $p = (X, I, O, \beta)$ and $q = (Y, I, O, \beta')$, $R \subseteq X \times Y$ is a concrete bisimulation if it satisfies the following properties.*

- *For $(x,y) \in R$, if $x \xrightarrow{i|v|o} x'$, there exists $y' \in Y$, such that $y \xrightarrow{i|v|o} y'$ and $(x', y') \in R$.*
- *For $(x,y) \in R$, if $y \xrightarrow{i|v|o} y'$, there exists $x' \in X$, such that $x \xrightarrow{i|v|o} x'$ and $(x', y') \in R$.*

Theorem 4. *Given two FMlA (X, I, O, α, e) and (Y, I, O, α', e'), R is a concrete bisimulation if and only if R is a $T_{I,O}$-bisimulation between their corresponding $T_{I,O}$-coalgebras.*

Proof. The proof of the result is immediate from the definition. □

Theorem 5. *Given two FUA (X, I, O, β) and (Y, I, O, β'), R is a concrete bisimulation if and only if R is an $F_{I,O}$-bisimulation between their corresponding $F_{I,O}$-coalgebras.*

Proof. The proof of the result is immediate from the definition. □

Since the core idea of fuzzy automata is fuzzing, the concrete bisimulation induced by coalgebraic bisimulation seems to be too strict. To find a more suitable characterization of bisimulation of fuzzy automata, we introduce the notion of approximate ϵ-bisimulation, which requires that membership values for states in an approximate ϵ-bisimulation of two transition branches should have a difference less than ϵ.

Definition 15 (ϵ-Bisimulation for FMIA). *Given two FMIA (X, I, O, α, e) and (Y, I, O, α', e'), a relation $R \subseteq X \times Y$ is an approximate ϵ-bisimulation ($\epsilon > 0$) if for all $(x, y) \in R$:*

- *If $x \xrightarrow[o,v_2]{i,v_1} x'$, there exists $y' \in Y$, such that $y \xrightarrow[o,u_2]{i,u_1} y'$, $|u_1 - v_1| \leq \epsilon$, $|u_2 - v_2| \leq \epsilon$ and $(x', y') \in R$.*
- *If $y \xrightarrow[o,u_2]{i,u_1} y'$, there exists $x' \in X$, such that $x \xrightarrow[o,v_2]{i,v_1} x'$, $|u_1 - v_1| \leq \epsilon$, $|u_2 - v_2| \leq \epsilon$ and $(x', y') \in R$.*

Example 4. *Consider two FMIA (X, I, O, α, e) and (Y, I, O, α', e'), where $X = \{x_1, x_2\}, Y = \{y_1, y_2\}, I = \{i\}, O = \{o\}, \alpha(x_1, i)(x_2) = 0.6, e(x_1, i)(o) = 0.4, \alpha'(y_1, i)(y_2) = 0.5, e'(y_1, i)(o)$ = 0.5. Then, $R = \{(x_1, y_1), (x_2, y_2)\}$ is an approximate 0.1-bisimulation.*

Definition 16 (ϵ-Bisimulation for FUA). *Given two FUA (X, I, O, β) and (Y, I, O, β'), a relation $R \subseteq X \times Y$ is an approximate ϵ-bisimulation ($\epsilon > 0$) if for all $(x, y) \in R$,*

- *If $x \xrightarrow{i|u|o} x'$, then there exists y' such that $y \xrightarrow{i|v|o} y'$, $|u - v| \leq \epsilon$ and $(x', y') \in R$;*
- *If $y \xrightarrow{i|v|o} y'$, then there exists x' such that $x \xrightarrow{i|u|o} x'$, $|u - v| \leq \epsilon$ and $(x', y') \in R$.*

Example 5. *Consider two FUA (X, I, O, β) and (Y, I, O, β'), where $X = \{x_1, x_2\}, Y = \{y_1, y_2\}$, $I = \{i\}, O = \{o\}, \beta(x_1, i)(x_2, o) = 0.8, \beta'(y_1, i)(y_2, o) = 0.7$. Then, $R = \{(x_1, y_1), (x_2, y_2)\}$ is an approximate 0.1-bisimulation.*

Proposition 1. *For approximate ϵ-bisimulation, we have*

1. *R is an approximate ϵ-bisimulation if and only if R^{-1} is an approximate ϵ-bisimulation.*
2. *If R_i is an approximate ϵ_i-bisimulation for $i = 1, 2$, then $R_1 \circ R_2$ is an approximate $(\epsilon_1 + \epsilon_2)$ -bisimulation.*
3. *If R_i is an approximate ϵ_i-bisimulation, then $\cup_i R_i$ is an approximate $max_i\{\epsilon_i\}$-bisimulation.*

Proof. The proof of the result is immediate from the definition. □

5. Composition for FMIA

A family of combinators for $B(- \times O)^I$-coalgebras where B is a monad, such as sequential composition (;), parallel ($\boxtimes$), choice ($\boxplus$) and concurrency ($\boxdot$) combinators were introduced in [32]. Therefore, the composition of FUA can be naturally instantiated. However, an FMIA assigns different membership values to the next state and the corresponding output, which should be separated for composition. With some abuse of notation, we construct sequential composition (;), parallel ($\boxtimes$), choice ($\boxplus$) and concurrency ($\boxdot$) combinators for FMIA. Consider three fuzzy Mealy automata p, q, r with the corresponding coalgebras

$$\begin{aligned}
[\![p]\!] &= (X_p, \langle \overline{\alpha_p}, \overline{e_p} \rangle : X_p \to \mathcal{Z}(X_p)^I \times \mathcal{Z}(O)^I) \\
[\![q]\!] &= (X_q, \langle \overline{\alpha_q}, \overline{e_q} \rangle : X_q \to \mathcal{Z}(X_q)^J \times \mathcal{Z}(R)^J) \qquad (\star) \\
[\![r]\!] &= (X_r, \langle \overline{\alpha_r}, \overline{e_r} \rangle : X_r \to \mathcal{Z}(X_r)^O \times \mathcal{Z}(R)^O).
\end{aligned}$$

Some standard isomorphisms in Set are used in the definitions of combinators:

$$\begin{aligned}
\mathsf{a} &: A \times B \times C \to A \times (B \times C) \\
\mathsf{s} &: A \times B \to B \times A \\
\mathsf{xr} &: A \times B \times C \to A \times C \times B \\
\mathsf{m} &: A \times B \times (C \times D) \to A \times C \times (B \times D) \\
\mathsf{dist} &: A \times (B + C) \to A \times B + A \times C
\end{aligned}$$

Furthermore, combinators $\mathtt{a}_+, \mathtt{s}_+, \mathtt{xr}_+, \mathtt{m}_+$ are the corresponding isomorphisms for sums in Set. Finally, the inverse of an isomorphism i is denoted by i^{-1}.

The sequential composition combinator ; requires the compatibility of interfaces. The sequential composition of p, r actually shares the data which is sent out from p. From a coalgebraic point of view, it is a $T_{I,R}$-coalgebra

$$[\![p;r]\!] = (X_p \times X_r, \langle \overline{\alpha_{p;r}}, \overline{e_{p;r}} \rangle)$$

where $\alpha_{p;r}$ is defined as:

$$X_p \times X_r \times I \xrightarrow{\mathtt{xr}} X_p \times I \times X_r \xrightarrow{\langle \alpha_p, e_p \rangle \times \mathsf{id}} \mathcal{Z}(X_p) \times \mathcal{Z}(O) \times X_r \xrightarrow{\mathtt{a} \circ \mathtt{xr}} \mathcal{Z}(X_p) \times (X_r \times \mathcal{Z}(O))$$
$$\xrightarrow{\mathsf{id} \times \sigma'_{X_r,O}} \mathcal{Z}(X_p) \times \mathcal{Z}(X_r \times O) \xrightarrow{\mathsf{id} \times \mathcal{Z}\alpha_r} \mathcal{Z}(X_p) \times \mathcal{Z}\mathcal{Z}(X_r) \xrightarrow{\gamma \circ (\mathsf{id} \times \mu)} \mathcal{Z}(X_p \times X_r)$$

and $e_{p;r}$ is defined as:

$$X_p \times X_r \times I \xrightarrow{\mathtt{xr}} X_p \times I \times X_r \xrightarrow{e_p \times \mathsf{id}} \mathcal{Z}(O) \times X_r \xrightarrow{\sigma'_{X_r,O} \circ \mathtt{s}} \mathcal{Z}(X_r \times O) \xrightarrow{\mathcal{Z}e_r} \mathcal{Z}\mathcal{Z}(R) \xrightarrow{\mu} \mathcal{Z}(R)$$

The parallel combinator $\boxtimes$ corresponds to synchronous product and composes two coalgebras into one with their inputs (outputs) merged together. The parallel $p \boxtimes q$ produces an output belonging to $O \times R$ after receiving an input belonging to $I \times J$. Coalgebraically, the semantics of the parallel combinator is a $T_{I \times J, O \times R}$-coalgebra

$$[\![p \boxtimes q]\!] = (X_p \times X_p, \langle \overline{\alpha_{p \boxtimes q}}, \overline{e_{p \boxtimes q}} \rangle)$$

where $\alpha_{p \boxtimes q}$ is defined as:

$$X_p \times X_q \times (I \times J) \xrightarrow{\mathtt{m}} X_p \times I \times (X_q \times J) \xrightarrow{\alpha_p \times \alpha_q} \mathcal{Z}(X_p) \times \mathcal{Z}(X_q) \xrightarrow{\gamma} \mathcal{Z}(X_p \times X_q)$$

and $e_{p \boxtimes q}$ is defined as

$$X_p \times X_q \times (I \times J) \xrightarrow{\mathtt{m}} X_p \times I \times (X_q \times J) \xrightarrow{e_p \times e_q} \mathcal{Z}(O) \times \mathcal{Z}(R) \xrightarrow{\gamma} \mathcal{Z}(O \times R)$$

The choice $p \boxplus q$ allows the environment to choose either to input a value of type I or one of type J, which will trigger the corresponding automata, producing the associated output. A formal definition is

$$[\![p \boxplus q]\!] = (X_p \times X_q, \langle \overline{\alpha_{p \boxplus q}}, \overline{e_{p \boxplus q}} \rangle)$$

where $\alpha_{p \boxplus q}$ is defined as

$$X_p \times X_q \times (I + J) \xrightarrow{\mathtt{dist}} X_p \times X_q \times I + X_p \times X_q \times J \xrightarrow{\mathtt{xr}+\mathtt{a}} X_p \times I \times X_q + X_p \times (X_q \times J)$$
$$\xrightarrow{\alpha_p \times \mathsf{id} + \mathsf{id} \times \alpha_q} \mathcal{Z}(X_p) \times X_q + X_p \times \mathcal{Z}(X_q) \xrightarrow{[\sigma_{X_p,X_q}, \sigma'_{X_p,X_q}]} \mathcal{Z}(X_p \times X_q)$$

and $e_{p \boxplus q}$ is defined as

$$X_p \times X_q \times (I + J) \xrightarrow{\mathtt{dist}} X_p \times X_q \times I + X_p \times X_q \times J \xrightarrow{\mathtt{xr}+\mathtt{a}} X_p \times I \times X_q + X_p \times (X_q \times J)$$
$$\xrightarrow{e_p \circ \pi_1 + e_q \circ \pi_2} \mathcal{Z}(O) + \mathcal{Z}(R) \xrightarrow{[\mathcal{Z}(\iota_1), \mathcal{Z}(\iota_2)]} \mathcal{Z}(O + R)$$

The concurrency combinator $\boxdot$ combines choice and parallel, in the sense that two fuzzy Mealy automata p and q can be executed depending on the input supplied. Let $I \boxdot J = I + J + I \times J$ and $O \boxdot R = O + R + O \times R$. The semantics of $\boxdot$ is given by

$$[\![p \boxdot q]\!] = (X_p \times X_q, \langle \overline{\alpha_{p \boxdot q}}, \overline{e_{p \boxdot q}} \rangle)$$

where $\alpha_{p \boxdot q}$ is defined as

$$X_p \times X_q \times (I \boxdot J) \xrightarrow{\texttt{dist}} X_p \times X_q \times (I + J) + X_p \times X_q \times (I \times J)$$
$$\xrightarrow{\alpha_{p \boxplus q} + \alpha_{p \boxtimes q}} \mathcal{Z}(X_p \times X_q) + \mathcal{Z}(X_p \times X_q) \xrightarrow{[\mathcal{Z}(\mathsf{id}), \mathcal{Z}(\mathsf{id})]} \mathcal{Z}(X_p \times X_q)$$

and $e_{p \boxdot q}$ is defined as

$$X_p \times X_q \times (I \boxdot J) \xrightarrow{\texttt{dist}} X_p \times X_q \times (I + J) + X_p \times X_q \times (I \times J)$$
$$\xrightarrow{e_{p \boxplus q} + e_{p \boxtimes q}} \mathcal{Z}(O + R) + \mathcal{Z}(O \times R) \xrightarrow{[\mathcal{Z}(\iota_1), \mathcal{Z}(\iota_2)]} \mathcal{Z}(O \boxdot R)$$

In coalgebra theory, it is [20] shown that the graph of a $T_{I,O}$-homomorphism is a $T_{I,O}$-bisimulation and the greatest $T_{I,O}$-bisimulation is an equivalence relationship $\sim$. Thus for two given FMlA p, q, if there exists a $T_{I,O}$-homomorphism between their corresponding coalgebras $[\![p]\!], [\![q]\!]$, we denote $p \sim q$.

Theorem 6. *For appropriately typed FMlA p, q, r, p', q',*

$$(p; q); r \sim p; (q; r)$$
$$(p \boxtimes p'); (q \boxtimes q') \sim (p; q) \boxtimes (p', q')$$
$$(p \boxplus p'); (q \boxplus q') \sim (p; q) \boxplus (p', q')$$
$$(p \boxdot p'); (q \boxdot q') \sim (p; q) \boxdot (p', q')$$

Proof. The proof proceeds by pointwise induction. For the first law, if we assume

$$\alpha_p(x_1, i)(x_1') = k_1, e_p(x_1, i)(j) = t_1$$
$$\alpha_q(x_2, j)(x_2') = k_2, e_q(x_2, j)(o) = t_2$$
$$\alpha_r(x_3, o)(x_3') = k_3, e_r(x_3, o)(h) = t_3$$

we can obtain

$$\alpha_{(p;q);r}(x_1, x_2, x_3)(i)(x_1', x_2', x_3')$$
$$-k_1 \otimes k_2 \otimes k_3 \otimes t_1 \otimes t_2$$
$$= \alpha_{p;(q;r)}(x_1, (x_2, x_3))(i)(x_1', (x_2', x_3'))$$
$$e_{(p;q);r}(x_1, x_2, x_3)(i)(h)$$
$$= t_1 \otimes t_2 \otimes t_3$$
$$= e_{p;(q;r)}(x_1, (x_2, x_3))(i)(h)$$

With these equations, it is easy to show a is a $T_{I,O}$-homomorphism from $[\![(p; q); r]\!]$ to $[\![p; (q; r)]\!]$. Other laws can be proved similarly. □

Connecting FMlA through isomorphisms leads to a bisimilarity up to an isomorphic rearranging of input types and output types. Let f, g be isomorphic rearrangements of input types and output types respectively. We use $p\{f, g\}$ to denote the FMlA after arranging the input and the output types in the FMlA p.

Theorem 7. *For appropriately typed FMIA p, q, r,*

$$\begin{aligned} p \boxtimes q &\sim (q \boxtimes p)\{\mathtt{s}, \mathtt{s}\} \\ p \boxplus q &\sim q \boxplus p\{\mathtt{s}_+, \mathtt{s}_+\} \\ p \boxdot q &\sim q \boxdot p\{\mathtt{s}_+ + \mathtt{s}, \mathtt{s}_+ + \mathtt{s}\} \\ (p \boxtimes q) \boxtimes r &\sim p \boxtimes (q \boxtimes r)\{\mathtt{a}, \mathtt{a}^{-1}\} \\ (p \boxplus q) \boxplus r &\sim p \boxplus (q \boxplus r)\{\mathtt{a}_+, \mathtt{a}_+^{-1}\} \\ (p \boxdot q) \boxdot r &\sim p \boxdot (q \boxdot r)\{\mathtt{a}_*, \mathtt{a}_*^{-1}\} \end{aligned}$$

where $\mathtt{a}_*$ is a natural isomorphism from $(A \boxdot B) \boxdot C$ to $A \boxdot (B \boxdot C)$ and its inverse is denoted by $\mathtt{a}_*^{-1}$.

Proof. Similar to Theorem 6. □

The two theorems demonstrate that our combinators are well defined. In the sequel, we compare them with the ones in [32] up to the natural transformation θ through a theorem and an example.

Theorem 8. *Given two FMIA p, q with the corresponding coalgebras in $(\star)$, the following equations holds.*

$$\begin{aligned} \theta(\langle \overline{\alpha_{p\boxtimes q}}, \overline{e_{p\boxtimes q}} \rangle) &= \theta(\langle \overline{\alpha_p}, \overline{e_p} \rangle) \boxtimes \theta(\langle \overline{\alpha_q}, \overline{e_q} \rangle) \\ \theta(\langle \overline{\alpha_{p\boxplus q}}, \overline{e_{p\boxplus q}} \rangle) &= \theta(\langle \overline{\alpha_p}, \overline{e_p} \rangle) \boxplus \theta(\langle \overline{\alpha_q}, \overline{e_q} \rangle) \\ \theta(\langle \overline{\alpha_{p\boxdot q}}, \overline{e_{p\boxdot q}} \rangle) &= \theta(\langle \overline{\alpha_p}, \overline{e_p} \rangle) \boxdot \theta(\langle \overline{\alpha_q}, \overline{e_q} \rangle) \end{aligned}$$

where $\boxtimes, \boxplus, \boxdot$ correspond to our combinators in the left side and the ones for composing $F_{I,O}$-coalgebras in [32] in the right side.

Proof. The proof proceeds by pointwise induction. For the first law, if we assume

$$\begin{aligned} \alpha_p(x_1, i)(x_1') &= k_1, e_p(x_1, i)(j) = t_1 \\ \alpha_q(x_2, j)(x_2') &= k_2, e_q(x_2, j)(o) = t_2 \end{aligned}$$

we obtain

$$\begin{aligned} &\theta(\langle \overline{\alpha_{p\boxtimes q}}, \overline{e_{p\boxtimes q}} \rangle)((x_1, x_2), i)((x_1', x_2'), o) \\ =&\overline{\alpha_{p\boxtimes q}}((x_1, x_2), i)(x_1', x_2') \otimes \overline{e_{p\boxtimes q}}((x_1, x_2), i)(o) \\ =&(k_1 \otimes k_2) \otimes (t_1 \otimes t_2) \\ =&(k_1 \otimes t_1) \otimes (k_2 \otimes t_2) \\ =&\theta(\langle \overline{\alpha_p}, \overline{e_p} \rangle)(x_1, i)(x_1', o) \otimes \theta(\langle \overline{\alpha_q}, \overline{e_q} \rangle)(x_2, i)(x_2', o) \\ =&\theta(\langle \overline{\alpha_p}, \overline{e_p} \rangle) \boxtimes \theta(\langle \overline{\alpha_q}, \overline{e_q} \rangle) \end{aligned}$$

Other laws can be proved similarly. □

Note that the case for the sequential composition combinator does not always hold. Actually, this depends on the complete residuated lattice used, since the state transition of the first component is considered twice, which can be demonstrated by the following example.

Example 6. *Recall the standard product algebra in Example 2. Consider two FMIA $p = (\{x_1, x_2\}, \{a\}, \{b\}, \alpha_p, e_p)$ and $r = (\{y_1, y_2\}, \{b\}, \{c\}, \alpha_r, e_r)$ where $\alpha_p(x_1, a)(x_2) = 0.4$, $e_p(x_1, a)(b) = 0.5$ and $\alpha_r(y_1, b)(y_2) = 0.8, e_r(y_1, b)(c) = 0.5$. Then we can obtain $[\![p; r]\!] = (U, \langle \overline{\alpha_{p;r}}, \overline{e_{p;r}} \rangle)$ where $U = \{(x_i, y_j) | i, j = 1, 2\}, \alpha_{p;r}((x_1, y_1), a)(x_2, y_2) = 0.4 \times 0.5 \times 0.8 = 0.16$ and $e_{p;r}((x_1, y_1), a)(c) = 0.5 \times 0.5 = 0.25$. Therefore*

$$\theta(\langle \overline{\alpha_{p;r}}, \overline{e_{p;r}} \rangle)((x_1, y_1), a)((x_2, y_2), c) = 0.16 \times 0.25 = 0.04.$$

However, $\theta(\langle\overline{\alpha_p},\overline{e_p}\rangle)(x_1,a)(x_2,b) = 0.4 \times 0.5 = 0.2$ *and* $\theta(\langle\overline{\alpha_r},\overline{e_r}\rangle)(x_1,a)(x_2,b) = 0.8 \times 0.5 = 0.4$. *Thus,*

$$\theta(\langle\overline{\alpha_p},\overline{e_p}\rangle);\theta(\langle\overline{\alpha_r},\overline{e_r}\rangle)((x_1,y_1),a)((x_2,y_2),c) = 0.2 \times 0.4 = 0.08.$$

Oppositely, if we consider the standard Gödel algebra, the two values will be both 0.4.

6. Case Study

In the sequel, we illustrate the use of fuzzy components by means of a concrete example. For simplicity, we consider an non-fuzzy input-output function and compose components with $F_{I,O}$-coalgebras. Consider the following example of a steam turbine.

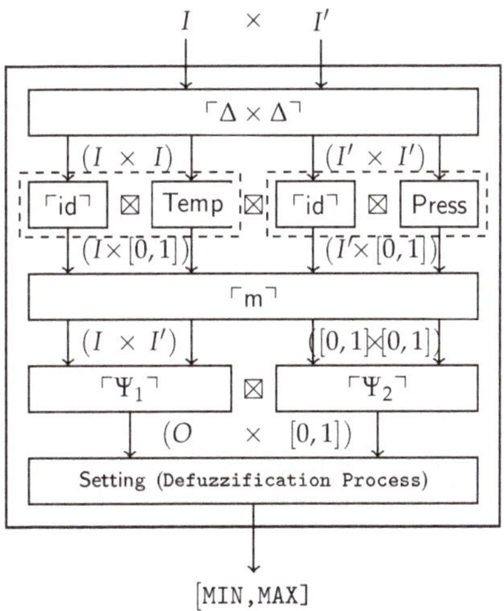

The system is composed of two fuzzification components Temp, Press and a defuzzification component Setting with corresponding membership functions illustrated in Figure 1. Note that Δ represents the copy operation.

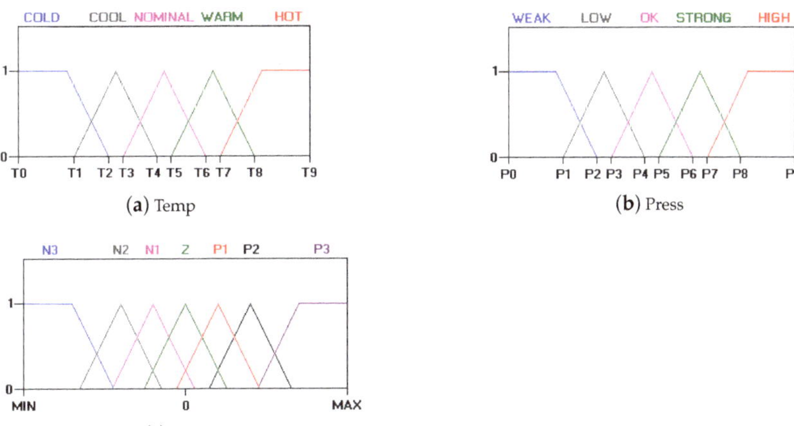

Figure 1. The graphs of membership functions.

In practice, the components Temp and Press execute in parallel. Each one will produce a membership value corresponding to the state and membership function after receiving a mode signal. After that, the minimum of the two output values will become the input of Setting. The membership function of the Setting component is determined by the following rules (for simplicity, only whose conditions with temperature COOL are displayed).

```
rule 1 : If temperature is COOL and pressure is WEAK then throttle is P3.
rule 2 : If temperature is COOL and pressure is LOW then throttle is P2.
rule 3 : If temperature is COOL and pressure is OK then throttle is Z.
rule 4 : If temperature is COOL and pressure is STRONG then throttle is N2.
rule 5 : If temperature is COOL and pressure is HIGH then throttle is N3.
  ...
```

The output functions are considered as non-fuzzy in this example.

(i) The coalgebraic semantic of component Temp

$$[\![\mathsf{Temp}]\!] = (T, \theta\langle\overline{\alpha_t}, \overline{e_t}\rangle) : T \to \mathcal{Z}(T \times [0,1])^I)$$

is actually an $F_{I,[0,1]}$-coalgebra. In this model, states are the temperature over $T = [T_0, T_9]$, inputs are operation modes over set I = {COLD, COOL, NORMAL, WARM, HOT} that are decided by users. The fuzzy transition function is constant on the temperature and given by $\alpha_t : T \times I \to [0,1]^T$ with

$\langle t, \mathtt{COLD}\rangle \mapsto \phi_{COL}$, $\langle t, \mathtt{COOL}\rangle \mapsto \phi_{COO}$, $\langle t, \mathtt{NORMAL}\rangle \mapsto \phi_{NOR}$, $\langle t, \mathtt{WARM}\rangle \mapsto \phi_{WAR}$, $\langle t, \mathtt{HOT}\rangle \mapsto \phi_{HOT}$

for all $t \in [T_0, T_9] \subseteq \mathbb{R}$. The output function $e_t : T \times I \to [0,1]$ is defined by $(t,i) \mapsto \mathsf{eval}(\alpha_t(t,i), t)$ where eval is an evaluation function. As a concrete example, suppose the fuzzy subset for the NORMAL mode is the function

$$\phi_{NOR}(t) = max\{0, \frac{2}{T_3 - T_6}(t - \frac{T_3 + T_6}{2}) + 1\}.$$

Then the membership value (output) over state $\frac{T_3+T_6}{2}$ under the mode NORMAL is $e_t(\frac{T_3+T_6}{2}, \mathtt{NORMAL}) = \mathsf{eval}(\phi_{NOR}, \frac{T_3+T_6}{2}) = 1$.

(ii) Press is a component whose state space P is given by the pressure in the steam turbine and inputs are over the set $I' = \{\mathtt{WEAK}, \mathtt{LOW}, \mathtt{OK}, \mathtt{STRONG}, \mathtt{HIGH}\}$, which represent the mode triggered by the users. The output of this component is the membership value corresponding to the current fuzzy state. The dynamics of this component is

$$[\![\mathsf{Press}]\!] = (P, \theta\langle\overline{\alpha_p}, \overline{e_p}\rangle) : P \to \mathcal{Z}(P \times [0,1])^{I'})$$

with the transition and output functions defined as $\alpha_p : P \times I' \to \mathcal{Z}(P)$:

$\langle p, \mathtt{WEAK}\rangle \mapsto \phi_{WEAK}$, $\langle p, \mathtt{LOW}\rangle \mapsto \phi_{LOW}$, $\langle p, \mathtt{OK}\rangle \mapsto \phi_{OK}$, $\langle p, \mathtt{STRONG}\rangle \mapsto \phi_{STRONG}$, $\langle p, \mathtt{HIGH}\rangle \mapsto \phi_{HIGH}$

for $p \in P$ and

$$o_p : P \times I' \to [0,1] : (p, i') \mapsto \mathsf{ev}(\alpha_p(p, i'), p).$$

(iii) The dynamics of Rule and And components are denoted by $\ulcorner\Psi_1\urcorner$ and $\ulcorner\Psi_2\urcorner$ where

$$\ulcorner\Psi_1\urcorner = (\mathbf{1}, \overline{\eta_{(\mathbf{1}\times O)} \cdot \langle \mathsf{id}, \Psi_1\rangle} : \mathbf{1} \to \mathcal{Z}(\mathbf{1} \times O)^{I \times I'}).$$

In this expression O is the output set determined by the output function, namely,

$$\Psi_1 : \mathbf{1} \times (I \times I') \to O$$

$$\Psi_1(\star,(i,i')) = \begin{cases} \texttt{P3} & i = \texttt{COOL} \wedge i' = \texttt{WEAK} \\ \texttt{P2} & i' = \texttt{COOL} \wedge \texttt{LOW} \\ \texttt{Z} & i' = \texttt{COOL} \wedge \texttt{OK} \\ \texttt{N2} & i' = \texttt{COOL} \wedge \texttt{STRONG} \\ \texttt{N3} & i' = \texttt{COOL} \wedge \texttt{HIGH} \\ \dots \end{cases}$$

$\mathbf{1} = \{*\}$ is the singleton set. The notation $\ulcorner f \urcorner$ is the representation of function $f : A \to B$, which is defined as a coalgebra $\ulcorner f \urcorner = (* \in \mathbf{1}, \overline{c}_{\ulcorner f \urcorner})$, where $c_{\ulcorner f \urcorner} = \mathbf{1} \times A \xrightarrow{\mathsf{id} \times f} \mathbf{1} \times B \xrightarrow{\eta_{(\mathbf{1} \times B)}} \mathcal{Z}(\mathbf{1} \times B)$. The definition of Ψ_2 is similar, given a pair of inputs of $[0,1]$, it outputs the minimum value of the two.

(iv) The last component Setting works as follows. Through the channel it interacts with Temp and Press. It receives the mode information and a membership value as the current state. The mode information determines which membership function is accessible for the component. Then the component outputs an area whose boundary consists of the horizontal axis and the graph of the membership function. Formally, this model is represented by a coalgebra

$$[\![\mathsf{Setting}]\!] = (D, \theta\langle \overline{\alpha_s}, \overline{e_s} \rangle) : D \to \mathcal{Z}(D \times \mathcal{P}(\mathbb{R}^2))^{O \times [0,1]})$$

where $D = [\texttt{MIN}, \texttt{MAX}]$ is an interval of real numbers. The output function is defined as $e_s(d,(o,r)) = \{(x,y) | 0 \le y \le min\{\alpha_s(x,(o,r)), r\}, x \in [\texttt{MIN}, \texttt{MAX}]\}$. Resorting to centroid defuzzification technique, the output stage processes combine areas and produce a control value, which will participate in the control of the system.

7. Conclusions and Future Work

The present work aims at addressing fuzzy automata from a coalgebraic perspective. Our starting point was studying the fuzzy-set monad further. We defined a left tensorial strength and a right tensorial strength, and proved it is a strong and commutative monad. With these properties, we modeled different types of fuzzy automata as coalgebraic models with the same transition structure. Based on these coalgebraic models, we defined the notions of fuzzy language bisimulation between fuzzy automata. Moreover, we developed some compositional combinators for fuzzy Mealy automata of two kinds: state transition and output function and compared it with the classical component calculi in [32]. Finally, through a case study, we discussed the application of our component calculi.

Besides these fundamental results, there are several topics left to explore. One is to define a notion of refinement [38] of fuzzy automata, to specify an inclusion relation of fuzzy behaviour. Fuzzy automata may involve complex behaviour such as non-deterministic transitions or branched transitions with probability [23,39]. Therefore another topic for future work is to develop more complex versions of fuzzy automata and analyze their behavior and discuss their properties, namely of the suitable notions of bisimulation as in [15,35,36].

Author Contributions: Conceptualization, M.S. and L.S.B.; methodology, A.L. and S.W.; formal analysis, A.L. and S.W.; investigation, A.L.; writing—original draft preparation, A.L. and S.W.; writing—review and editing, M.S. and L.S.B. All authors have read and agreed to the published version of the manuscript.

Funding: This work has been supported by the Guangdong Science and Technology Department (Grant No. 2018B010107004) and the National Natural Science Foundation of China under grant No. 61772038, 61532019 and 61272160. L.S.B. was supported by the ERDF—European Regional Development Fund through the Operational Programme for Competitiveness and Internationalisation-COMPETE 2020 Programme and by National Funds through the Portuguese funding agency, FCT, within project KLEE - POCI-01-0145-FEDER-030947.

Institutional Review Board Statement: Not applicable.

Informed Consent Statement: Not applicable.

Data Availability Statement: Not applicable.

Acknowledgments: This work is also supported by Hiroshima University. Many thanks to the reviewers and editors.

Conflicts of Interest: The authors declare no conflict of interest.

References

1. Tanaka, K. *An Introduction to Fuzzy Logic for Practical Applications*; Springer: Berlin/Heidelberg, Germany, 1997.
2. Zadeh, L.A. Soft Computing and Fuzzy Logic. *IEEE Softw.* **1994**, *11*, 48–56. [CrossRef]
3. Syropoulos, A.; Grammenos, T. *Fuzzy Computation;* A Modern Introduction to Fuzzy Mathematics; John Wiley & Sons: Hoboken, NJ, USA, 2020; pp. 191–214. [CrossRef]
4. Wu, H.; Gu, X.; Zhen, L. Fuzzy Principal Component Analysis Model on Evaluating Innovation Service Capability. *Sci. Program.* **2020**, *2020*, 8834901. [CrossRef]
5. Böhme, M.; Pham, V.; Roychoudhury, A. Coverage-Based Greybox Fuzzing as Markov Chain. *IEEE Trans. Softw. Eng.* **2019**, *45*, 489–506. [CrossRef]
6. Simon, D.J. Introduction to Fuzzy Control. In *Embedded Systems Programming;* Electrical Engineering & Computer Science Faculty Publications: Cambridge, MA, USA, 2003; Volume 16, pp. 55–56.
7. Doostfatemeh, M.; Kremer, S.C. New directions in fuzzy automata. *Int. J. Approx. Reason.* **2005**, *38*, 175–214. [CrossRef]
8. Chaudhari, S.R.; Desai, A.S. On fuzzy Mealy and Moore machines. *Bull. Pure Appl. Math* **2010**, *4*, 375–384.
9. Mordeson, J.N.; Nair, P.S. Fuzzy Mealy machines. *Kybernetes* **1966**, *25*, 18–33. [CrossRef]
10. Li, Y.; Pedrycz, W. The equivalence between fuzzy Mealy and fuzzy Moore machines. *Soft Comput.* **2006**, *10*, 953–959. [CrossRef]
11. Todinca, D.; Sora, I.; Butoianu, D.; Precup, R. A Novel Method to Compute the Membership Value of the States of Fuzzy Automata. In Proceedings of the 2018 IEEE 12th International Symposium on Applied Computational Intelligence and Informatics (SACI), Timisoara, Romania, 17–19 May 2018; pp. 107–112. [CrossRef]
12. Pan, H.; Li, Y.; Cao, Y.; Li, P. Nondeterministic fuzzy automata with membership values in complete residuated lattices. *Int. J. Approx. Reason.* **2017**, *82*, 22–38. [CrossRef]
13. Tiwari, S.P.; Pal, P. On a category of deterministic fuzzy automata. In *11th Conference of the European Society for Fuzzy Logic and Technology (EUSFLAT 2019);* Atlantis Studies in Uncertainty Modelling; Atlantis Press: Paris, France, 2019; Volume 1. [CrossRef]
14. Mockor, J. Monads and a common framework for fuzzy type automata. *Int. J. Gen. Syst.* **2019**, *48*, 406–442. [CrossRef]
15. Singh, A.K.; Tiwari, S.P. Fuzzy Regular Languages Based on Residuated Lattice. *New Math. Nat. Comput.* **2020**, *16*, 363–376. [CrossRef]
16. Yang, C.; Li, Y. Approximate bisimulations and state reduction of fuzzy automata under fuzzy similarity measures. *Fuzzy Sets Syst.* **2020**, *391*, 72–95. [CrossRef]
17. Yang, C.; Li, Y. ϵ-Bisimulation Relations for Fuzzy Automata. *IEEE Trans. Fuzzy Syst.* **2018**, *26*, 2017–2029. [CrossRef]
18. Yang, C.; Li, Y. Approximate bisimulation relations for fuzzy automata. *Soft Comput.* **2018**, *22*, 4535–4547. [CrossRef]
19. Rutten, J.J.M.M. Automata and coinduction (an exercise in coalgebra). In *International Conference on Concurrency Theory, Proceedings of the CONCUR 1998: CONCUR'98 Concurrency Theory, Nice, France, 8–11 September 1998;* Springer: Berlin/Heidelberg, Germany, 1998; Volume 1466, pp. 194–218.
20. Rutten, J.J.M.M. Universal coalgebra: A theory of systems. *Theor. Comput. Sci.* **2000**, *249*, 3–80. [CrossRef]
21. Jacobs, B. *Introduction to Coalgebra: Towards Mathematics of States and Observation;* Cambridge Tracts in Theoretical Computer Science; Cambridge University Press: Cambridge, UK, 2016; Volume 59.
22. Silva, A.; Bonchi, F.; Bonsangue, M.M.; Rutten, J.J.M.M. Generalizing determinization from automata to coalgebras. *Log. Methods Comput. Sci.* **2013**, *9*. [CrossRef]
23. Sokolova, A. Coalgebraic Analysis of Probabilistic Systems. Ph.D. Thesis, Technische Universiteit Eindhoven, Eindhoven, The Netherlands, 2005.
24. Neves, R.; Barbosa, L.S. Hybrid Automata as Coalgebras. In *International Colloquium on Theoretical Aspects of Computing, Proceedings of the ICTAC 2016: Theoretical Aspects of Computing, Taipei, Taiwan, 24–31 October 2016;* Lecture Notes in Computer Science; Springer: Berlin/Heidelberg, Germany, 2016; Volume 9965, pp. 385–402. [CrossRef]
25. Neves, R.; Barbosa, L.S. Languages and models for hybrid automata: A coalgebraic perspective. *Theor. Comput. Sci.* **2018**, *744*, 113–142. [CrossRef]
26. Liu, A.; Sun, M. A Coalgebraic Semantics Framework for Quantum Systems. In *International Conference on Formal Engineering Methods, Proceedings of the ICFEM 2019: Formal Methods and Software Engineering, Shenzhen, China, 5–9 November 2019;* Lecture Notes in Computer Science; Springer: Berlin/Heidelberg, Germany, 2019; Volume 11852, pp. 387–402. [CrossRef]
27. Feng, Y.; Duan, R.; Ying, M. Bisimulation for Quantum Processes. *ACM Trans. Program. Lang. Syst.* **2012**, *34*, 1–43. [CrossRef]
28. Larsen, K.G.; Skou, A. Bisimulation through probabilistic testing. *Inf. Comput.* **1991**, *94*, 1–28. [CrossRef]

29. Haghverdi, E.; Tabuada, P.; Pappas, G.J. Bisimulation Relations for Dynamical and Control Systems. *Electr. Notes Theor. Comput. Sci.* **2002**, *69*, 120–136. [CrossRef]
30. Jacobs, B. Invariants, Bisimulations and the Correctness of Coalgebraic Refinements. In *International Conference on Algebraic Methodology and Software Technology, Proceedings of the AMAST 1997: Algebraic Methodology and Software Technology, Sydney, Australia, 13–17 December 1997*; Lecture Notes in Computer Science; Springer: Berlin/Heidelberg, Germany, 1997; Volume 1349, pp. 276–291. [CrossRef]
31. Venema, Y. Algebras and coalgebras. In *Handbook of Modal Logic*; Studies in Logic and Practical Reasoning; Elsevier B.V.: Amsterdam, The Netherlands, 2007; Volume 3, pp. 331–426. [CrossRef]
32. Barbosa, L.S. Components as Coalgebras. Ph.D. Thesis, Universidade do Minho, Braga, Portugal, 2001.
33. Guilherme, R.J.P. A Coalgebraic Approach to Fuzzy Automata. Ph.D. Thesis, Universidade Nova De Lisboa, Lisbon, Portugal, 2016.
34. Wu, H.; Chen, Y. Coalgebras for Fuzzy Transition Systems. *Electron. Notes Theor. Comput. Sci.* **2014**, *301*, 91–101. [CrossRef]
35. Wu, H.; Chen, Y.; Bu, T.; Deng, Y. Algorithmic and logical characterizations of bisimulations for non-deterministic fuzzy transition systems. *Fuzzy Sets Syst.* **2018**, *333*, 106–123. [CrossRef]
36. Wu, H.; Chen, T.; Han, T.; Chen, Y. Bisimulations for fuzzy transition systems revisited. *Int. J. Approx. Reason.* **2018**, *99*, 1–11. [CrossRef]
37. Nikravesh, M.; Kacprzyk, J.; Zadeh, L.A. *Forging New Frontiers: Fuzzy Pioneers I*; University of California: Berkeley, CA, USA, 2007.
38. Meng, S.; Barbosa, L.S. Components as coalgebras: The refinement dimension. *Theor. Comput. Sci.* **2006**, *351*, 276–294. [CrossRef]
39. Narasimha, M.; Cleaveland, R.; Iyer, S.P. The role of observations in probabilistic open systems. *Electr. Notes Theor. Comput. Sci.* **1999**, *25*, 133–144. [CrossRef]

MDPI

Article

Mutated Specification-Based Test Data Generation with a Genetic Algorithm †

Rong Wang [1,*], Yuji Sato [1] and Shaoying Liu [2]

1 Department of Computer Science, Hosei University, Tokyo 184-8584, Japan; yuji@hosei.ac.jp
2 Graduate School of Advanced Science and Engineering, Hiroshima University, Hiroshima 739-8511, Japan; sliu@hiroshima-u.ac.jp
* Correspondence: rong.wang.99@stu.hosei.ac.jp

† This paper is an extended version of our paper published in 2019 IEEE Congress on Evolutionary Computation (CEC), Wellington, New Zealand, 10–13 June 2019.

Abstract: Specification-based testing methods generate test data without the knowledge of the structure of the program. However, the quality of these test data are not well ensured to detect bugs when non-functional changes are introduced to the program. To generate test data effectively, we propose a new method that combines formal specifications with the genetic algorithm (GA). In this method, formal specifications are reformed by GA in order to be used to generate input values that can kill as many mutants of the target program as possible. Two classic examples are presented to demonstrate how the method works. The result shows that the proposed method can help effectively generate test cases to kill the program mutants, which contributes to the further maintenance of software.

Keywords: test data generation; genetic algorithm; specification-based testing; regression testing; mutation testing

Citation: Wang, R.; Sato, Y.; Liu, S. Mutated Specification-Based Test Data Generation with a Genetic Algorithm. *Mathematics* **2021**, *9*, 331. https://doi.org/10.3390/math9040331

Academic Editor: David Greiner

Received: 31 December 2020
Accepted: 4 February 2021
Published: 7 February 2021

1. Introduction

Regression testing is an important technique to ensure that previously tested software still performs in the same way after it is changed or integrated with other software [1–3]. In general, changes to software are mainly concerned with the efficiency enhancement, robustness improvement, and configuration changes, but these changes should not result in big alternation of the functionality defined in the specification of the software. Therefore, specification-based testing (SBT) methods can be effectively used in regression testing.

SBT is characterized by test data being generated from the specification without concerning the structure of the corresponding program and test results being analyzed based on the specification [4–8]. Formal specifications may allow the test data generation and test result analysis to be done rigorously, systematically, and even automatically in many cases [9–12]. In our work, we mainly use formal specifications in pre- and post-conditions, such as Vienna Development Method (VDM) [13], a formal method that has been developed over past years [14,15], and Structured-Object-Oriented Formal Language (SOFL) [16], which has the potential of practical use in industry and serves as a solid foundation to develop a method of *functional scenario-based test data generation* [17].

However, in spite of considerable progresses having been made, it is still not easy for SBT to generate various test data only from specifications to detect different bugs that are contained in the program. This is because different features and effects of the program output cannot be controlled and triggered by only input data suites that satisfy some parts of the specification (some constraints over only input variables). Consequently, many faulty program paths would not be detected and thus the bug detection would be likely to fail in some cases. For the existing SBT, one of the major deficiencies is that the test data generation only considers the constraints over input variables in the formal specification, without making use of the constraints over outputs before the execution of a program.

To overcome the shortcomings of the existing SBT methods, we propose a new method for test data generation in this paper. The proposed method introduces dummy variables into some specified constraints in the specification, and makes use of the constraints over both input and output variables to guide the test data generation, in contrast to the conventional SBT methods that concerns only constraints over input variables. This method features the combination of three techniques, SBT, mutation testing, and GA. It is to obtain the enhanced (mutated) formal specifications by using a GA so that input data generated from these specifications are more likely to kill different kinds of mutants of the target program under test. The expected effect of the test data generated in this way is to detect various bugs probably occurring in the program that is being developed or improved.

The work in this paper is an extension for our previous work in [18]. Comparing with the previous work, we use more different kinds of mutants (16 types against previously 10) for each program in our case study to gain better mutated functional scenarios that are capable of detecting more bugs. In addition, we give additional experimental results of the proposed approach that uses no dummy variables, and point out the importance of introducing dummy variables. In addition, we carefully conduct more experiments with the dummy variables that are introduced to reform different parts of original specification, the equality and inequality relations. Based on that, we enriched the analysis for the effect of different ways of introducing dummy variables on bug detection.

The remaining part of the paper is organized as follows. Section 2 briefly introduces the existing work related to our approach. Section 3 illustrates how to transform formal specifications to chromosomes as well as the corresponding genetic manipulations in GA. Section 4 describes the main procedures of obtaining desirable reformed specifications for test data generation by integrating GA. Section 5 presents two classic cases to demonstrate the feasibility and efficiency of the approach. Finally, Section 6 concludes the paper and points out future research directions.

2. Related Work

In this section, we introduce several advanced techniques that relate to our methodology for test data generation.

Data flow analysis, a technique for computing the def-use associations for the control flow graph (CFG) of a program, are often used to develop different strategies for test data generation over a long history [19–21]. Many research works have proposed promising methods for automatic test data generation that integrates the data flow analysis with the heuristics, such as GAs [22–24], particle swarm optimization [25], and ant colony optimization [26]. Different from these techniques, our approach conducts the testing under the circumstance where the source code of thirty-party library under test cannot be accessed. Thus, these techniques rely too much on the knowledge and analysis of internal design (or code structure), but our approach focuses on generating test data without analyzing the source code when applying the GA to the formal specification. In addition, with respect to the usage of the GA, our approach uses the GA to search for the optimal mutated specifications that are later used for input data generation, while the techniques mentioned above use the GA to directly search for good input data.

The techniques of mutation-based test data generation [27,28] are used to select a set of "good" test data by executing designed incorrect versions of an original program with a great number of test data from the domain. Test data are selected if it can cause unintended behaviors for a certain number of incorrect versions. These techniques mainly concentrate on designing appropriate mutation operators to introduce small modifications for different kinds of programming languages such as Java [29], C# [30], and C++ [31]. The incorrect versions, also called program mutants, are created by inserting the mutant operators into the original program. Compared with these techniques, our method selects a set of "good" mutated specifications as a seed for further test data generation by using not only program mutants but also the mutated specifications with the GA.

The SBT techniques, some of them integrated with heuristic search strategies, have been well developed to cope with different kinds of specifications, such as SOFL [9,16], Alloy [32,33], protocol specifications [34,35], and Object Constraint Language (OCL) specification [36,37]. Among these specifications, we take an interest in the formal specification of pre-post style like SOFL and Alloy. On the one hand, the SBT techniques for both SOFL and Alloy generate test data only from the pre-condition and use the post-condition as an oracle to check if the outcome is correct. On the other hand, the SBT with SOFL still needs to be improved since a data suite generated only from the original SOFL specification is not sufficient enough to trigger different kinds of bad behaviors of programs. On the contrary, our approach uses both the pre- and post- conditions to generate input data, as well as selects the optimal mutated specifications to enhance the bug detection.

3. GA with Mutated Specification

We first briefly introduce the basics of GA and then discuss how to obtain reformed specifications.

3.1. Description for GA

GA is a heuristic search method inspired from evolutionary biology and was first proposed by John Holland [38]. In general, a GA is involved in an iterative process with three steps: (i) create an initial population of solutions (called individuals) represented by a pre-defined chromosome that are typically encoded the solutions to a problem; (ii) in the existing population, select a group of individuals by a specified *selection* strategy based on a *fitness* function, and generate the next population from applying two key genetic operators, *crossover* and *mutation* to those selected individuals; (iii) repeat step (ii) until the remaining individuals in the generation are good enough according to both the fitness function and the stop criteria.

Since GA works well in finding optimal solutions for nonlinear problems and the specifications of pre-post style could be easily transformed to chromosomes with few efforts, we employ GA to find the best mutated specifications in this paper. Later, we will first describe how to transform the original formal specifications into a chromosome, and then carefully describe the evolution in step (ii) in detail for our specific goal: to obtain all the mutated functional scenarios from the specification, each a constraint over only input variables.

3.2. Mutation Testing

Mutation testing, also called program mutation [39], is used to design test cases and evaluates the quality of existing testing techniques. In mutation testing, some small modifications are injected into the original program. Each mutated version is called *program mutant* and test data are regarded as the good one if it kills the program mutants, that is, it makes the behavior of program mutants different from that of the original program.

In our approach, both programs and the specifications are mutated. The program mutants are used to evaluate the quality of mutated specifications. We search for good mutated specifications that can be used to effectively generate test data for bug detection.

3.3. Mutated Specifications

We use SOFL as the formal notation for specifications in this paper. There are two reasons: one is that SOFL, as a formal notation, is more comprehensible than other formal notations since it uses the comprehensible condition data flow diagrams for system structure as well as pre- and post- conditions for defining individual operations in the system. Another reason is that SOFL is familiar to us and its use in industry has been increasing [40].

In SOFL, the *defining condition* describes the constraints over input and output variables after a method in the program performs. Generally, the defining condition is not used for directly generating input data in most of existing techniques because the values of outputs in defining condition are unknown to us before the execution of the program. We consider the defining conditions as an important factor for test data generation from which the test data are sensitive to bad behaviors of the program.

Since defining conditions describe how output variables relate to input variables, they are often used to check whether an execution of the program is correct or not, rather than being used to directly generate input values. For a program, it is usually difficult to directly generate input values that satisfy a defining condition without knowing the corresponding output values. For instance, suppose input variable x and output variable y satisfy the defining condition $(x * y > x + y)$, we cannot generate input x from $(x * y > x + y)$ due to the unknown output y. Thus, usually $(x * y > x + y)$ is not used to help generate the input values but can be used to check the result of executing the program with input x.

Nevertheless, by assigning appropriate values to the output variables in the defining conditions, we can get some useful mutated specifications that can be used to directly generate input values. For the defining condition $(x * y > x + y)$ mentioned above, input data generated from $(x * 2 > x + 2)$ (when $y = 2$) may be more likely to trigger bugs than from $(0 > x)$ (when $y = 0$). Currently, it cannot be determined without further checking. However, $(x * 2 > x + 2)$ is definitely better than $(x * 1 > x + 1)$ (when $y = 1$) because the latter is always false and cannot be used to generate test data.

Our work mainly concentrates on developing a way to find appropriate output values for the specification. These output values are then used to build the mutated specifications that are the constraints over only input variables. Then, the mutated specifications can be directly used to generate input values in regression testing. To achieve that, we employ GA to seek such appropriate values of outputs from the defining condition.

Moreover, to obtain mutated specifications that are more powerful in bug detection, some extension is considered for reforming defining conditions before applying GA. In this extension, we make a slight change in defining conditions so as to induce the generated test data that satisfy those reformed ones to trigger as many bad behaviors of the program as possible.

In our method, *mutated specifications* are made from after applying GA to the original specification. More precisely, the mutated specifications can be obtained by following *two rules*:

1. Reforming the original specification by introducing dummy variables into defining conditions so that test data that satisfy those reformed ones can trigger bad behaviors of the program;
2. Finding appropriate concrete values through GA and assigning them to the output and dummy variables that occur in the reformed specification.

Our goal is to obtain a new version of the specification from which the test data suite can be generated to trigger as many bugs as possible in the program. Next, we will define the chromosome forms for the reformed specification, as well as describe the crossover operator and mutation operator. Then, we apply the GA for gaining the suitable mutated specifications that can do well in bug detection.

We define the form of chromosomes for a condition data flow diagram (CDFD) that is part of the SOFL language.

A condition data flow diagram (CDFD) is a directed graph that specifies how processes work together to provide functional behaviors [41]. Every process has its own pre- and post-conditions. For instance, Figure 1 displays a small CDFD that consists of two processes A and B where process A first consumes two input variables x and y and produces output z, and then process B consumes z and produces w.

The two separately defined processes A and B may not be automatically combined into a bigger process C since we can not always infer $C\_pre(x, y) \wedge C\_post(x, y, w)$ just from $A\_pre(x, y) \wedge A\_post(x, y, z) \wedge B\_pre(z) \wedge B\_post(z, w)$ unless we know the expression $z = Expr(x, y)$ in $A\_post(x, y, z)$, since, in that case, we can easily replace z with $Expr(x, y)$ and derive the following predicate expression:

$$C\_pre(x,y) \wedge C\_post(x,y,w) = A\_pre(x,y) \wedge A\_post(x,y,Expr(x,y)) \wedge B\_pre(Expr(x,y)) \wedge B\_post(Expr(x,y),w).$$

However, the intermediate variables between two processes like variable z can not always be replaced in real CDFDs. Therefore, our discussion on test data generation from specifications focuses on a single process.

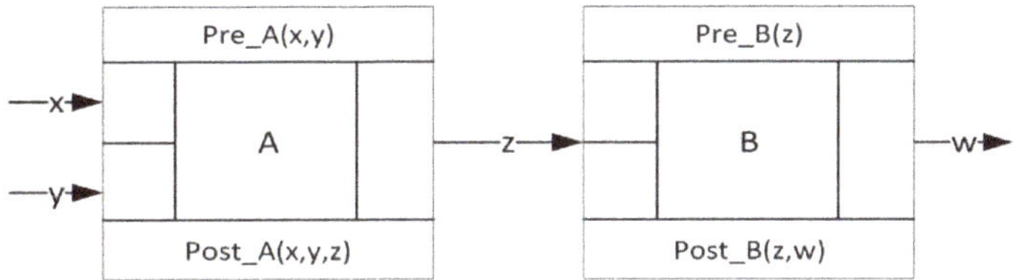

Figure 1. The process A and process B.

3.4. Chromosome Formation

In this approach, the specification is converted into an equivalent expression called *functional scenario form* (FSF).

Definition 1. *A FSF of process is the disjunction of functional scenarios:* $\vee_{i=1}^{n}(T_i \wedge D_i) := S_{pre} \wedge (\vee_{i=1}^{n}(G_i \wedge D_i))(i = 1, \cdots, N)$ *where* $T_i = S_{pre} \wedge G_i$ *is called a test condition, which is the conjunction of the pre-condition* S_{pre} *and the guard-condition* G_i*; and* D_i *is a predicate called a defining condition.*

The pre-condition S_{pre} of process S is a constraint on the input, and it contains only input variables. A guard condition G_i is part of the post-condition but contains no output variables. A defining condition D_i is also part of the post-condition but contains at least one output variable. The functional scenario $T_i \wedge D_i$ describes a single specific functional behavior: when test condition T_i is true, the output of the operation is defined using defining condition D_i. In this paper, we assume that any FSF $\vee_{i=1}^{n}(T_i \wedge D_i)$ of process S is complete, which means that any input satisfying S_{pre} must make $\vee_{i=1}^{n} T_i$ true.

Each functional scenario defines an independent function of the operation: when the test condition holds on the input variables, the output variables will be defined by the defining condition. Currently, test data generation from a functional scenario only takes the test condition into account meanwhile leaving the defining condition untouched [9,42,43].

Now, we explain how to make a slight extension to change the form of defining conditions so as to allow bad behaviors to occur. To obtain a more flexible and useful reformed specification, we introduce *dummy variables*, $d_i (i = 1, \cdots, c)$, to the relationship of inputs and outputs from the defining condition. Then, we build an *output vector* from both the *dummy variables* and output variables.

Definition 2. *An output vector* o' *is a vector constructed by output variables and dummy variables:* $o' = (o_1, \cdots, o_n, d_1, \cdots, d_c)$*, where* o_i $(i = 1, \cdots, n)$ *are output variables, and* d_i $(i = 1, \cdots, c)$ *are dummy integer variables.*

For a relation $(f(inputs, outputs) \bigtriangleup 0)$ (where $\bigtriangleup$ is a operator such as $=, >, < \ldots$) in the defining condition D_i, by introducing dummy variables d_1 and d_2, we construct an inequality $d_1 <= f(inputs, outputs) <= d_2$ and replace the relation f with this new inequality in D_i. Then, the output vector is formed as $o' = (o_1, \cdots, o_n, d_1, d_2)$. In our work, we mainly make such change to only equality relations.

We change an equality relation to such an inequality relation because an equality relation is quite a strict condition that would drastically narrow down the exploration of input values for a single functional scenario when output values are determined by GA.

Therefore, dummy variables need to be introduced for equality. For the inequality relation in the specification, dummy variables are not introduced to them because, compared with equality relation, inequality relation is not a too strict condition for the generation of input values. Thus, these kind of relations are used to preserve some original features of specifications. In addition, the experimental results in Section 5 also indicate that additional dummy variables for inequality cannot help considerably improve the quality of the mutated specifications.

Definition 3. *A chromosome* $[T_i \wedge D_i]_{o'}$ $(i = 1, \cdots, N)$ *is a reformed functional scenario* $T_i \wedge D_i$, *where some dummy variables are introduced to* D_i. *An individual (a mutated specification) is a constraint over symbolic inputs, established from the chromosome* $[T_i \wedge D_i]_{o'}$ *by assigning concrete values to the output vector* $o' = (o_1, \cdots, o_n, d_1, \cdots, d_c)$. *A population is a group of such individuals. For convenience, the output vector* o' *is also called d-chromo, and each element of* o' *is called a genetic.*

From this definition, a d-chromo o' with concrete values determines an individual formula $[T_i \wedge D_i]_{o'}$ that is a constraint on symbolic inputs. Such an individual is a reformed specification that can be used to generate test data for the program. In order to obtain good individuals to generate test data that are useful for bug detection, we apply the genetic manipulation to a group of individuals $[T_i \wedge D_i]_{o'}$ and find the appropriate d-chromo o'. Each individual will be scored by evaluating the quality of the test data that are generated from it.

3.5. Genetic Manipulations and Selection

The genetic manipulation refers to the change of genetic structure in biology, but, in the GA, it indicates that a "child" solution is produced from a pair of "parent" solutions by using genetic operators like crossover and mutation.

In the existing population, a pair of individuals (solutions) are selected as parents to perform the *crossover* operator to produce their offspring. More specifically, as illustrated in Figure 2, first select two individuals $[T_i \wedge D_i]_{o'_1}$ and $[T_i \wedge D_i]_{o'_2}$ from the current population as parents and get their d-chromos o'_1 and o'_2, then swap each two genetics of the two d-chromos with possibility p $(0 < p < 1)$ to obtain two new individuals.

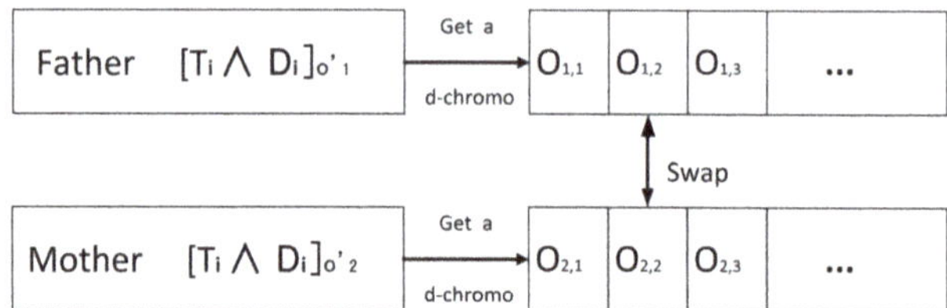

Figure 2. Crossover operator.

To perform the *mutation* operator, each genetic of an individual is mutated with possibility q $(0 < q < 1)$, as displayed in Figure 3. More clearly, for one individual $[T_i \wedge D_i]_{o'}$ with its d-chromo $o' = (o_1, \cdots, o_n, d_1, \cdots, d_c)$, each genetic of it has the possibility q to be mutated:

$o'_i := o'_i + \triangle$, where $\triangle$ is a different scalar of small value.

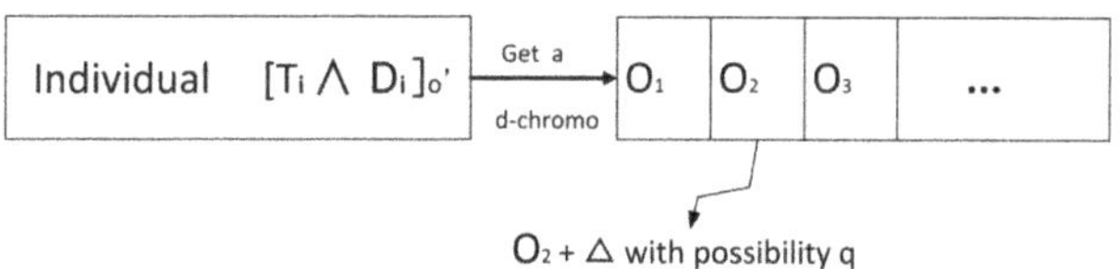

Figure 3. Mutation operator.

Fitness function *Grade* is used to evaluate an individual (a solution) $[T_i \wedge D_i]_{o'}$ by assigning a fitness value. Let $Datas = data\_suite\_from([T_i \wedge D_i]_{o'})$ where $data\_suite\_from$ generates a suite of input data from $[T_i \wedge D_i]_{o'}$ by using a constraint solver. Let $N\_kill_{i,o'} = (k_1, \ldots, k_m)$ where k_j is the number of test data that have been generated from $[T_i \wedge D_i]_{o'}$ and have killed the program mutant mu_j as well. A test case that kills a program mutant indicates that it fails based on the original specifications after it is executed by the program mutant. We consider both the killing rate of program mutants *Kill_rate*, and the killing rate of a data suite as important factors to compute the grade for $[T_i \wedge D_i]_{o'}$:

$$Grade([T_i \wedge D_i]_{o'}) = \frac{Kill\_rate(N\_kill_{i,o'}) \cdot Sum(N\_kill_{i,o'})}{(m \cdot (length(Datas) + 1))} \tag{1}$$

$$where \begin{cases} Kill\_rate(N\_kill_{i,o'}) = \dfrac{\Sigma_{j=1}^{m} I(k_j > 0)}{length(N\_kill_{i,o'})}, \\ I(k_j > 0) = \begin{cases} 1 & k_j > 0 \\ 0 & k_j \leq 0 \end{cases} \end{cases} \tag{2}$$

The factor $\Sigma_{j=1}^{m} I(k_j > 0)$ in *Kill_rate* is intended to encourage each mutated functional scenario to generate a test data suite that can kill as many different kinds of program mutants as possible. The factor $Sum(N\_kill_{i,o'})$ as a part of the numerator in *Grade* would inspire every mutated functional scenario to generate a test data suite where most test data are effective enough to kill as many program mutants as possible.

For a given chromosome $[T_i \wedge D_i]_{o'}$, its individual with appropriate d-chromo $o'_{i,best}$ is regarded as the best if and only if this individual possesses the highest value of *Grade* in the whole population. GA is to find such best individuals for these chromosomes $[T_i \wedge D_i]_{o'}$ $(i = 1, \cdots, N)$.

After all the individuals from the current population are evaluated, GA would select most of the best ones to form a new population for the next generation. This process is called *selection*. In our approach, we evaluate all the individuals and sort them by descending, then select individuals in the top 50 percent of the current population to breed the next generation.

As we can see, GA is used to find the best individuals separately for each chromosomes $[T_i \wedge D_i]_{o'}$ $(i = 1, \cdots, N)$. In order to evaluate all the best individuals represented by different chromosomes, the final formula of evaluation is made as follows:

$$Grade(\vee_{i=1}^{n} [T_i \wedge D_i]_{o'_{i,best}}) = \frac{Kill\_rate(N\_kill) \cdot Sum(N\_kill)}{(m \cdot length(Datas))} \tag{3}$$

$$where \begin{cases} N\_kill = \Sigma_{i=1}^{n} N\_kill_{i,o'_{i,best}}, \\ Datas = \{data\_suite\_from([T_i \wedge D_i]_{o'_{i,best}})\}_i \end{cases} \tag{4}$$

We use the final formula to find all the mutated functional scenarios that together hit the highest final grade (i.e., do the best in bug detection), each mutated one with well-tuned values for dummy variables and output variables. Additionally, this final grade is also used for comparison between our approach and other techniques. In the case study, our method is compared with the conventional specification-based method with respect to test data generation for bug detection.

4. Algorithm Summary

Our approach that incorporates GA accomplishes the goal of obtaining the mutated specifications by taking three key steps:

1. Inject faults into the original program to obtain a set of program mutants;
2. Use reformed specification $[T_i \wedge D_i]_{o'}$ as seed chromosomes. Each chromosome corresponds to a group of individuals that are generated by assigning concrete values to the output vector in the chromosome;
3. Apply GA to each chromosome and select the best individuals (the best mutated specifications). According to the original specifications, determine whether or not a test case from a mutated specification (a constraint over inputs) kills the program mutants.

Figure 4 displays the whole evolution process of GA. In the first round of evolution, a group of individuals are initialized and evaluated. Then, the best individuals in the top k ($k = 50\%$ in this paper) of the group are selected to perform both crossover and mutation operators to produce their offspring for the next round. In the next round, all of the individuals are evaluated and the top k of them again prepare to breed a new generation by performing genetic manipulations. The population iteratively evolves in this process until there has been no improvement in the population or it reaches the predefined maximum number of generations.

In the mutation testing, we use Z3 [44], a widely used satisfiability modulo theories (SMT) solver, as our constraint solver to generate the data suite for each individual formula (i.e., each mutated functional scenario). The generation for a data suite takes three steps: (1) use Z3 to generate a test data that satisfies the logical formula, (2) exclude all the test data obtained previously from the logical formula; (3) go to step (1) to obtain another piece of test data unless enough test data are obtained. Each individual formula is evaluated by the fitness function that measures the quality of the test data suite.

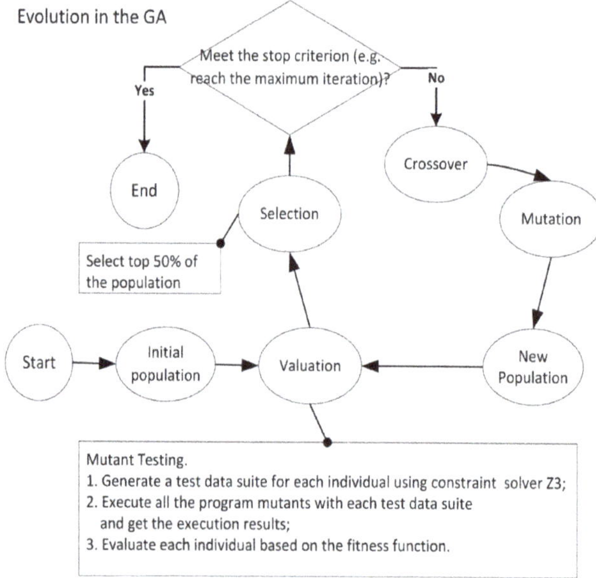

Figure 4. The evolution in GA.

We give the pseudo-code of the algorithm in Appendix A.

5. Case Study

In this section, we apply GA to two classic examples to demonstrate the effectiveness of the proposed method. The original specifications are used as test oracles for determining whether the outputs are correct or not during the evaluation of individuals.

We compare our method with the conventional method, called *original specification-based method*, which directly generates input data from the original specifications by using Z3. In the original specification-based method, neither dummy variables nor defining condition are used to generate input values, since the defining condition contains output variables with unknown values. The input data are directly generated from only test conditions (pre-condition and guard-condition over only input variables) by using Z3.

Sixteen program mutants are prepared for each program in the way that the injected faults will not cause execution crash and infinite loops since we only focus on the functional bugs in this paper. Both methods generate a test suite of the same size 20 every time to execute these program mutants in the evaluation process.

5.1. Case Study 1: Mod

In this program, process *Mod* is to find the quotient q and remainder r from dividing y by x. For *Mod*, we give its formal specification in SOFL and the implementation in Python.

The formal specification of *Mod* is:

process Mod (y: int, x: int) r: int, q: int
pre $x \neq 0$
post $x > 0 \wedge y \neq 0 \wedge y = q * x + r \wedge Abs(r) < x \wedge xr \geq 0 \vee$
$x < 0 \wedge y \neq 0 \wedge y = q * x + r \wedge Abs(r) < -x \wedge xr \geq 0 \vee$
$y = 0 \wedge q = 0 \wedge r = 0$
end_process

Its implementation in Python is:

```
def Abs(x):
if x>=0:
return x
else:
return -x
def mod(y, x):
r = y;
q = 0;
if y!=0:
if x*y > 0:
while Abs(x) <= Abs(r):
r = r - x
q = q + 1

else:
while x*r < 0:
r = r + x
q = q - 1
return r, q
```

In this specification, *Abs* is a function for calculating the absolute value of its input. To shorten the explanation of each step, assume *Abs* is an inline executable predicate. Both $-7\ mod\ 5 = 3$ and $-7\ mod\ 5 = -2$ satisfy the classic definition $y = q * x + r \wedge Abs(r) < Abs(x)$. To avoid the ambiguity, the specification of *Mod* puts an additional condition $xr \geq 0$ in order to get only one result of $-7\ mod\ 5 = 3$.

In the specification, the pre-condition, guard-conditions, and defining conditions are listed as:

$S_{pre} := x \neq 0;$
$G_1 := x > 0 \wedge y \neq 0; D_1 := y = q * x + r \wedge Abs(r) < x \wedge xr \geq 0;$
$G_2 := x < 0 \wedge y \neq 0; D_2 := y = q * x + r \wedge Abs(r) < -x \wedge xr \geq 0;$
$G_3 := x > 0 \wedge y = 0; D_3 := q = 0 \wedge r = 0.$

We can obtain the functional scenarios $T_i \wedge D_i := S_{pre} \wedge G_i \wedge D_i$ as follows:
$T_1 \wedge D_1 := x > 0 \wedge y \neq 0 \wedge y = q * x + r \wedge Abs(r) < x \wedge xr \geq 0;$
$T_2 \wedge D_2 := x < 0 \wedge y \neq 0 \wedge y = q * x + r \wedge Abs(r) < -x \wedge xr \geq 0;$
$T_3 \wedge D_3 := x \neq 0 \wedge y = 0 \wedge q = 0 \wedge r = 0.$

For $T_3 \wedge D_3$, the input x and y are not related to the output q and r, so we do not need to apply GA to it. Since there is an equality $y = q * x + r$ in which inputs and outputs are related, we introduce two dummy variables d_1 and d_2. The chromosomes of *Mod* are displayed in Table 1.

Table 1. Chromosome forms for functional scenarios of process Mod.

Chromosome	D-Chromo	Dummy Vars
$[T_1 \wedge D_1]_{o'} : x > 0 \wedge y \neq 0 \wedge$ $d_1 \leq q * x + r - y \leq d_2 \wedge$ $Abs(r) < x \wedge xr \geq 0$	$o' =$ (q, r, d_1, d_2)	d_1, d_2
$[T_2 \wedge D_2]_{o'} : x < 0 \wedge y \neq 0 \wedge$ $d_1 \leq q * x + r - y \leq d_2 \wedge$ $Abs(r) < -x \wedge xr \geq 0$	$o' =$ (q, r, d_1, d_2)	d_1, d_2

Apply Algorithm A1 to these chromosomes. The results are displayed in Table 2.

Table 2. Results for process Mod after applying GA.

The Best Individual of Chromosome	Grade
$[T_1 \wedge D_1]_{o'}$ $o' = (q, r, d_1, d_2)$ $o'_{1,best} = (-4, 0, -6, -6)$	0.58
$[T_2 \wedge D_2]_{o'}$ $o' = (q, r, d_1, d_2)$ $o'_{2,best} = (-7, 0, 9, 13)$	0.55
$Total : \vee_{i=1}^{2} [T_i \wedge D_i]_{o'_{i,best}}$	0.59

To illustrate the effectiveness of data generation from the mutated specifications, Table 3 displays the results of the conventional method that generates the test data directly from the original specifications. For the original specification, we generate test data only from the test condition T_i consisting of both pre-condition S_{pre} and guard-condition G_i meanwhile ignoring the defining condition D_i because the defining condition D_i involves unknown output variables that can not directly help to generate test data.

Table 3. Results for process Mod with original specifications.

Original Specification	Grade
$T_1 : x > 0 \wedge y \neq 0$	0.32
$T_2 : x < 0 \wedge y \neq 0$	0.38
$Total : \vee_{i=1}^{2} T_i$	0.37

For the proposed method, the final *Kill_rate* of $\vee_{i=1}^{2}[T_i \wedge D_i]_{o'_{i,best}}$ is 100%, the same as the conventional method. It means that every program mutant has been killed by at least one piece of test data. The corresponding final *Grade* is 0.59, larger than the *Grade* of 0.37 with the original specification-based method, indicating that the test data generated from $\vee_{i=1}^{2}[T_i \wedge D_i]_{o'_{i,best}}$ are of high quality that are more likely to kill all the program mutants. The result suggests that it is plausible to use these best individuals of chromosomes to make four mutated specifications for test case generation in the further maintenance of the original program.

Comparing the reformed specifications with the original ones in Figure 5, we can find the *Grade* of reformed ones that are always larger than that of original ones. It means that the data suite generated from the mutated specifications is more likely to pinpoint bugs than that of original ones, although both of them share the same *Kill_rate* of 100%.

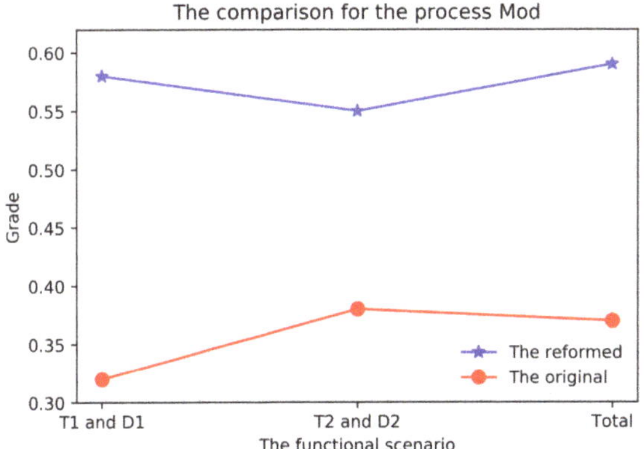

Figure 5. The grade of the mutated and original.

The Effect with Dummy Variables

We conduct additional experiments to figure out how dummy variables introduced into the different parts of defining conditions would affect the quality of the obtained mutated specifications. We make three versions of modifications to our approach as follows:

1. Version *V1*: Introducing dummy variables into only inequality relation;
2. Version *V2*: Introducing dummy variables into both equality and inequality relation;
3. Version *V3*: Putting no dummy variables in defining conditions.

For convenience, the approach with no modification, that is, with dummy variables for only equality relation, is called Version *V0*.

The previous experimental result for *V0* and the original, as well as the results from after applying variations of the approach *V1*,*V2*, and *V3* to process *Mod*, are together displayed in Figure 6.

According to Figure 6, three approaches with dummy variables *V0*, *V1*, and *V2* gain higher *Grade*s than the approach without dummy variables *V3*, and even *V3* seems to behave better than the conventional method. There are no significant differences between the evaluation of *V0* and *V2*. However, *V2* would occupy more computation resources than *V0* due to the consideration of more dummy variables. It seems that *V1* gains a little better final *Grade* than *V0*, though its *Grade* for each single mutated functional scenario is not good enough.

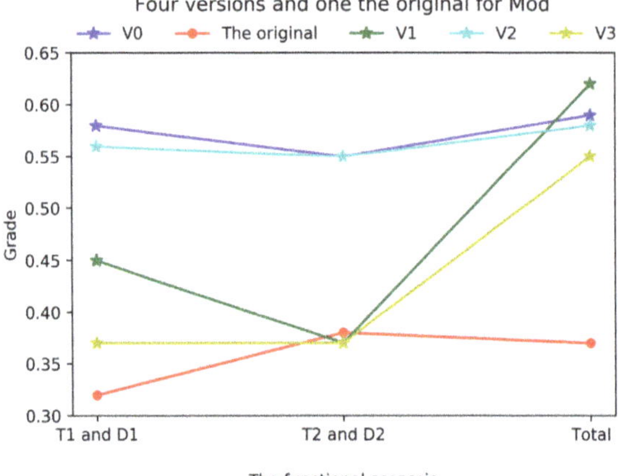

Figure 6. Results by four versions and the original for Mod.

In addition, by using an approach without dummy variables *V1* and *V3*, every obtained single mutated functional scenario demonstrates the strong capability to kill some specific program mutants while leaving other program mutants not killed, though the combination of all the functional scenarios in *V1* can reach 100% total *Kill_rate* while, for *V3*, the total *Kill_rate* unfortunately remains in 87.5%. This result demonstrates the importance of introducing dummy variables into equality relation in order to accomplish both good single and total *Grade*s and *Kill_rate*s.

In summary, it is necessary to introduce dummy variables into equality relations, and the additional dummy variables for inequality relation cannot significantly improve the proposed approach.

5.2. Case Study 2: Gcd

Process *gcd* is to compute the greatest common divisor of two inputs by using Stein's algorithm.

The formal specifications of *gcd* is:

process gcd (x: int, y: int) r: int
pre $x \geq 0 \wedge y \geq 0$
post $x > 0 \wedge y > 0 \wedge x \geq y \wedge r = gcd(y, x\%y) \vee$
$x > 0 \wedge y > 0 \wedge x < y \wedge r = gcd(y, y\%x) \vee$
$y = 0 \wedge r = x \vee$
$x = 0 \wedge r = y$
end_process

The implementation of process gcd in Python is:

```
def gcd(x, y):
if x < y:
x, y = y, x
if (0 == y):
return x
if x % 2 == 0 and y % 2 == 0:
return 2 * gcd(x//2, y//2)
if x % 2 == 0:
return gcd(x // 2, y)
if y % 2 == 0:
return gcd(x, y // 2)
return gcd((x - y) // 2, y)
```

Process *gcd* is a recursive process and its post-condition contains itself, so it is difficult to generate data from this kind of post-condition. We transform the original post-condition to the following ones:

$T_1 \wedge D_1 := x > 0 \wedge y > 0 \wedge x \geq y \wedge x\%r = 0 \wedge y\%r = 0 \wedge x\%y\%r = 0;$
$T_2 \wedge D_2 := x > 0 \wedge y > 0 \wedge x < y \wedge x\%r = 0 \wedge y\%r = 0 \wedge y\%x\%r = 0;$
$T_3 \wedge D_3 := x \geq 0 \wedge y = 0 \wedge r = x;$
$T_4 \wedge D_4 := y \geq 0 \wedge x = 0 \wedge r = y.$

Table 4 shows the chromosomes of process *gcd*.

Apply the algorithm to all of the chromosomes; in the meantime, make use of the original post-condition to determine whether the outputs of codes are correct or not. The results are displayed in Table 5.

The final *Kill_rate* of $\vee_{i=1}^{4}[T_i \wedge D_i]_{o'_{i,best}}$ is 100%. The corresponding *Grade* is 0.46, which means roughly 46 percent of test data that are randomly generated from $\vee_{i=1}^{4}[T_i \wedge D_i]_{o'_{i,best}}$ can kill all the program mutants.

Table 4. Chromosome forms for functional scenarios of process gcd.

Chromosome	D-Chromo	Dummy Vars
$[T_1 \wedge D_1]_{o'} : x \geq y \wedge (d_1 \leq x\%r \leq d_2) \wedge (d_3 \leq y\%r \leq d_4) \wedge (d_5 \leq x\%y\%r \leq d_6)$	$o' = (r, d_1, d_2, d_3, d_4, d_5, d_6)$	$d_1, d_2, d_3, d_4, d_5, d_6$
$[T_2 \wedge D_2]_{o'} : x < y \wedge (d_1 \leq x\%r \leq d_2) \wedge (d_3 \leq y\%r \leq d_4) \wedge (d_5 \leq y\%x\%r \leq d_6)$	$o' = (r, d_1, d_2, d_3, d_4, d_5, d_6)$	$d_1, d_2, d_3, d_4, d_5, d_6$
$[T_3 \wedge D_3]_{o'} : y = 0 \wedge (d_1 \leq x - r \leq d_2)$	$o' = (r, d_1, d_2)$	d_1, d_2
$[T_4 \wedge D_4]_{o'} : x = 0 \wedge (d_1 \leq y - r \leq d_2)$	$o' = (r, d_1, d_2)$	d_1, d_2

Table 5. Results for process gcd after applying GA.

The Best Individual of Chromosome	Grade
$[T_1 \wedge D_1]_{o'}$ $o'_{1,best} = (0, 1, 10, 10, 10, 3, 6)$	0.72
$[T_2 \wedge D_2]_{o'}$ $o'_{2,best} = (8, 4, 6, 2, 2, 0, 8)$	0.58
$[T_3 \wedge D_3]_{o'}$ $o'_{3,best} = (5, -1, 20)$	0.0037
$[T_4 \wedge D_4]_{o'}$ $o'_{4,best} = (4, -3, 16)$	0.0037
$Total : \vee_{i=1}^{4}[T_i \wedge D_i]_{o'_{i,best}}$	0.46

Conversely, the result for applying the method that generates test data directly from the original specification is displayed in Table 6.

Table 6. Results for process gcd with original specifications.

Original Specification	Grade
$T_1 : x > 0 \wedge y > 0 \wedge x \geq y$	0.29
$T_2 : x > 0 \wedge y > 0 \wedge x < y$	0.49
$T_3 : x \geq 0 \wedge y = 0$	0.0035
$T_4 : y \geq 0 \wedge x = 0$	0.0035
$Total : \vee_{i=1}^{2} T_i$	0.25

Comparing the reformed specifications with the original ones in Figure 7, we can find that the first two reformed ones $[T_1 \wedge D_1]_{o'}$ and $[T_2 \wedge D_2]_{o'}$ have very high values of *Grade*, 0.72 and 0.58, respectively, higher than 0.29 and 0.49 with the original specifications. In addition, the *Kill_rate* of a sole $[T_i \wedge D_i]_{o'}$ $(i = 1,2)$ is 94%, indicating that the test data generated from the first two reformed specifications are likely to pinpoint most bugs probably occurring in the program. Only a few program mutants (6% of total), with some faults that directly relates to the last two functional scenarios $T_3 \wedge D_3$ and $T_4 \wedge D_4$ (where $x = 0$ or $y = 0$), cannot be killed by the test data generated from either the first two reformed specifications or the first two original specifications. Due to the very simple forms and the limited functionality of the last two functional scenarios, there is no improvement of test data generation using our method against the original ones.

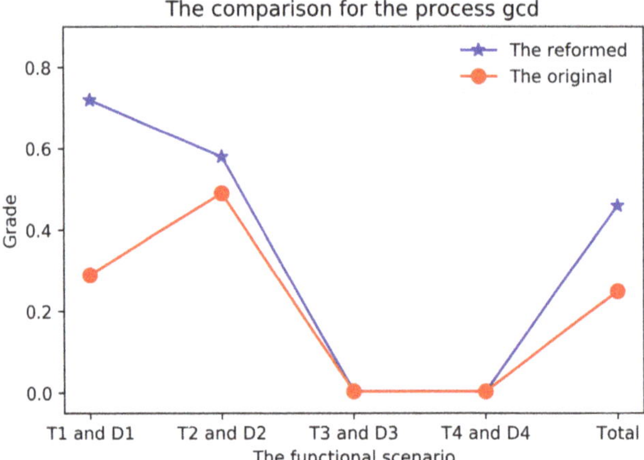

Figure 7. The grade of the reformed and original.

The results from both classic examples demonstrate that the input data generated from the mutated specifications are more likely to kill the mutants of programs than that from the original specifications.

The Effect without Dummy Variables

Like what we have done for process *Mod*, we conduct additional experiments with the approach without using dummy variable V3 since *gcd* only has equality relations in the defining conditions. The results are shown in Figure 8.

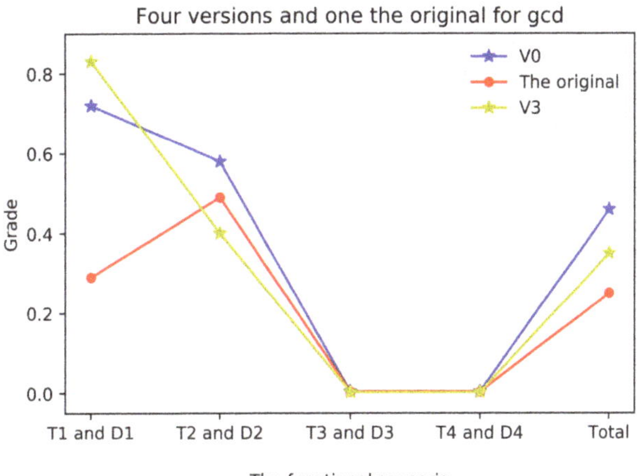

Figure 8. Results by four versions and the original for gcd.

The experimental results are similar to that in *Mod*. $V0$ performs better than $V3$. $V3$ still encounters the problem that every single mutated functional scenario is not able to kill all the program mutants. It shows that the test data generated from those strict equality relations are less likely to trigger some bad behaviors of program.

5.3. Complexity of Our Approach

We present an abstract analysis of the complexity for our approach. Generally, a GA complexity is on the order of $O(g * n * m)$ without the effect of the fitness function, where g is the number of generations, n is the population size, and m is the number of functional scenarios. Since our approach uses a fitness function involved in the mutation testing, we should take both the program execution time and the data suite generation time into consideration.

As the speed of the constraint solver to solve an individual formula (to generate a test suite) depends on the complexity of the functional scenarios (logical formulas) whose complexity cannot be easily determined, we associate the cost of using the constraint solver for a singular individual with the number of input variables in, the number of output variables out, and the number of dummy variables d. In addition, the number of dummy variables relies on the number of equality relation in each functional scenario, which varies in different kinds of programs. We simply assume that each functional scenario has at least one equality relation. Thus, the complexity of using the constraint solver for each individual is $O(in + out + 2 * d * m)$. Moreover, the complexity of all the executions for program mutants is approximately $O(mu * sui)$ with mu the number of program mutants and sui the size of test data suite. Finally, considering the complexity of the GA together with the mutation testing, the complexity for our approach is

$$O(g * n * ((in + out + 2 * d * m) * (mu * sui)) * m).$$

6. Conclusions

We propose a new method for effective test data generation based on mutated pre-post style formal specifications. The method is characterized by the integration of the functional scenario-based testing, a genetic algorithm and the mutation testing. In the approach, by assigning appropriate values to the unknown output and dummy variables to the variations for the original specifications, we can obtain useful mutated specifications that are sensitive to small syntactic structural changes of program codes.

We have also carried out two classic cases to evaluate the performance of our method. The results of case studies demonstrate that, for a complicated functional scenario, the proposed approach is capable of effectively generating useful test data to kill as many program mutants as possible, which outperforms the conventional data generation method.

In spite of the advantages of our method as mentioned above, there are also some limitations and disadvantages in the application of our method. First, the proposed method can only work on arithmetical relationships between inputs and outputs in which outputs affect the generation of inputs. Second, as the GA usually iterates many times and executes all the program mutants for every iteration, the cost would not be low. However, if we have enough computing resources for applying our method, it might be worth taking time to obtain good reformed specifications for the further maintenance of software.

In order to cope well with complex real programs, some additional extensions can be made in our approach. Firstly, by using the character encoding standard like US-ASCII [45], we can convert a String to a byte array so that the relationship that contains string variables can also be manipulated by our method. Moreover, since many research works exist concerning about the techniques of encoding complex data [46–48] that may occur in specifications like images and videos, it is possible to transform these specifications into appropriate arithmetical relationships so that our approach can be used in such cases. Secondly, although there exist specifications where the input and output variables are not specified by some explicit arithmetical equality relation, our method would still be applicable. Because instead of directly using these specifications, we can design some mutated arithmetical relationships (in form of inequality) of input and output that can not only approximate to the real properties of program but also leave open the possibility of occurrence of unexpected behaviors. Thirdly, when testing a big complex system, we can decompose it into a set of subroutines and focus on testing small procedures one by one using our approach. Thus, there is no need to repeatedly executing the whole system with our algorithm.

In future work, we will focus on enhancing the capability of this method to deal with more kinds of relationships between inputs and outputs where the values of outputs may not directly determine the inputs. We will conduct more experiments to ensure that our method can be well used in different kinds of programs.

Author Contributions: Conceptualization, R.W., S.L., and Y.S.; methodology, R.W. and Y.S.; investigation, R.W.; resources, R.W.; data curation, R.W.; writing—original draft preparation, R.W.; writing—review and editing, S.L. and Y.S.; visualization, R.W.; supervision, S.L. and Y.S.; project administration, S.L. and Y.S.; funding acquisition, S.L. and Y.S. All authors have read and agreed to the published version of the manuscript.

Funding: This work was funded by JSPS KAKENHI Grant No. 26240008.

Conflicts of Interest: The authors declare no conflict of interest.

Appendix A

In Algorithm A1, the function *one_step* creates a new population with fitness values from the previous population through applying crossover and mutation operations; and the function *do_valuation* assigns fitness values calculated by the function *Grade* to all of the individuals by using the feedback of testing program mutants.

Algorithm A1 GA to obtain mutated specifications.

Inputs: the functional scenarios from the specification: $T_i \wedge D_i$
Individuals: $o' = (o_{1,} \cdots, o_n, d_{1,\dots,} d_m)$ with concrete values
Outputs: the reformed specification $[T_i \wedge D_i]_{o'}$

```
run(){
   result = list()
   for [T_i ∧ D_i]_o' in functional scenarios:
       spec = T_i ∧ D_i
       population = initial(o')
       while(not enough iterations){
            one_step(spec)
       }
       best_individual =
                select_best_individual(population)
       reformed_specification = (spec, best_individual)
       result.append(reformed_specification)
}
one_step(spec) {
   # This function selects top 50% of the current population
   population = keep_good_individuals(population)
   do:
       father, mother = random_select_two(population)
       child1, child2 = crossover_operation(father,mother)
       child1, child2 = mutation_operation(child1,child2)
       population.put(child1,child2)
   until population increases enough
   do_valuation(population,spec)
}
do_valuation(population,spec){
   for individual in population:
       datas = data_suite_from(individual,spec)
       statistic_sum = kill_program_mutants(datas)
       individual.value = Grade(statistic_sum)
}
```

References

1. Wong, W.E.; Horgan, J.R.; London, S.; Agrawal, H. A study of effective regression testing in practice. In Proceedings of the Eighth International Symposium on Software Reliability Engineering, Albuquerque, NM, USA, 2–5 November 1997; pp. 264–274.
2. Leung, H.K.; White, L. Insights into regression testing (software testing). In Proceedings of the Conference on Software Maintenance, Miami, FL, USA, 16–19 October 1989; pp. 60–69.
3. Kazmi, R.; Jawawi, D.N.; Mohamad, R.; Ghani, I. Effective regression test case selection: A systematic literature review. *ACM Comput. Surv. (CSUR)* **2017**, *50*, 1–32. [CrossRef]
4. Stocks, P.; Carrington, D. A framework for specification-based testing. *IEEE Trans. Softw. Eng.* **1996**, 777–793. [CrossRef]
5. Richardson, D.; O'Malley, O.; Tittle, C. *Approaches to Specification-Based Testing*; ACM: New York, NY, USA, 1989; Volume 14.
6. Khurshid, S.; Marinov, D. TestEra: Specification-based testing of Java programs using SAT. *Autom. Softw. Eng.* **2004**, *11*, 403–434. [CrossRef]
7. Hierons, R.M.; Bogdanov, K.; Bowen, J.P.; Cleaveland, R.; Derrick, J.; Dick, J.; Gheorghe, M.; Harman, M.; Kapoor, K.; Krause, P.; et al. Using formal specifications to support testing. *ACM Comput. Surv. (CSUR)* **2009**, *41*, 1–76. [CrossRef]
8. Dokhanchi, A.; Hoxha, B.; Fainekos, G. Formal requirement debugging for testing and verification of cyber-physical systems. *ACM Trans. Embed. Comput. Syst. (TECS)* **2017**, *17*, 1–26. [CrossRef]
9. Offutt, A.J.; Liu, S. Generating Test Data from SOFL Specifications. *J. Syst. Softw.* **1999**, *49*, 49–62. [CrossRef]
10. Dick, J.; Faivre, A. Automating the generation and sequencing of test cases from model-based specifications. In Proceedings of the International Symposium of Formal Methods Europe, Odense, Denmark, 19–23 April 1993; Springer: Berlin/Heidelberg, Germany, 1993; pp. 268–284.

11. Ed-Douibi, H.; Izquierdo, J.L.C.; Cabot, J. Automatic generation of test cases for REST APIs: A specification-based approach. In Proceedings of the 2018 IEEE 22nd International Enterprise Distributed Object Computing Conference (EDOC), IEEE, Stockholm, Sweden, 16–19 October 2018; pp. 181–190.
12. Alrawashed, T.A.; Almomani, A.; Althunibat, A.; Tamimi, A. An Automated Approach to Generate Test Cases From Use Case Description Model. *Comput. Model. Eng. Sci.* **2019**, *119*, 409–425. [CrossRef]
13. Jones, C.B. *Systematic Software Development Using VDM*; Citeseer: Princeton, NJ, USA, 1990; Volume 2.
14. Larsen, P.G.; Battle, N.; Ferreira, M.; Fitzgerald, J.; Lausdahl, K.; Verhoef, M. The overture initiative integrating tools for VDM. *ACM SIGSOFT Softw. Eng. Notes* **2010**, *35*, 1–6. [CrossRef]
15. Tran-Jørgensen, P.W.; Nilsson, R.S.; Lausdahl, K. Enhancing Testing of VDM-SL models. In Proceedings of the 16th Overture Workshop, Oxford, UK, 14 July 2018; pp. 7–22.
16. Liu, S. *Formal Engineering for Industrial Software Development: Using the SOFL Method*; Springer Science & Business Media: Berlin, Germany, 2013.
17. Liu, S.; Nakajima, S. Combining Specification Testing, Correctness Proof, and Inspection for Program Verification in Practice. In Proceedings of the 3rd International Workshop on SOFL + MSVL (SOFL+MSVL 2013), LNCS 8332, Queenstown, New Zealand, 29 October 2013; Springer: Queenstown, New Zealand, 2013; pp. 3–16.
18. Wang, R.; Sato, Y.; Liu, S. Specification-based Test Case Generation with Genetic Algorithm. In Proceedings of the 2019 IEEE Congress on Evolutionary Computation (CEC), Wellington, New Zealand, 10–13 June 2019; pp. 1382–1389.
19. Rapps, S.; Weyuker, E.J. Selecting software test data using data flow information. *IEEE Trans. Softw. Eng.* **1985**, 367–375. [CrossRef]
20. Weyuker, E.J. More experience with data flow testing. *IEEE Trans. Softw. Eng.* **1993**, *19*, 912–919. [CrossRef]
21. Khedker, U.; Sanyal, A.; Sathe, B. *Data Flow Analysis: Theory and Practice*; CRC Press: Boca Raton, FL, USA, 2017.
22. Pargas, R.P.; Harrold, M.J.; Peck, R.R. Test-data generation using genetic algorithms. *Softw. Test. Verif. Reliab.* **1999**, *9*, 263–282. [CrossRef]
23. Girgis, M.R. Automatic Test Data Generation for Data Flow Testing Using a Genetic Algorithm. *J. UCS* **2005**, *11*, 898–915.
24. Girgis, M.R.; Ghiduk, A.S.; Abd-Elkawy, E.H. Automatic generation of data flow test paths using a genetic algorithm. *Int. J. Comput. Appl.* **2014**, *89*, 29–36.
25. Nayak, N.; Mohapatra, D.P. Automatic test data generation for data flow testing using particle swarm optimization. In Proceedings of the International Conference on Contemporary Computing, Noida, India, 9–11 August 2010; Springer: Cham, Switzerland, 2010; pp. 1–12.
26. Biswas, S.; Kaiser, M.S.; Mamun, S. Applying ant colony optimization in software testing to generate prioritized optimal path and test data. In Proceedings of the 2015 International Conference on Electrical Engineering and Information Communication Technology (ICEEICT), IEEE, Dhaka, Bangladesh, 21–23 May 2015; pp. 1–6.
27. Harman, M.; Jia, Y.; Langdon, W.B. Strong higher order mutation-based test data generation. In Proceedings of the 19th ACM SIGSOFT Symposium and the 13th European Conference on Foundations of Software Engineering, ACM, Szeged, Hungary, 5–9 September 2011; pp. 212–222.
28. Papadakis, M.; Kintis, M.; Zhang, J.; Jia, Y.; Le Traon, Y.; Harman, M. Mutation testing advances: An analysis and survey. In *Advances in Computers*; Elsevier: Amsterdam, The Netherlands, 2019; Volume 112, pp. 275–378.
29. Ma, Y.S.; Offutt, J.; Kwon, Y.R. MuJava: An automated class mutation system. *Softw. Test. Verif. Reliab.* **2005**, *15*, 97–133. [CrossRef]
30. Derezinska, A.; Kowalski, K. Object-oriented mutation applied in common intermediate language programs originated from c. In Proceedings of the 2011 IEEE Fourth International Conference on Software Testing, Verification and Validation Workshops, Berlin, Germany, 21–25 March 2011; pp. 342–350.
31. Delgado-Pérez, P.; Medina-Bulo, I.; Palomo-Lozano, F.; García-Domínguez, A.; Domínguez-Jiménez, J.J. Assessment of class mutation operators for C++ with the MuCPP mutation system. *Inf. Softw. Technol.* **2017**, *81*, 169–184. [CrossRef]
32. Jackson, D. Alloy: A lightweight object modelling notation. *ACM Trans. Softw. Eng. Methodol. (TOSEM)* **2002**, *11*, 256–290. [CrossRef]
33. Sullivan, A.; Wang, K.; Zaeem, R.N.; Khurshid, S. Automated test generation and mutation testing for Alloy. In Proceedings of the 2017 IEEE International Conference on Software Testing, Verification and Validation (ICST), Tokyo, Japan, 13–17 March 2017; pp. 264–275.
34. Martins, E.; Sabião, S.B.; Ambrosio, A.M. ConData: A tool for automating specification-based test case generation for communication systems. *Softw. Qual. J.* **1999**, *8*, 303–320. [CrossRef]
35. McMillan, K.L.; Zuck, L.D. Formal specification and testing of QUIC. In *Proceedings of the ACM Special Interest Group on Data Communication*; ACM: New York, NY, USA, 2019; pp. 227–240.
36. Ali, S.; Iqbal, M.Z.; Arcuri, A.; Briand, L.C. Generating test data from OCL constraints with search techniques. *IEEE Trans. Softw. Eng.* **2013**, *39*, 1376–1402. [CrossRef]
37. Jalila, A.; Mala, D.J. Automated optimal test data generation for OCL specification using harmony search algorithm. *Int. J. Bus. Intell. Data Min.* **2020**, *16*, 231–259. [CrossRef]
38. Holland, J.H. *Adaptation in Natural and Artificial Systems: An Introductory Analysis with Applications to Biology, Control, and Artificial Intelligence*; MIT Press: Cambridge, MA, USA, 1992.

39. DeMillo, R.A.; Lipton, R.J.; Sayward, F.G. Hints on test data selection: Help for the practicing programmer. *Computer* **1978**, *11*, 34–41. [CrossRef]
40. Luo, J.; Liu, S.; Wang, Y.; Zhou, T. Applying SOFL to a railway interlocking system in industry. In Proceedings of the International Workshop on Structured Object-Oriented Formal Language and Method, Tokyo, Japan, 15 November 2016; Springer: Cham, Switzerland, 2016; pp. 160–177.
41. Liu, S. *Formal Engineering for Industrial Software Development Using the SOFL Method*; Springer: Berlin, Germany, 2004; ISBN 3-540-20602-7.
42. Sen, K. Concolic testing. In Proceedings of the Twenty-Second IEEE/ACM International Conference on Automated Software Engineering, ACM, Atlanta, GA, USA, 5–9 November 2007; pp. 571–572.
43. Sato, Y.; Sugihara, T. Automatic Generation of Specification-Based Test Cases by Applying Genetic Algorithms in Reinforcement Learning. In Proceedings of the International Workshop on Structured Object-Oriented Formal Language and Method, Paris, France, 6 November 2015; Springer: Cham, Switzerland, 2015; pp. 59–71.
44. De Moura, L.; Bjørner, N. Z3: An efficient SMT solver. In Proceedings of the International conference on Tools and Algorithms for the Construction and Analysis of Systems, Budapest, Hungary, 29 March–6 April 2008; Springer: Cham, Switzerland, 2008; pp. 337–340.
45. Mackenzie, C.E. *Coded-Character Sets: History and Development*; Addison-Wesley Longman Publishing Co., Inc.: Boston, MA, USA, 1980.
46. Basavaprasad, B.; Ravi, M. A study on the importance of image processing and its applications. *IJRET Int. J. Res. Eng. Technol.* **2014**, *3*, 1.
47. Barannik, V.; Podlesny, S.; Tarasenko, D.; Barannik, D.; Kulitsa, O. The video stream encoding method in infocommunication systems. In Proceedings of the 14th International Conference on Advanced Trends in Radioelecrtronics, Telecommunications and Computer Engineering (TCSET), IEEE, Lviv-Slavske, Ukraine, 20–24 February 2018; pp. 538–541.
48. Hur, T.; Bang, J.; Huynh-The, T.; Lee, J.; Kim, J.I.; Lee, S. Iss2Image: A novel signal-encoding technique for CNN-based human activity recognition. *Sensors* **2018**, *18*, 3910. [CrossRef] [PubMed]

 *mathematics*

Article

A Divide and Conquer Approach to Eventual Model Checking

Moe Nandi Aung [1,†], Yati Phyo [2,†], Canh Minh Do [2,†] and Kazuhiro Ogata [2,*,†]

1 Faculty of Information Science, University of Information Technology (UIT), Hlaing Township, Yangon PO 11052, Myanmar; moenandiaung@uit.edu.mm
2 School of Information Science, Japan Advanced Institite of Science and Technology (JAIST), Nomi, Ishikawa 923-1292, Japan; yatiphyo@jaist.ac.jp (Y.P.); canhdominh@jaist.ac.jp (C.M.D.)
* Correspondence: ogata@jaist.ac.jp
† These authors contributed equally to this work.

Abstract: The paper proposes a new technique to mitigate the state of explosion in model checking. The technique is called a divide and conquer approach to eventual model checking. As indicated by the name, the technique is dedicated to eventual properties. The technique divides an original eventual model checking problem into multiple smaller model checking problems and tackles each smaller one. We prove a theorem that the multiple smaller model checking problems are equivalent to the original eventual model checking problem. We conducted a case study that demonstrates the power of the proposed technique.

Keywords: eventual property; model checking; Maude

Citation: Aung, M.N.; Phyo, Y.; Do, C.M.; Ogata, K. A Divide and Conquer Approach to Eventual Model Checking. *Mathematics* **2021**, *9*, 368. https://doi.org/10.3390/math9040368

Academic Editor: Tadashi Dohi

Received: 17 January 2021
Accepted: 8 February 2021
Published: 12 February 2021

Publisher's Note: MDPI stays neutral with regard to jurisdictional clai-ms in published maps and institutio-nal affiliations.

1. Introduction

Model checking is an attractive and promising formal verification technique because it is possible to automatically conduct model checking experiments once good concise formal models are made. It has also been used in industries, especially hardware industries. There are still some challenges to tackle in model checking, one of which is the state explosion, the most annoying one. Many techniques to mitigate the state explosion have been devised, such as symbolic model checking [1] and SAT-based bounded model checking (BMC) [2], where SAT stands for Boolean satisfiability problem. As those existing techniques are not enough to deal with the state explosion, it is still worth tackling the issue.

Moe Nandi Aung et al. [3] tried to check that an autonomous vehicle intersection control protocol [4] enjoyed some desired properties, where there were 13 vehicles, and encountered the notorious state space explosion, making it impossible to conduct the model checking experiments. Note that it was possible to conduct the model checking experiments for a case wherein there were five vehicles. One property is the starvation freedom property that can be expressed as an eventual property. An informal description of the starvation freedom property is that every vehicle will pass the intersection concerned. The case motivated us to come up with the technique proposed in the present paper.

The present paper proposes a divide and conquer approach to eventual model checking. The technique splits the reachable state space from each initial state into $L+1$ layers, where $L \geq 1$, generating multiple smaller sub-state spaces, dividing the original eventual mode checking problem into multiple smaller model checking problems and tackling each smaller one. As the name indicates, the technique proposed in the present paper is dedicated to eventual properties. Many important software requirements can be expressed as eventual properties. For example, halting is one important requirement many programs should enjoy. Halting can be expressed as an eventual property. We prove a theorem that the multiple smaller model checking problems are equivalent to the original eventual model checking problem. We conducted a case study that demonstrates the power of the proposed technique. Maude [5] was used as the formal specification language and Maude LTL (linear temporal logic) model checker was used as the model checker.

The model checking algorithm adopted by Maude LTL model checker is the same as the one used by SPIN [6], which is one of the most popular model checkers for model checking software systems. It has been reported that Maude LTL model checker is comparable with SPIN with respect to model checking running performance. This implies that whenever Maude LTL model checker encounters the state space explosion problem, making it impossible to conduct model checking experiments, SPIN does so as well, and so do most existing model checkers. The proposed technique aims at mitigating the state space explosion problem and we demonstrate that it can mitigate the problem through a case study. We are allowed to use Maude as a formal specification language for systems under model checking. Maude is extremely expressive because it is one direct descendant of and OBJ language family, such as OBJ3 [7] and CafeOBJ [8]. Inductively-defined data structures, associative and/or commutative binary operators, etc., can be used in systems' specifications under model checking with the Maude LTL model checker. Inductively-defined data structures and associative and/or commutative binary operators cannot be used in systems' specifications under model checking for most existing model checkers, such as SPIN and NuSMV [9]. This is mainly why we used the Maude LTL model checker. Those who are more interested in the flavor of the Maude LTL model checker are recommended to see the paper [10] in which the Maude LTL model checker is intensively compared with the Symbolic Analysis Laboratory (SAL) [11], a collection of model checkers.

The remaining part of the paper is organized as follows. Section 2 explains some preliminaries, such as Kripke structures and LTL. Section 3 uses a simple example to outline the proposed technique. Section 4 describes the theoretical part of the proposed technique. Section 5 describes the proposed technique. Section 6 reports on a case study. Section 7 mentions some existing related work. Section 8 concludes the paper and suggests some future directions.

2. Preliminaries

This section describes some preliminaries needed to read the technical contents of the paper. We give the definitions of Kripke structures, the syntax of LTL formulas and the semantics of LTL formulas. We need infinite sequences of states (called paths of Kripke structure) to define the semantics of LTL formulas. We introduce several notations or symbols for paths, sets of paths and satisfaction relations, where satisfaction relations are the essence of the semantics of LTL formulas. We prepared tables for those notations or symbols. We use the symbol $\triangleq$ as "if and only if" or "be defined as."

Definition 1 (Kripke structures). *A Kripke structure* $\boldsymbol{K} \triangleq \langle \boldsymbol{S}, \boldsymbol{I}, \boldsymbol{T}, \boldsymbol{A}, \boldsymbol{L} \rangle$ *consists of a set* $\boldsymbol{S}$ *of states, a set* $\boldsymbol{I} \subseteq \boldsymbol{S}$ *of initial states, a left-total binary relation* $\boldsymbol{T} \subseteq \boldsymbol{S} \times \boldsymbol{S}$ *over states, a set* $\boldsymbol{A}$ *of atomic propositions and a labeling function* $\boldsymbol{L}$ *whose type is* $\boldsymbol{S} \rightarrow 2^{\boldsymbol{A}}$*. An element* $(s, s') \in \boldsymbol{T}$ *is called a (state) transition from* s *to* s' *and may be written as* $s \rightarrow_{\boldsymbol{K}} s'$*.*

$\boldsymbol{S}$ does not need to be finite. The set $\boldsymbol{R}$ of reachable states is inductively defined as follows: $\boldsymbol{I} \subseteq \boldsymbol{R}$ and if $s \in \boldsymbol{R}$ and $(s, s') \in \boldsymbol{T}$, then $s' \in \boldsymbol{R}$. We suppose that $\boldsymbol{R}$ is finite. $\boldsymbol{K}$ in $s \rightarrow_{\boldsymbol{K}} s'$ may be omitted if it is clear from the context.

An infinite sequence of states is a sequence that consists of states infinitely many times, where infinitely many copies of some states may occur. Let $s_0, s_1, \ldots, s_i, s_{i+1}, \ldots$ be an infinite sequence of states, where s_0 is the top element (called 0th element), s_1 is the next element (called 1st element) and s_i is the ith element. As we suppose that $\boldsymbol{R}$ is finite, if $s_0 \in \boldsymbol{R}$, then $s_0, s_1, \ldots, s_i, s_{i+1}, \ldots$ only consists of bounded number of different states, although infinitely many copies of some states occur. As usual, let ∞ be used to denote the infinity.

An infinite sequence $s_0, s_1, \ldots, s_i, s_{i+1}, \ldots$ of states is called a path of $\boldsymbol{K}$ if and only if for any natural number i, $(s_i, s_{i+1}) \in \boldsymbol{T}$. Let π be $s_0, s_1, \ldots, s_i, s_{i+1}, \ldots$ and some notations are defined as follows:

$$\begin{aligned}
&\pi(i) \triangleq s_i \\
&\pi^i \triangleq s_i, s_{i+1}, \ldots \\
&\pi_i \triangleq s_0, s_1, \ldots, s_i, s_i, \ldots \\
&\pi_\infty \triangleq \pi \\
&\pi^{(i,j)} \triangleq \begin{cases} s_i, s_{i+1}, \ldots, s_j, s_j, \ldots & \textbf{if } i \leq j \\ s_i, s_i, \ldots & \textbf{otherwise} \end{cases} \\
&\pi^{(i,\infty)} \triangleq \pi^i \\
&\pi^i_j \triangleq \pi^{(i,j)}
\end{aligned}$$

where i and j are any natural numbers. Note that $\pi^{(0,j)} = \pi_j$. Note that $\pi_i(k) = \pi(k)$ if $k = 0, \ldots, i$ and $\pi_i(k) = \pi(i)$ if $k > i$. Note that $\pi^{(i,j)}(k) = \pi(i+k)$ if $i \leq j$ and $k = 0, \ldots, m$, where $j = i + m$, $\pi^{(i,j)}(k) = \pi(j)$ if $i \leq j$ and $k > j$ and $\pi^{(i,j)}(k) = \pi(i)$ if $i > j$ and k is a natural number. A path π of $\boldsymbol{K}$ is called a computation of $\boldsymbol{K}$ if and only if $\pi(0) \in \boldsymbol{I}$.

Let $\boldsymbol{P_K}$ be the set of all paths of $\boldsymbol{K}$. Let $\boldsymbol{P}_{(\boldsymbol{K},s)}$ be $\{\pi \mid \pi \in \boldsymbol{P_K}, \pi(0) = s\}$, where $s \in \boldsymbol{S}$. Let $\boldsymbol{P}^b_{(\boldsymbol{K},s)}$ be $\{\pi_b \mid \pi \in \boldsymbol{P}_{(\boldsymbol{K},s)}\}$, where $s \in \boldsymbol{S}$ and b is a natural number. Note that $\boldsymbol{P}^\infty_{(\boldsymbol{K},s)}$ is $\boldsymbol{P}_{(\boldsymbol{K},s)}$. If $\boldsymbol{R}$ is finite and $s \in \boldsymbol{R}$, then $\boldsymbol{P}_{(\boldsymbol{K},s)}$ is finite and so is $\boldsymbol{P}^b_{(\boldsymbol{K},s)}$.

Definition 2 (Syntax of LTL). *The syntax of linear temporal logic (LTL) is as follows:*

$$\varphi ::= a \mid \top \mid \neg\varphi \mid \varphi \vee \varphi \mid \bigcirc\, \varphi \mid \varphi\, \mathcal{U}\, \varphi$$

where $a \in \boldsymbol{A}$.

Definition 3 (Semantics of LTL). *For any Kripke structure* $\boldsymbol{K}$*, any path* π *of* $\boldsymbol{K}$ *and any LTL formula* φ*,* $\boldsymbol{K}, \pi \models \varphi$ *is inductively defined as follows:*

- $\boldsymbol{K}, \pi \models a$ *if and only if* $a \in \pi(0)$
- $\boldsymbol{K}, \pi \models \top$
- $\boldsymbol{K}, \pi \models \neg\varphi_1$ *if and only if* $\boldsymbol{K}, \pi \not\models \varphi_1$
- $\boldsymbol{K}, \pi \models \varphi_1 \vee \varphi_2$ *if and only if* $\boldsymbol{K}, \pi \models \varphi_1$ *and/or* $\boldsymbol{K}, \pi \models \varphi_2$
- $\boldsymbol{K}, \pi \models \bigcirc\, \varphi_1$ *if and only if* $\boldsymbol{K}, \pi^1 \models \varphi_1$
- $\boldsymbol{K}, \pi \models \varphi_1\, \mathcal{U}\, \varphi_2$ *if and only if there exists a natural number* i *such that* $\boldsymbol{K}, \pi^i \models \varphi_2$ *and for each natural number* $j < i$*,* $\boldsymbol{K}, \pi^j \models \varphi_1$

where φ_1 *and* φ_2 *are LTL formulas. Then,* $\boldsymbol{K} \models \varphi$ *if and only if* $\boldsymbol{K}, \pi \models \varphi$ *for all computations* π *of* $\boldsymbol{K}$*.*

$\bot \triangleq \neg\top$ and some other connectives are defined as follows: $\varphi_1 \wedge \varphi_2 \triangleq \neg((\neg\varphi_1) \vee (\neg\varphi_2))$, $\varphi_1 \Rightarrow \varphi_2 \triangleq (\neg\varphi_1) \vee \varphi_2$, $\varphi_1 \Leftrightarrow \varphi_2 \triangleq (\varphi_1 \Rightarrow \varphi_2) \wedge (\varphi_2 \Rightarrow \varphi_1)$, $\Diamond\varphi_1 \triangleq \top\, \mathcal{U}\, \varphi_1$, $\Box\varphi_1 \triangleq \neg(\Diamond\neg\varphi_1)$ and $\varphi_1 \rightsquigarrow \varphi_2 \triangleq \Box(\varphi_1 \Rightarrow \Diamond\varphi_2)$. $\bigcirc$, $\mathcal{U}$, $\Diamond$, $\Box$ and $\rightsquigarrow$ are called next, until, eventually, always and leads-to temporal connectives, respectively. Although it is unnecessary to directly define the semantics for $\Diamond$, $\Box$ and $\rightsquigarrow$, we can define it as follows:

- $\boldsymbol{K}, \pi \models \Diamond\varphi_1$ if and only if there exists a natural number i such that $\boldsymbol{K}, \pi^i \models \varphi_1$
- $\boldsymbol{K}, \pi \models \Box\varphi_1$ if and only if for all natural numbers i, $\boldsymbol{K}, \pi^i \models \varphi_1$
- $\boldsymbol{K}, \pi \models \varphi_1 \rightsquigarrow \varphi_2$ if and only if for each natural number i such that $\boldsymbol{K}, \pi^i \models \varphi_1$, there exists a natural number $j \geq i$ such that $\boldsymbol{K}, \pi^j \models \varphi_2$.

Definition 4 (State propositions). *State propositions are LTL formulas such that they do not have any temporal connectives.*

Proposition 1. *Let $\boldsymbol{K}$ be any Kripke structure. If φ is any state proposition, then $(\boldsymbol{K}, \pi \models \varphi) \Leftrightarrow (\boldsymbol{K}, \pi' \models \varphi)$ for any paths π and π' of $\boldsymbol{K}$ such that $\pi(0) = \pi'(0)$.*

Proof. The first state $\pi(0)$ decides if $\boldsymbol{K}, \pi \models \varphi$ holds. □

Eventual properties are those that are expressed in the form of $\Diamond\, \varphi$, where φ is an LTL formula. In this paper, furthermore, we give the constraint to φ: φ is a state proposition.

Let $\boldsymbol{K}, s \models \varphi$, where $s \in \boldsymbol{S}$, be $\boldsymbol{K}, \pi \models \varphi$ for all $\pi \in \boldsymbol{P}_{(\boldsymbol{K},s)}$. Note that $\boldsymbol{K}, s \models \varphi$ for all $s \in \boldsymbol{I}$ is equivalent to $\boldsymbol{K} \models \varphi$. Let $\boldsymbol{K}, s, b \models \varphi$, where $s \in \boldsymbol{S}$ and b is a natural number or ∞, be $\boldsymbol{K}, \pi \models \varphi$ for all $\pi \in \boldsymbol{P}^b_{(\boldsymbol{K},s)}$. Note that $\boldsymbol{K}, s, \infty \models \varphi$ is $\boldsymbol{K}, s \models \varphi$.

Some logical connectives are abused for $\boldsymbol{K}, \pi \models \varphi$ as follows:

- $(\boldsymbol{K}, \pi \models \varphi) \wedge (\boldsymbol{K}', \pi' \models \varphi') \triangleq \boldsymbol{K}, \pi \models \varphi$ and $\boldsymbol{K}', \pi' \models \varphi'$
- $(\boldsymbol{K}, \pi \models \varphi) \vee (\boldsymbol{K}', \pi' \models \varphi') \triangleq \boldsymbol{K}, \pi \models \varphi$ and/or $\boldsymbol{K}', \pi' \models \varphi'$
- $(\boldsymbol{K}, \pi \models \varphi) \Rightarrow (\boldsymbol{K}', \pi' \models \varphi') \triangleq$ if $\boldsymbol{K}, \pi \models \varphi$, then $\boldsymbol{K}', \pi' \models \varphi'$
- $(\boldsymbol{K}, \pi \models \varphi) \Leftrightarrow (\boldsymbol{K}', \pi' \models \varphi') \triangleq \boldsymbol{K}, \pi \models \varphi$ if and only if $\boldsymbol{K}', \pi' \models \varphi'$

We summarize some notations or symbols used in the paper in the three tables: Tables 1–3. Table 1 describes notations or symbols for paths. Table 2 describes notations or symbols for sets of paths. Table 3 describes notations or symbols for satisfaction relations.

Table 1. Descriptions of path notations (or symbols), where i and j are natural numbers.

Symbol	Description
π	a path; an infinite sequence $s_0, s_1, \ldots, s_i, s_{i+1}, \ldots$ of states such that $s_i \rightarrow_K s_{i+1}$ for each i; if s_o is an initial state, it is called a computation
$\pi(i)$	the ith state s_i in π
π^i	the postfix $s_i, s_{i+1}, \ldots$ obtained by deleting the first i states $s_0, s_1, \ldots, s_{i-1}$ from π
π_i	$s_0, s_1, \ldots, s_i, s_i, \ldots$ constructed by first extracting the prefix $s_0, s_1, \ldots, s_i$, the first $i+1$ states from π and then adding s_i, the final state of the prefix, to the prefix at the end infinitely many times
π_∞	$s_0, s_1, \ldots, s_i, s_{i+1}, \ldots$, the same as π
$\pi^{(i,j)}$	if $i \leq j$, then $s_i, \ldots, s_j, s_j, \ldots$, the same as $(\pi^i)_{j-i}$; otherwise, $s_i, s_i, \ldots$, the infinite sequence in which only s_i occurs infinitely many times
$\pi^{(i,\infty)}$	$s_i, s_{i+1}, \ldots$, the same as π^i
π^i_j	the same as $\pi^{(i,j)}$

Table 2. Descriptions of path-set notations (or symbols), where b is a natural number.

Symbol	Description
$\boldsymbol{P}_{\boldsymbol{K}}$	the set of all paths of $\boldsymbol{K}$
$\boldsymbol{P}_{(\boldsymbol{K},s)}$	the set of all paths π of $\boldsymbol{K}$ such that $\pi(0)$, the 0th state of the path π, is s
$\boldsymbol{P}^b_{(\boldsymbol{K},s)}$	the set of all paths π^b such that $\pi \in \boldsymbol{P}_{(\boldsymbol{K},s)}$
$\boldsymbol{P}^\infty_{(\boldsymbol{K},s)}$	the same as $\boldsymbol{P}_{(\boldsymbol{K},s)}$

Table 3. Descriptions of satisfaction relation $\models$ notations (or symbols), where b is a natural number.

Symbol	Description
$\boldsymbol{K}, \pi \models \varphi$	an LTL formula φ holds for a path π of $\boldsymbol{K}$
$\boldsymbol{K} \models \varphi$	an LTL formula φ holds for all computations of $\boldsymbol{K}$
$\boldsymbol{K}, s \models \varphi$	an LTL formula φ holds for all paths in $\boldsymbol{P}_{(\boldsymbol{K},s)}$
$\boldsymbol{K}, s, b \models \varphi$	an LTL formula φ holds for all paths in $\boldsymbol{P}^b_{(\boldsymbol{K},s)}$
$\boldsymbol{K}, s, \infty \models \varphi$	the same as $\boldsymbol{K}, s \models \varphi$

3. Outline of the Proposed Technique

Let us outline the proposed technique with a simple system (or Kripke structure) called SimpSys as depicted in Figure 1 so that you can intuitively comprehend the technique. SimpSys has four states s_0, s_1, s_2 and s_3, where s_0 is the only initial state. There are seven transitions depicted as arrows in Figure 1. Let us consider three atomic propositions init, middle and final. The labeling function is defined as depicted in Figure 1. For example, middle holds in s_1 and s_2 and does not in s_0 and s_3. Let us take $\lozenge$ final as a property concerned. We can straightforwardly check that SimpSys satisfies $\lozenge$ final, namely SimpSys $\models$ $\lozenge$ final, and then do not need to use the proposed technique for this model checking experiment. We, however, use this simple model checking experiment to sketch the technique.

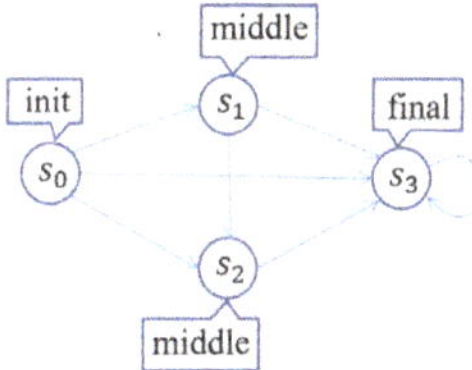

Figure 1. A simple system called SimpSys.

The left part of Figure 2 shows the computation tree made from the reachable states such that its root is the initial state s_0. Let us split the computation tree into two layers such that the first layer depth is 1. Note that it is unnecessary to specify the second (or the final) layer depth. The first layer has one sub-state space such that its initial state is s_0 as shown in the right part of Figure 2. The second layer has three sub-state spaces such that their initial states are s_1, s_2 and s_3, respectively. We first conduct the model checking experiment that $\lozenge$ final holds for the sub-state space in the first layer. There are two counterexamples: (1) $s_0, s_1, s_1, \ldots$ and (2) $s_0, s_2, s_2, \ldots$, where s_1 and s_2 are called counterexample states. As $\lozenge$ final holds for $s_1, s_3, s_3, \ldots$, we do not need to conduct the model checking experiment that $\lozenge$ final holds for the sub-state space whose initial state is s_3 in the second layer. It suffices to conduct the model checking experiments that $\lozenge$ final holds for the two sub-state spaces whose initial states are s_1 and s_2, respectively. There are no counterexamples for the two model checking experiments and then we can conclude that SimpSys satisfies $\lozenge$ final.

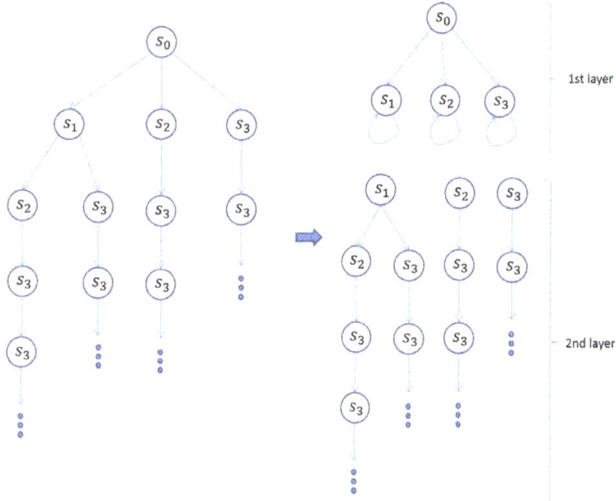

Figure 2. Two-layer division of the SimpSys reachable state space.

This is how the proposed technique works. For this simple example, the number of different states in each sub-state space is the same as or almost the same as the number of different states in the original state space. If the number of each sub-state space is much smaller than the number of the original state space, then even though it is impossible to conduct a model checking experiment for the original reachable state space because of the state space explosion, it may be possible to conduct the model checking experiment for each sub-state space. This is how the proposed technique mitigates the state space explosion problem.

4. Multiple Layer Division of Eventual Model Checking

This section describes the theoretical contribution of the paper. An overview of the proposed technique is as follows: an eventual model checking problem is divided into multiple smaller model checking problems and each smaller model checking problem is tackled so as to tackle the original eventual model checking experiment. We need to guarantee that tackling each smaller model checking problem is equivalent to tackling the original eventual model checking problem. We prove a theorem for it.

We prove that an eventual model checking problem for a Kripke structure $\boldsymbol{K}$ and a path π of $\boldsymbol{K}$ is equivalent to $L+1$ eventual model checking problems for $\boldsymbol{K}$ and $L+1$ paths of $\boldsymbol{K}$, where $L \geq 1$ and the $L+1$ paths are obtained by splitting π into $L+1$ parts. The $L+1$ paths are $\pi^{(d(0),d(1))}$ $(= \pi_{d(0)}), \ldots, \pi^{(d(l),d(l+1))}, \ldots, \pi^{(d(L),d(L+1))}$ $(= \pi^{d(L)})$. Please see Figure 3.

We first tackle the case in which L is 1.

Lemma 1 (Two-layer division of $\Diamond$). *Let φ be any state proposition of $\boldsymbol{K}$. For any natural number k, $(\boldsymbol{K}, \pi \models \Diamond\varphi) \Leftrightarrow ((\boldsymbol{K}, \pi_k \models \Diamond\varphi) \vee ((\boldsymbol{K}, \pi_k \not\models \Diamond\varphi) \Rightarrow (\boldsymbol{K}, \pi^k \models \Diamond\varphi)))$. (We could use $(\boldsymbol{K}, \pi_k \models \Diamond\varphi) \vee (\boldsymbol{K}, \pi^k \models \Diamond\varphi)$ instead of $(\boldsymbol{K}, \pi_k \models \Diamond\varphi) \vee ((\boldsymbol{K}, \pi_k \not\models \Diamond\varphi) \Rightarrow (\boldsymbol{K}, \pi^k \models \Diamond\varphi))$ because they are equivalent).*

Proof. (1) Case "only if" ($\Rightarrow$): There must be i such that $\boldsymbol{K}, \pi^i \models \varphi$. If $i \leq k$, $\boldsymbol{K}, \pi_k^i \models \varphi$ from Proposition 1 because φ is a state proposition. Thus, $\boldsymbol{K}, \pi_k \models \Diamond\varphi$. Otherwise, $\boldsymbol{K}, \pi_k \not\models \Diamond\varphi$. However, $i > k$ and $\boldsymbol{K}, \pi^i \models \varphi$. Hence, $\boldsymbol{K}, \pi^k \models \Diamond\varphi$. (2) Case "if" ($\Leftarrow$): If $\boldsymbol{K}, \pi_k \models \Diamond\varphi$, there must be i such that $i \leq k$ and $\boldsymbol{K}, \pi_k^i \models \varphi$. As φ is a state proposition, $\boldsymbol{K}, \pi^i \models \varphi$ from Proposition 1 and then $\boldsymbol{K}, \pi \models \Diamond\varphi$. If $\boldsymbol{K}, \pi_k \not\models \Diamond\varphi$, then there must be j such that $j > k$ and $\boldsymbol{K}, \pi^j \models \varphi$. Thus, $\boldsymbol{K}, \pi \models \Diamond\varphi$. □

Lemma 1 makes it possible to divide the original model checking problem $\boldsymbol{K}, \pi \models \Diamond\varphi$ into two model checking problems $\boldsymbol{K}, \pi_k \models \Diamond\varphi$ and $\boldsymbol{K}, \pi^k \models \Diamond\varphi$. We only need to tackle $\boldsymbol{K}, \pi^k \models \Diamond\varphi$ unless $\boldsymbol{K}, \pi_k \models \Diamond\varphi$ holds.

Definition 5 (Eventually$_L$). *Let L be any non-zero natural number, k be any natural number and d be any function such that $d(0)$ is 0, $d(x)$ is a natural number for $x = 1, \ldots, L$ and $d(L+1)$ is ∞.*

1. $0 \leq k < L-1$

$$\begin{aligned}&\text{Eventually}_L(\boldsymbol{K}, \pi, \varphi, k)\\ &\triangleq\ (\boldsymbol{K}, \pi^{(d(k),d(k+1))} \models \Diamond\varphi) \vee [(\boldsymbol{K}, \pi^{(d(k),d(k+1))} \not\models \Diamond\varphi) \Rightarrow \text{Eventually}_L(\boldsymbol{K}, \pi, \varphi, k+1)].\end{aligned}$$

2. $k = L-1$

$$\begin{aligned}&\text{Eventually}_L(\boldsymbol{K}, \pi, \varphi, k)\\ &\triangleq\ (\boldsymbol{K}, \pi^{(d(k),d(k+1))} \models \Diamond\varphi) \vee [(\boldsymbol{K}, \pi^{(d(k),d(k+1))} \not\models \Diamond\varphi) \Rightarrow (\boldsymbol{K}, \pi^{(d(k+1),d(k+2))} \models \Diamond\varphi)]\end{aligned}$$

.

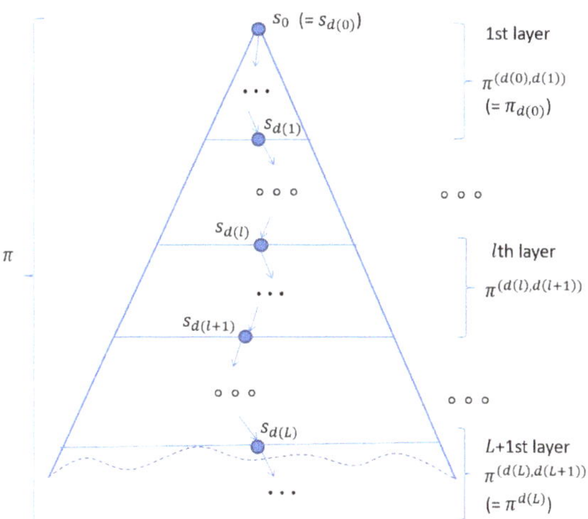

Figure 3. $L+1$ layer division of the reachable state space.

Theorem 1 ($L+1$ layer division of $\Diamond$)**.** *Let L be any non-zero natural number. Let $d(0)$ be 0, $d(x)$ be any natural number for $x = 1, \ldots, L$ and $d(L+1)$ be ∞. Let φ be any state proposition of $\boldsymbol{K}$. Then,*

$$(\boldsymbol{K}, \pi \models \Diamond\varphi) \Leftrightarrow \text{Eventually}_L(\boldsymbol{K}, \pi, \varphi, 0)$$

Proof. By induction on L.

- Base case ($L = 1$): It follows from Lemma 1.
- Induction case ($L = l+1$): We prove the following:

$$(\boldsymbol{K}, \pi \models \Diamond\varphi) \Leftrightarrow \text{Eventually}_{l+1}(\boldsymbol{K}, \pi, \varphi, 0)$$

Let d_{l+1} be d used in $\text{Eventually}_{l+1}(\boldsymbol{K}, \pi, \varphi, 0)$ such that $d_{l+1}(0) = 0$, $d_{l+1}(i)$ is an arbitrary natural number for $i = 1, \ldots, l+1$ and $d_{l+1}(l+2) = \infty$. The induction hypothesis is as follows:

$$(\boldsymbol{K}, \pi \models \Diamond\varphi) \Leftrightarrow \text{Eventually}_l(\boldsymbol{K}, \pi, \varphi, 0)$$

Let d_l be d used in $\text{Eventually}_l(\boldsymbol{K}, \pi, \varphi, 0)$ such that $d_l(0) = 0$, $d_l(i)$ is an arbitrary natural number for $i = 1, \ldots, l$ and $d_l(l+1) = \infty$. As $d_{l+1}(i)$ is an arbitrary natural number for $i = 1, \ldots, l+1$, we suppose that $d_{l+1}(1) = d_l(1)$ and $d_{l+1}(i+1) = d_l(i)$ for $i = 1, \ldots, l$. As π is any path of $\boldsymbol{K}$, π can be replaced with $\pi^{d_l(1)}$. If so, we have the following as an instance of the induction hypothesis:

$$(\boldsymbol{K}, \pi^{d_l(1)} \models \Diamond\varphi) \Leftrightarrow \text{Eventually}_l(\boldsymbol{K}, \pi^{d_l(1)}, \varphi, 0)$$

From Definition 5, $\text{Eventually}_l(\boldsymbol{K}, \pi^{d_l(1)}, \varphi, 0)$ is $\text{Eventually}_{l+1}(\boldsymbol{K}, \pi, \varphi, 1)$ because $d_l(0) = d_{l+1}(0) = 0$, $d_l(1) = d_{l+1}(1)$ and $d_l(i) = d_{l+1}(i+1)$ for $i = 1, \ldots, l$ and $d_l(l+1) = d_{l+1}(l+2) = \infty$. Therefore, the induction hypothesis instance can be rephrased as follows:

$$(\boldsymbol{K}, \pi^{d_{l+1}(1)} \models \Diamond\varphi) \Leftrightarrow \text{Eventually}_{l+1}(\boldsymbol{K}, \pi, \varphi, 1)$$

From Definition 5, $\text{Eventually}_{l+1}(\boldsymbol{K}, \pi, \varphi, 0)$ is

$$(\boldsymbol{K}, \pi^{(d_{l+1}(0), d_{l+1}(1))} \models \Diamond\varphi) \vee [(\boldsymbol{K}, \pi^{(d_{l+1}(0), d_{l+1}(1))} \not\models \Diamond\varphi) \Rightarrow \text{Eventually}_{l+1}(\boldsymbol{K}, \pi, \varphi, 1)]$$

which is

$$(\boldsymbol{K}, \pi^{(d_{l+1}(0),d_{l+1}(1))} \models \Diamond\varphi) \vee [(\boldsymbol{K}, \pi^{(d_{l+1}(0),d_{l+1}(1))} \not\models \Diamond\varphi) \Rightarrow (\boldsymbol{K}, \pi^{d_{l+1}(1)} \models \Diamond\varphi)]$$

because of the induction hypothesis instance. From Lemma 1, this is equivalent to $\boldsymbol{K}, \pi \models \Diamond\varphi$.

□

Theorem 1 makes it possible to divide the original model checking problem $\boldsymbol{K}, \pi \models \Diamond\varphi$ into $L+1$ model checking problems $\boldsymbol{K}, \pi^{(d(0),d(1))} \models \Diamond\varphi, \ldots, \boldsymbol{K}, \pi^{(d(i-1),d(i))} \models \Diamond\varphi$, $\boldsymbol{K}, \pi^{(d(i),d(i+1))} \models \Diamond\varphi, \ldots, \boldsymbol{K}, \pi^{(d(L),d(L+1))} \models \Diamond\varphi$. We only need to tackle $\boldsymbol{K}, \pi^{(d(i),d(i+1))} \models \Diamond\varphi$ if all of $\boldsymbol{K}, \pi^{(d(0),d(1))} \models \Diamond\varphi, \ldots, \boldsymbol{K}, \pi^{(d(i-1),d(i))} \models \Diamond\varphi$ do not hold.

5. A Divide and Conquer Approach to an Eventual Model Checking Algorithm

This section describes an algorithm that carries out the proposed technique. The algorithm takes as inputs a Kripke structure $\boldsymbol{K}$, a state proposition φ, a non-zero natural number L and a function d such that $d(x)$ is a natural number for $x = 1, \ldots, L$, where $d(x)$ is the depth of layer x; and returns as an output success if $\boldsymbol{K} \models \Diamond\varphi$ holds and failure otherwise.

An algorithm can be constructed based on Theorem 1, which is shown as Algorithm 1. For each initial state $s_0 \in \boldsymbol{K}$, unfolding s_0 by using $\boldsymbol{T}$ such that each node except for s_0 has exactly one incoming edge, an infinite tree whose root is s_0 is made. The infinite tree may have multiple copies of some states. Such an infinite tree can be divided into $L+1$ layers, as shown in Figure 3, where L is a non-zero natural number. Although there does not actually exist layer 0, it is convenient to just suppose that we have layer 0. Therefore, let us suppose that there is virtually layer 0 and s_o is located at the bottom of layer 0. Let n_l be the number of states located at the bottom of layer $l = 0, 1, \ldots, L$ and then there are n_l sub-state spaces in layer $l+1$. In this way, the reachable state space from s_0 is divided into multiple smaller sub-state spaces. As $\boldsymbol{R}$ is finite, the number of different states in each layer and in each sub-state space is finite. Theorem 1 makes it possible to check $\boldsymbol{K} \models \Diamond\,\varphi$ in a stratified way in that for each layer $l \in \{1, \ldots, L+1\}$ we can check $\boldsymbol{K}, s, d(l) \models \Diamond\,\varphi$ for each $s \in \{\pi(d(l-1)) \mid \pi \in \boldsymbol{P}^{d(l-1)}_{(\boldsymbol{K},s_0)}\}$, where $d(0)$ is 0, $d(x)$ is a non-zero natural number for $x = 1, \ldots, L$ and $d(L+1)$ is ∞.

ES and ES' are variables to which sets of states are set. Each iteration of the outermost loop in Algorithm 1, which conducts the model checking experiment in layer $l = 1, \ldots, L+1$. ES, is the set of states located at the bottom of layer $l = 0, 1, \ldots L$ and ES' is the empty set before the model checking experiments conducted in the $l+1$st iteration. If $\boldsymbol{K}, \pi \not\models \Diamond\varphi$ for $\pi \in \boldsymbol{P}^{d(l)}_{(\boldsymbol{K},s)}$, then $\pi(d(l))$ is added to ES'. ES' is set to ES at the end of each iteration. If ES is empty at the beginning of an iteration, Success is returned, meaning that $\boldsymbol{K} \models \Diamond\varphi$ holds. After the outermost loop, we check whether ES is empty. If so, Success is returned, and otherwise, Failure is returned.

Although Algorithm 1 does not construct a counterexample when failure is returned, it could be constructed. For each $l \in \{0, 1, \ldots, L\}$, $\boldsymbol{ES}_l$ is prepared. As elements of $\boldsymbol{ES}_l$, pairs (s, s') are used, where s is a state in $\boldsymbol{S}$ or a dummy state denoted δ-stt that is different from any state in $\boldsymbol{S}$, s' is a state in $\boldsymbol{S}$ and s' is reachable from s if $s \in \boldsymbol{S}$. The assignment at line 6 should be revised as follows:

$\boldsymbol{ES}_l \leftarrow \emptyset$

The assignment at line 10 should be revised as follows:

$\boldsymbol{ES}_l \leftarrow \boldsymbol{ES}_l \cup \{(s, \pi(d(l)))\}$

The assignment at line 14 should be revised as follows:

$ES \leftarrow \{s \mid (s, s') \in \boldsymbol{ES}_l\}$

$\boldsymbol{ES}_0$ is set to $\{(\delta\text{-stt}, s) \mid s \in \boldsymbol{I}\}$. We could then construct a counterexample, when failure is returned, by searching through $\boldsymbol{ES}_L, \ldots, \boldsymbol{ES}_1$ and $\boldsymbol{ES}_0$.

Algorithm 1: A divide and conquer approach to eventual model checking.

```
input : K—a Kripke structure
        φ—a state proposition
        L—a non-zero natural number
        d—a function such that d(x) is a natural number for x = 1,...,L, where
        d(x) is the depth of layer x
output: Success (K ⊨ ◇φ) or Failure (K ⊭ ◇φ)
ES ← I
forall l ∈ {1,...,L+1} do
    if ES = ∅ then
        return Success
    end
    ES' ← ∅
    forall s ∈ ES do
        forall π ∈ P^{d(l)}_{(K,s)} do
            if K, π ⊭ ◇φ then
                ES' ← ES' ∪ {π(d(l))}
            end
        end
    end
    ES ← ES'
end
if ES = ∅ then
    return Success
end
else
    return Failure
end
```

6. A Case Study

Many systems' requirements can be expressed as eventual properties. Termination or halting is one important requirement that many programs should satisfy, which can be expressed as an eventual property. The starvation freedom property that should be satisfied by systems, such as an autonomous vehicle intersection control protocol [4], can be expressed as an eventual property. Some communication protocols, such as Alternating Bit Protocol (ABP) and the sliding window protocol used in Transmission Control Protocol (TCP), guarantee that all data sent by a sender are delivered to a receiver without any data loss and duplication. The requirement can be expressed as an eventual property.

We use a mutual exclusion protocol as an example in the case study. The requirement we take into account is that the protocol guarantees that a process can enter the critical section, doing some tasks there, leaving the section and reaching a final position (or terminating). The requirement can be expressed as an eventual property. The mutual exclusion protocol is called Qlock, an abstract version of the Dijkstra binary semaphore in that an atomic queue of process IDs is used.

In the rest of the section, we first describe Qlock, how to formally specify Qlock and the property concerned in Maude and how to model check the eventual property with the proposed technique. Let us note that when there are 10 processes that participate in Qlock, it is impossible to complete the model checking experiment with Maude LTL model checker, while it is possible to do so with the proposed technique. We finally summarize the case study.

6.1. Qlock

We report on a case study that demonstrates the power of the proposed technique. The case study used a mutual exclusion protocol called Qlock whose pseudo-code for each process p can be described as follows:

"Start Section"
ss : enq($queue, p$);
ws : **repeat until** top($queue$) $= p$;
"Critical Section"
cs : deq($queue$);
fs : ...
"Finish Section"

where $queue$ is an atomic queue of process IDs shared by all processes participating in Qlock. enq($queue, p$) atomically puts a process ID p into $queue$ at bottom. top($queue$) atomically returns the top element of $queue$. deq($queue$) atomically deletes the top element of $queue$. If $queue$ is empty, deq($queue$) does nothing. $queue$ is initially empty. Each process p is supposed to be located at one of the four locations ss (start section), ws (waiting section), cs (critical section) and fs (finish section), and is initially located at ss. Let us suppose that each process p stays fs once it gets there, implying that it enters the critical section at most once.

The property to be checked in this case study is that a process will eventually get to fs. The property can be formalized as an eventual property. When there were 10 processes, it did not complete the model check with the Maude LTL model checker running on a computer that carried a 2.10 GHz microprocessor and 8 GB main memory because of the state space explosion.

6.2. Formal Specification

We describe how to formally specify Qlock in Maude. A state is expressed as a braced soup of observable components, where observable components are name–value pairs and soups are associative–commutative collections. When there are n processes, the initial state of Qlock is as follows:

```
{(queue: empq) (pc[p1]: ss) ... (pc[pn]: ss) (cnt: n)}
```

where `(queue: empq)` is an observable component saying that the shared queue is empty, `(pc[pi]: ss)` is an observable component saying that process pi is in the ss and `(cnt: n)` is an observable component whose value is a natural number n. The role of `(cnt: n)` will be described later.

Transitions are described in terms of rewrite rules. The transitions of Qlock are specified as follows:

```
rl [start] : {(queue: Q) (pc[I]: ss) OCs} => {(queue: (Q | I)) (pc[I]: ws) OCs} .
rl [wait] : {(queue: (I | Q)) (pc[I]: ws) OCs}
 => {(queue: (I | Q)) (pc[I]: cs) OCs} .
rl [exit] : {(queue: Q) (pc[I]: cs) (cnt: N) OCs}
 => {(queue: deq(Q)) (pc[I]: fs) (cnt: dec(N)) OCs} .
rl [fin] : {(cnt: 0) OCs} => {(cnt: 0) OCs} .
```

where `Q` is a variable of queues, `I` is a variable of process IDs, `OCs` is a variable of observable component soups and `N` is a variable of natural numbers. `I | Q` denotes a non-empty queue such that `I` is the top and `Q` is the remaining part of the queue. `deq(Q)` returns the empty queue if `Q` is empty and what is obtained by deleting the top from `Q` otherwise. `dec(N)` returns 0 if `N` is 0 and the predecessor number of `N` otherwise.

`start`, `wait`, `exit` and `fin` are the labels given to the four rules, respectively. Rule `start` says that if process `I` is in `ss`, then it puts its ID into `Q` at end and moves to `ws`. Rule wait says that if process I is in ws and the top of the shared queue is I, then I enters cs. Rule exit says that if process I is in cs, then it deletes the top from the shared queue, decrements the natural number N stored in (cnt: N) and moves to fs. Rule fin says that if the natural number N stored in (cnt: N) is 0, a self-transition $s \rightarrow_K s$ occurs. Rule fin is used to make the transitions total. The natural number N stored in (cnt: N) is the number of processes that have not yet reached fs. Use of it and rule fin make it unnecessary to use any fairness assumptions to model check an eventual property.

Let us consider one atomic proposition inFs1. inFs1 holds in a state if and only if the state matches {(pc[p1]: fs) OCs}, namely, that process p1 is in fs.

6.3. Model Checking with the Proposed Technique

It quickly completes to model check $\Diamond$ inFs1 for Qlock when there are five processes, finding no counterexample. It is, however, impossible to model check the same property for Qlock when there are 10 processes. We then use Algorithm 1 to tackle the latter case, where $L = 1$ and $d(1) = 3$.

We use one more observable component (depth: d), where d is a natural number, to work on the first layer. The initial state turns into the following:

```
{(queue: empq) (pc[p1]: ss) ... (pc[p10)]: ss) (cnt: 10) (depth: 0)}
```

The rules turn into the following:

```
crl [start] : {(queue: Q) (pc[I]: ss) (depth: D) OCs}
 => {(queue: (Q | I)) (pc[I]: ws) (depth: (D + 1)) OCs}
 if D < Bound .
crl [wait] : {(queue: (I | Q)) (pc[I]: ws) (depth: D) OCs}
 => {(queue: (I | Q)) (pc[I]: cs) (depth: (D + 1)) OCs}
 if D < Bound .
crl [exit] : {(queue: Q) (pc[I]: cs) (cnt: N)(depth: D) OCs}
 => {(queue: deq(Q)) (pc[I]: fs) (cnt: dec(N)) (depth: (D + 1)) OCs}
 if D < Bound .
crl [fin] : {(cnt: 0) (depth: D) OCs} => {(cnt: 0) (depth: (D + 1)) OCs}
 if D < Bound .
crl [stutter] :{(depth: D) OCs} => {(depth: D) OCs} if D >= Bound .
```

where D is a variable of natural numbers and Bound is 3. Rule stutter has been added to make each state at depth three have a transition to itself. The revised version of rule start says that if D is less than Bound and process I is in ss, then I puts its ID into Q at end and moves to ws and D is incremented. The other revised rules can be interpreted likewise. When we model checked $\Diamond$ inFs1 for the revised specification of Qlock, we found a counterexample that is a finite state sequence starting from the initial state and leading to a state loop that consists of one state that is as follows:

```
{(queue: (p1 | p2 | p3)) (cnt: 10) (depth: 3) (pc[p1]: ws)
 (pc[p2]: ws) (pc[p3]: ws) (pc[p4]: ss) (pc[p5]: ss) (pc[p6]: ss)
 (pc[p7]: ss) (pc[p8]: ss) (pc[p9]: ss) (pc[p10]: ss)}
```

We needed to find all counterexamples and then revise the definition of inFs1 such that inFs1 holds in the state as well. When we model checked the same property for the revised specification, we found another counterexample. This process was repeated until no more counterexamples were found. We totally found 819 counterexamples and 819 counterexample states at depth three.

We gathered all states at depth three from the initial state, which totaled 820 states, including the 819 states found in the last step. There was one state at depth three such that process p1 was located at fs. For each of the 819 states as an initial state, we model checked $\Diamond$ inFs1 for the original specification of Qlock, finding no counterexample. Therefore, we can conclude that it completed model check $\lozenge$ `inFs1` for Qlock when there were 10 processes, finding no counterexample. It took about 44 h to conduct the model checking experiments for the second layer and it took less than 200 ms to conduct each model checking experiment for the first layer. As there were 819 counterexamples for $\lozenge$ `inFs1` in the first layer, we needed to conduct 820 model checking experiments for the first layer.

6.4. Summary of the Case Study

The proposed divide and conquer approach to eventual model checking makes it possible to successfully conduct the model checking experiment $\lozenge$ `inFs1` for Qlock when there are 10 processes and each process enters the critical section at most once, which cannot be otherwise tackled by the computer used in the case study. The specifications in Maude used in the case study are available at the webpage (http://www.jaist.ac.jp/~ogata/code/dca2emc/).

7. Related Work

The state space explosion problem is one of the biggest challenges in model checking. Many techniques to mitigate it have been proposed so far. Among them are partial order reduction [12], symmetry reduction [13], abstraction [14–16], abstract logical model checking [17] and SAT-based bounded model checking (BMC) [2]. The proposed divide and conquer approach to eventual model checking is a new technique to mitigate the problem when model checking eventual properties. The second, third and fourth authors of the present paper proposed a ($L+1$-layer) divide and conquer approach to leads-to model checking [18]. The technique proposed in the present paper can be regarded as an extension of the one described in the paper [18] to eventual properties.

Clarke et al. summarized several techniques that address the state space explosion problem in model checking [19]. One of them is SAT-based BMC. SAT-based BMC is used in industries, especially hardware industries. BMC can find a flaw located within some reasonably shallow depth k from each initial state but cannot prove that systems whose (reachable) state space is enormous (including infinite-state systems) enjoy the desired properties. Some extensions have been made to SAT-based BMC so that we can prove that such systems enjoy the desired properties. One extension is k-induction [20,21]. k-induction is a combination of mathematical induction and SAT/SMT-based BMC, where SMT stands for SAT modulo theories. The bounded state space from each initial state up to depth k is tackled with BMC, which is regarded as the base case. For each state sequence $s_0, s_1, \ldots, s_k$, where s_0 is an arbitrary state, such that a property concerned is not broken in each state s_i for $i = 0, 1, \ldots, k$, it is checked that the property is not broken in all successor states s_{k+1} of s_k, which is done with an SAT/SMT solver and regarded as the induction case. If an SMT solver is used, infinite-state systems, for example, in which integers are used, could be handled. Our proposed technique can be regarded as another extension of BMC, although we do not use any SAT/SMT solvers.

SAT/SMT-based BMC has been extended to model check concurrent programs [22]. Given a concurrent (or multithreaded) program P together with two parameters u and r that are the loop unwinding bound and the number of round-robin schedules, respectively, an intermediate bounded program P_u is first generated by unwinding all loops and inlining all function calls in P with u as a bound, except for those used for creating threads, and then P_u is transformed into a sequential program $Q_{u,r}$ that simulates all behaviors of P_u within r round-robin schedules. $Q_{u,r}$ is then transformed into a propositional formula, which is converted into an equisatisfiable CNF formula that can be analyzed by an SAT/SMT solver. This way to model check multithreaded programs can be parallelized by decomposing the set of execution traces of a concurrent program into symbolic subsets and analyzing the set of execution traces in parallel [23]. Instead of generating a single formula from P via $Q_{u,r}$, multiple propositional sub-formulas are generated. Each sub-formula corresponds to a different symbolic partition of the execution traces of P and can be checked for satisfiability independently from the others. The approaches to BMC of multithreaded programs seem able to deal with safety properties only, while our tool is able to deal with leads-to properties, a class of liveness properties. Another difference between their approach and our approach is that the target of our approach is designs of concurrent/distributed systems, while the one of theirs is concurrent programs.

Barnat et al. [24] surveyed some recent advancements of parallel model checking algorithms for LTL. Graph search algorithms need to be redesigned to make the best use of multi-core and/or multi-processor architectures. Parallel model checkers based on such parallel model checking algorithms have been developed, among which are DiVinE 3.0 [25], Garakabu2 [26,27] and a multicore extension of SPIN [28]. In the technique proposed in the present paper, there are generally multiple sub-state spaces in each layer, and model checking experiments for these sub-state spaces are totally independent from each other. Furthermore, model checking experiments for many sub-state spaces in different layers are independent. It is possible to conduct such model checking experiments in parallel. Therefore, it is possible to parallelize Algorithm 1, which never requires us to redesign any graph search algorithms and makes it possible to use any existing LTL model checker, such as Maude LTL model checker.

To tackle a large system that cannot be handled by an exhaustive verification mode, SPIN has a bit-state verification mode that may not exhaustively search the entire reachable state space of a large system, but can achieve a higher coverage of large state spaces by using a few bits of memory per state stored. The larger a system under verification becomes, the higher chances the SPIN bit-state verification mode may overlook flaws lurking in the system. To overcome such situations, swarm verification [29] has been proposed. The key ideas of swam verification are parallelism and search diversity. For each of the multiple different search strategies, one instance of bit-state verification is conducted. These instances are totally independent and can be conducted in parallel. Different search strategies traverse different portions of the entire reachable state space, making it more likely to achieve higher coverage of the entire reachable state space and find flaws lurking in a large system if any. An implementation of swarm verification on GPUs, called Grapple [30], has also been developed. Although the technique proposed in the present paper splits the reachable state space from each initial state into multiple layers, generating multiple sub-state spaces, it exhaustively searches each sub-state space with the Maude LTL model checker. It may be worth adopting the swarm verification idea into our technique such that swarm verification is conducted for each sub-state space instead of exhaustive search, which may make it possible to quickly find a flaw lurking in a large system.

One hot theme in research on methods to formally verify liveness properties including program termination is liveness-to-safety reductions. Biere et al. [31] have proposed a technique that formally verifies that finite-state systems satisfy liveness properties by showing the absence of fair cycles in every execution and coined the term "liveness-to-safety reduction" to refer to the technique. The technique can be extended to what is called "parameterized systems" in which the state space is infinite but actually finite for every system instance [32]. Padon et al. [33] have further extended "liveness-to-safety reduction" to systems such that processes can be dynamically created and each process state space is infinite so that they can formally verify that such systems enjoy liveness properties under fairness assumptions. Their technique basically reduces a infinite-state system liveness formal verification problem under fairness to a infinite-state system safety formal verification problem that can be expressed in first-order logic. The latter problem can be solved by existing first-order theorem provers, such as IC3 [34,35] and VAMPIRE [36]. The technique proposed in the present paper does not take into account fairness assumptions. We need to use fairness assumptions to model check liveness properties, including eventual ones from time to time. We might adopt the idea used in the Padon et al.'s liveness-to-safety reduction technique. To our knowledge, the liveness-to-safety reduction technique has not been parallelized. Our approach to eventual model checking might make it possible to parallelize the liveness-to-safety reduction technique.

8. Conclusions

We have proposed a new technique to mitigate the state explosion in model checking. The technique is dedicated to eventual properties. It divides an eventual model checking problem into multiple smaller model checking problems and tackles each smaller one. We have proved that the multiple smaller model checking problems are equivalent to the original eventual model checking problem. We have reported on a case study demonstrating the power of the proposed technique.

There are several things left to do as our future research. One piece of future work for us will be to develop a tool supporting the proposed technique. We will use Maude as an implementing language with its reflective programming (meta-programming) facilities to develop the tool that will do all necessary modifications to systems specifications (or systems models) so that human users do not need to change systems specifications to use the divide and conquer approach to eventual properties. It was impossible to conduct the model checking experiment with Maude LTL model checker; the autonomous vehicle intersection control protocol [4] enjoys the starvation freedom property when there are 13 vehicles with the tool supporting the proposed technique. The starvation freedom property can be expressed as an eventual property. Another piece of future work will be to complete the model checking experiment with the tool supporting the proposed technique. To complete the model checking experiment, we may need to make the best use of up-to-date multi-core/processor architectures. To this end, we need to parallelize Algorithm 1 and the tool supporting the proposed technique. Therefore, yet another piece of future work may be to evolve the tool into a parallel version that can make best use of up-to-date multi-core/processor architectures.

Author Contributions: Conceptualization, methodology, software, investigation and formal analysis, M.N.A., Y.P., C.M.D. and K.O.; project administration and funding acquisition, K.O. All authors have read and agreed to the published version of the manuscript.

Funding: This research was partially funded by JSPS KAKENHI Grant Number JP19H04082.

Institutional Review Board Statement: Not applicable.

Informed Consent Statement: Not applicable.

Data Availability Statement: The specifications in Maude used in the case study are available at the webpage http://www.jaist.ac.jp/~ogata/code/dca2emc/ (accessed on 16 January 2021).

Acknowledgments: The authors would like to thank the anonymous reviewers who carefully read an earlier version of the paper and gave them valuable comments without which they were not able to complete the present paper.

Conflicts of Interest: The authors declare no conflict of interest.

References

1. Burch, J.R.; Clarke, E.M.; McMillan, K.L.; Dill, D.L.; Hwang, L.J. Symbolic Model Checking: 10^{20} States and Beyond. *Inf. Comput.* **1992**, *98*, 142–170. [CrossRef]
2. Clarke, E.M.; Biere, A.; Raimi, R.; Zhu, Y. Bounded Model Checking Using Satisfiability Solving. *Form. Methods Syst. Des.* **2001**, *19*, 7–34. [CrossRef]
3. Aung, M.N.; Phyo, Y.; Ogata, K. Formal Specification and Model Checking of the Lim-Jeong-Park-Lee Autonomous Vehicle Intersection Control Protocol. In Proceedings of the 31st International Conference on Software Engineering and Knowledge Engineering, SEKE 2019, Lisbon, Portugal, 10–12 July 2019; pp. 159–208. [CrossRef]
4. Lim, J.; Jeong, Y.; Park, D.; Lee, H. An efficient distributed mutual exclusion algorithm for intersection traffic control. *J. Supercomput.* **2018**, *74*, 1090–1107. [CrossRef]
5. Clavel, M.; Durán, F.; Eker, S.; Lincoln, P.; Martí-Oliet, N.; Meseguer, J.; Talcott, C. *All About Maude—A High-Performance Logical Framework: How to Specify, Program and Verify Systems in Rewriting Logic*; Lecture Notes in Computer Science (LNCS); Springer: Berlin/Heidelberg, Germany, 2007; Volume 4350. [CrossRef]
6. Holzmann, G.J. *The SPIN Model Checker—Primer and Reference Manual*; Addison-Wesley: Reading, MA, USA, 2004.

7. Goguen, J.A.; Kirchner, C.; Kirchner, H.; Mégrelis, A.; Meseguer, J.; Winkler, T.C. An Introduction to OBJ 3. In Proceedings of the Conditional Term Rewriting Systems, 1st International Workshop, Orsay, France, 8–10 July 1987; Lecture Notes in Computer Science; Kaplan, S., Jouannaud, J., Eds.; Springer: Berlin/Heidelberg, Germany, 1987; Volume 308, pp. 258–263. [CrossRef]
8. Diaconescu, R.; Futatsugi, K. *Cafeobj Report—The Language, Proof Techniques, and Methodologies for Object-Oriented Algebraic Specification*; AMAST Series in Computing; World Scientific: Singapore, 1998; Volume 6. [CrossRef]
9. Cimatti, A.; Clarke, E.M.; Giunchiglia, E.; Giunchiglia, F.; Pistore, M.; Roveri, M.; Sebastiani, R.; Tacchella, A. NuSMV 2: An OpenSource Tool for Symbolic Model Checking. In Proceedings of the Computer Aided Verification, 14th International Conference, CAV 2002, Copenhagen, Denmark, 27–31 July 2002; Lecture Notes in Computer Science; Brinksma, E., Larsen, K.G., Eds.; Springer: Berlin/Heidelberg, Germany, 2002; Volume 2404, pp. 359–364. [CrossRef]
10. Ogata, K.; Futatsugi, K. Comparison of Maude and SAL by Conducting Case Studies Model Checking a Distributed Algorithm. *IEICE Trans. Fundam. Electron. Commun. Comput. Sci.* **2007**, *90*, 1690–1703. [CrossRef]
11. de Moura, L.M.; Owre, S.; Rueß, H.; Rushby, J.M.; Shankar, N.; Sorea, M.; Tiwari, A. SAL 2. Computer Aided Verification. In Proceedings of the 16th International Conference, CAV 2004, Boston, MA, USA, 13–17 July 2004; Lecture Notes in Computer Science; Alur, R., Peled, D.A., Eds.; Springer: Berlin/Heidelberg, Germany, 2004; Volume 3114, pp. 496–500. [CrossRef]
12. Clarke, E.M.; Grumberg, O.; Minea, M.; Peled, D.A. State Space Reduction Using Partial Order Techniques. *Int. J. Softw. Tools Technol. Transf.* **1999**, *2*, 279–287. [CrossRef]
13. Clarke, E.M.; Emerson, E.A.; Jha, S.; Sistla, A.P. Symmetry Reductions in Model Checking. In Proceedings of the CAV 1998, Vancouver, BC, Canada, 28 June–2 July 1998; Lecture Notes in Computer Science; Springer: Vancouver, BC, Canada, 1998; Volume 1427, pp. 147–158. [CrossRef]
14. Clarke, E.M.; Grumberg, O.; Long, D.E. Model Checking and Abstraction. *ACM Trans. Program. Lang. Syst.* **1994**, *16*, 1512–1542. [CrossRef]
15. Clarke, E.M.; Grumberg, O.; Jha, S.; Lu, Y.; Veith, H. Counterexample-guided abstraction refinement for symbolic model checking. *J. ACM* **2003**, *50*, 752–794. [CrossRef]
16. Meseguer, J.; Palomino, M.; Martí-Oliet, N. Equational abstractions. *Theor. Comput. Sci.* **2008**, *403*, 239–264. [CrossRef]
17. Bae, K.; Escobar, S.; Meseguer, J. Abstract Logical Model Checking of Infinite-State Systems Using Narrowing. In Proceedings of the RTA 2013, Eindhoven, The Netherlands, 24–26 June 2013; LIPIcs; Schloss Dagstuhl–Leibniz-Zentrum fuer Informatik: Eindhoven, The Netherlands, 2013; Volume 21, pp. 81–96. [CrossRef]
18. Phyo, Y.; Minh, C.D.; Ogata, K. A Divideeventual model checking Conquer Approach to Leads-to Model Checking. *Comput. J.* **2021**, [CrossRef]
19. Clarke, E.M.; Klieber, W.; Novácek, M.; Zuliani, P. Model Checking and the State Explosion Problem. In *LASER Summer School 2011*; Lecture Notes in Computer Science; Springer: Elba Island, Italy, 2011; Volume 7682, pp. 1–30. [CrossRef]
20. Sheeran, M.; Singh, S.; Stålmarck, G. Checking Safety Properties Using Induction and a SAT-Solver. In Proceedings of the FMCAD, Austin, TX, USA, 1–3 November 2000; Lecture Notes in Computer Science; Springer: Austin, TX, USA, 2000; Volume 1954, pp. 108–125. [CrossRef]
21. de Moura, L.M.; Rueß, H.; Sorea, M. Bounded Model Checking and Induction: From Refutation to Verification. In Proceedings of the CAV 2003, Boulder, CO, USA, 8–12 July 2003; Lecture Notes in Computer Science; Springer: Boulder, CO, USA, 2003; Volume 2725, pp. 14–26. [CrossRef]
22. Inverso, O.; Tomasco, E.; Fischer, B.; Torre, S.L.; Parlato, G. Bounded Model Checking of Multi-threaded C Programs via Lazy Sequentialization. In Proceedings of the Computer Aided Verification—26th International Conference, CAV 2014, Held as Part of the Vienna Summer of Logic, VSL 2014, Vienna, Austria, 18–22 July 2014; Lecture Notes in Computer Science; Biere, A., Bloem, R., Eds.; Springer: Berlin/Heidelberg, Germany, 2014; Volume 8559, pp. 585–602. [CrossRef]
23. Inverso, O.; Trubiani, C. Parallel and distributed bounded model checking of multi-threaded programs. In Proceedings of the PPoPP '20: 25th ACM SIGPLAN Symposium on Principles and Practice of Parallel Programming, San Diego, CA, USA, 22–26 February 2020; Gupta, R., Shen, X., Eds.; ACM: New York, NY, USA, 2020; pp. 202–216. [CrossRef]
24. Barnat, J.; Bloemen, V.; Duret-Lutz, A.; Laarman, A.; Petrucci, L.; van de Pol, J.; Renault, E. Parallel Model Checking Algorithms for Linear-Time Temporal Logic. In *Handbook of Parallel Constraint Reasoning*; Springer: Berlin/Heidelberg, Germany, 2018; pp. 457–507. [CrossRef]
25. Barnat, J.; Brim, L.; Havel, V.; Havlícek, J.; Kriho, J.; Lenco, M.; Rockai, P.; Still, V.; Weiser, J. DiVinE 3.0—An Explicit-State Model Checker for Multithreaded C & C++ Programs. In *CAV 2013*; LNCS; Springer: Berlin/Heidelberg, Germany, 2013; Volume 8044, pp. 863–868. [CrossRef]
26. Kong, W.; Liu, L.; Ando, T.; Yatsu, H.; Hisazumi, K.; Fukuda, A. Facilitating Multicore Bounded Model Checking with Stateless Explicit-State Exploration. *Comput. J.* **2015**, *58*, 2824–2840. [CrossRef]
27. Kong, W.; Hou, G.; Hu, X.; Ando, T.; Hisazumi, K.; Fukuda, A. Garakabu2: An SMT-based bounded model checker for HSTM designs in ZIPC. *J. Inf. Sec. Appl.* **2016**, *31*, 61–74. [CrossRef]
28. Holzmann, G.J.; Bosnacki, D. The Design of a Multicore Extension of the SPIN Model Checker. *IEEE Trans. Softw. Eng.* **2007**, *33*, 659–674. [CrossRef]
29. Holzmann, G.J.; Joshi, R.; Groce, A. Swarm Verification Techniques. *IEEE Trans. Softw. Eng.* **2011**, *37*, 845–857. [CrossRef]

30. DeFrancisco, R.; Cho, S.; Ferdman, M.; Smolka, S.A. Swarm model checking on the GPU. *Int. J. Softw. Tools Technol. Transf.* **2020**, *22*, 583–599. [CrossRef]
31. Biere, A.; Artho, C.; Schuppan, V. Liveness Checking as Safety Checking. *Electron. Notes Theor. Comput. Sci.* **2002**, *66*, 160–177. [CrossRef]
32. Pnueli, A.; Shahar, E. Liveness and Acceleration in Parameterized Verification. In Proceedings of the Computer Aided Verification, 12th International Conference, CAV 2000, Chicago, IL, USA, 15–19 July 2000; Lecture Notes in Computer Science; Emerson, E.A., Sistla, A.P., Eds.; Springer: Berlin/Heidelberg, Germany, 2000; Volume 1855, pp. 328–343. [CrossRef]
33. Padon, O.; Hoenicke, J.; Losa, G.; Podelski, A.; Sagiv, M.; Shoham, S. Reducing liveness to safety in first-order logic. *Proc. ACM Program. Lang.* **2018**, *2*, 1–33. [CrossRef]
34. Bradley, A.R. Understanding IC3. In Proceedings of the Theory and Applications of Satisfiability Testing—SAT 2012—15th International Conference, Trento, Italy, 17–20 June 2012; Lecture Notes in Computer Science; Cimatti, A., Sebastiani, R., Eds.; Springer: Berlin/Heidelberg, Germany, 2012; Volume 7317, pp. 1–14. [CrossRef]
35. Bradley, A.R. IC3 and beyond: Incremental, Inductive Verification. In Proceedings of the Computer Aided Verification—24th International Conference, CAV 2012, Berkeley, CA, USA, 7–13 July 2012; Lecture Notes in Computer Science; Madhusudan, P., Seshia, S.A., Eds.; Springer: Berlin/Heidelberg, Germany, 2012; Volume 7358, p. 4. [CrossRef]
36. Riazanov, A.; Voronkov, A. The design and implementation of VAMPIRE. *AI Commun.* **2002**, *15*, 91–110.

Article

A Metamorphic Testing Approach for Assessing Question Answering Systems

Kaiyi Tu, Mingyue Jiang * and Zuohua Ding

School of Information Science and Technology, Zhejiang Sci-Tech University, Hangzhou 310018, China; 201930605023@mails.zstu.edu.cn (K.T.); zuohuading@zstu.edu.cn (Z.D.)
* Correspondence: mjiang@zstu.edu.cn

Abstract: Question Answering (QA) enables the machine to understand and answer questions posed in natural language, which has emerged as a powerful tool in various domains. However, QA is a challenging task and there is an increasing concern about its quality. In this paper, we propose to apply the technique of metamorphic testing (MT) to evaluate QA systems from the users' perspectives, in order to help the users to better understand the capabilities of these systems and then to select appropriate QA systems for their specific needs. Two typical categories of QA systems, namely, the textual QA (TQA) and visual QA (VQA), are studied, and a total number of 17 metamorphic relations (MRs) are identified for them. These MRs respectively focus on some characteristics of different aspects of QA. We further apply MT to four QA systems (including two APIs from the AllenNLP platform, one API from the Transformers platform, and one API from CloudCV) by using all of the MRs. Our experimental results demonstrate the capabilities of the four subject QA systems from various aspects, revealing their strengths and weaknesses. These results further suggest that MT can be an effective method for assessing QA systems.

Keywords: textual question answering; visual question answering; metamorphic testing; metamorphic relations; quality assessment

Citation: Tu, K.; Jiang, M., Ding Z. A Metamorphic Testing Approach for Assessing Question Answering Systems. *Mathematics* **2021**, *9*, 726. https://doi.org/10.3390/math9070726

Academic Editors: Vassilis C. Gerogiannis and Tadashi Dohi

Received: 8 February 2021
Accepted: 25 March 2021
Published: 28 March 2021

Publisher's Note: MDPI stays neutral with regard to jurisdictional clai-ms in published maps and institutio-nal affiliations.

1. Introduction

Question answering (QA) [1,2] focuses on returning right answers to given questions. Among various QA systems, the textual question answering (TQA) and visual question answering (VQA) represent a typical paradigm that enables the machine to answer a question in natural language by referring to the given contents (i.e., text or image). As shown in Figure 1, TQA [3] focuses on answering a question about a passage of text, which is also known as an NLP task of machine reading comprehension; while VQA [4] focuses on answering a question based on an image, which leverages techniques from the domains of NLP and computer vision. Both TQA and VQA have various potential applications. For example, TQA has been widely adopted by conversational agents [5] and customer service support [6]; VQA has a broad range of applications in the autonomous agents and virtual assistants [7]. On the other hand, a large number of neural network models have been created for implementing both TQA and VQA. For instances, BiDAF [8], BERT [9], RoBERTa [10] for TQA, and ViLBERT [11] for VQA.

Due to the importance and popularity of QA, it it critical to properly assess QA systems in order to demonstrate their capabilities and limitations. QA systems are commonly evaluated by a test dataset. However, the dataset may not necessarily be representative of the real world. Due to this, various different approaches have been proposed and applied to evaluate QA systems, revealing a series of problems concerning different aspects. Jia et al. [12] proposed an adversarial evaluation scheme to investigate whether QA can answer questions about passages containing adversarially inserted sentences, and their experimental results revealed that the QA models under investigation had poor performance. Divyansh et al. [13] investigated popular QA benchmarks and then revealed that

TQA might ignore the passage of text when answering questions. Mudrakarta et al. [14] proposed to apply the notion of attribution to generate adversarial questions, based on which it was observed that QA systems often ignored important terms in questions. On the other hand, recent studies investigated the robustness of QA systems [15,16] and further proposed strategies for improving their robustness [17].

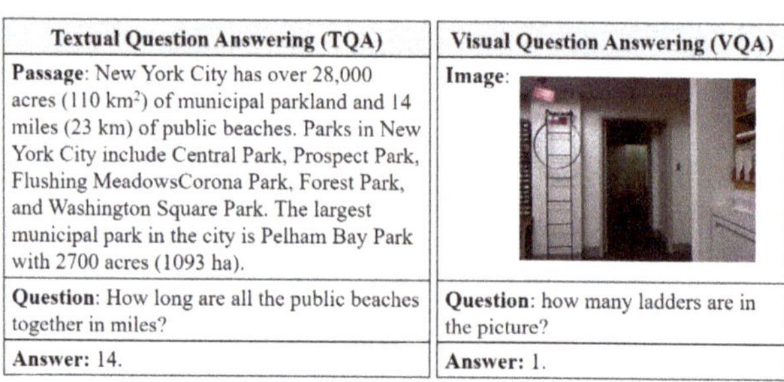

Figure 1. Textual question answering (TQA) and visual question answering (VQA): TQA answers a question with reference to a passage, while VQA answers a question with respect to an image.

This study focuses on assessing TQA and VQA systems from the users' perspective in order to reveal to which degree QA systems satisfy the users' expectations. This kind of assessment is helpful for the users to better understand QA systems such that they are able to select appropriate QA systems for their specific needs. To this end, we propose to adopt the technique of metamorphic testing (MT). MT is a property based testing technique, which has shown promising effectiveness in various software engineering activities, such as testing [18], fault localization [19], and program repair [20,21]. The key component of MT is metamorphic relations (MRs), which encode system properties via the relationship among multiple related inputs and outputs. MT is originally applied for software verification. In recent year, it has been successfully extended to software validation and system comprehension [22,23].

In this study, we identify a total number of 17 MRs for QA systems. These MRs respectively focus on different aspects of TQA and VQA, which can help the users to understand the capability of TQA and VQA systems from different perspectives, and can also provide guidances for the users to select appropriate systems to satisfy their specific needs. We conduct experiments by employing four QA systems (two TQA APIs provided by AllenNLP [24] and Transformers [25], and two VQA APIs provided by AllenNLP and CouldCV) using all of the MRs, demonstrating the capabilities and limitations of the QA systems under investigation. To summarize, the paper makes three major contributions.

- We proposed to apply the technique of metamorphic testing to assess QA systems from the users' perspectives, and presented 17 MRs by considering different aspects of QA systems.
- We conducted experiments on four common QA systems (two TQA systems and two VQA systems), demonstrating the feasibility and effectiveness of MT in assessing QA systems.
- We conducted comparison analysis among subject QA systems to reveal their capabilities of understanding and processing the input data, and also demonstrated how the analysis results can help the user to select appropriate QA system for their specific needs.

The remainder of the paper is organized as follows. Section 2 introduces the technique of metamorphic testing. Section 3 clarifies the overall approach, and Section 4 presents a list of MRs identified for QA systems. Our experimental setup is introduced in Section 5, and the experimental results are presented and analyzed in Section 6. Section 7 discusses related work, and Section 8 concludes the present study.

2. Metamorphic Testing

Metamorphic testing (MT) [26,27] is a property based testing technique. MT proposes to describe the necessary properties of the target system through the relationships among inputs and outputs of multiple executions. Such properties are expressed by metamorphic relation (MRs). Specifically, an MR describes how to construct the follow-up input from the given input (which is known as the source input), and also encodes the relationship among the source and follow-up outputs (namely, the outputs for the source and follow-up inputs respectively). As an example for illustration, consider the program *Max* that implements the algorithm of finding the maximum value among two input values. An MR for Max can be "Suppose that the source input is $t_s = (x, y)$, where x and y can be arbitrary numeric values, and the follow-up input t_f is constructed by swapping the two input values of t_s (that is, $t_f = (y, x)$). As a result, the source and follow-up outputs are expected to be identical".

Generally, MRs can be identified by referring to the system's requirements or based on the users' expectations on the system. Given an MR and a set of its source inputs (which can be generated by arbitrary strategies), MT can be conducted as below. At first, the corresponding follow-up inputs are constructed based on the source inputs according to the MR. After that, for every group of source and follow-up inputs, MT respectively runs the target program on both source and follow-up inputs, yielding the source and follow-up outputs. MT finally checks each group of source and follow-up inputs and outputs against the relevant MR to see whether or not the MR is violated. Any group of source and follow-up inputs with which the program violates the MR is regarded to incur an MR violation. Specifically, an MR violation is an indicator of the existence of defects in the target system if the relevant MR is identified with reference to the system's requirements. Nevertheless, an MR violation reveals either the existence of defects or the the discrepancies between the system behavior and the users expectations if the MR is identified with respect to the users' expected characteristics of the system.

Different from traditional testing techniques that check the correctness of the output of individual inputs, MT checks the satisfaction of MRs on individual groups of source and follow-up executions. Because of this, MT can be conducted without using oracles, and has been applied for software verification and validation [18,22] as well as for helping users to understand the system behaviors [23]. It is also noted that after MRs are identified, the whole procedure of MT can be easily automated.

3. Methodology

This study proposes to apply MT to evaluate QA systems by considering different users' requirements. An overview of the approach is presented in Figure 2. Given a set of source inputs (namely, passage-question pairs for TQA and image-question pairs for VQA) and a list of MRs, a corresponding set of perturbed passage-question pairs and image-question pairs are generated, which are respectively the follow-up inputs for TQA and VQA. By executing the TQA and VQA systems with source and follow-up inputs that are relevant to individual MRs, their source and follow-up answers are collected. Since both TQA and VQA provide a phrase or a sentence as an output answer, we conduct semantic similarity analysis on groups of original and follow-up answers with respect to the relevant MR to determine the testing result. At last, for each MR and every TQA and VQA system under investigation, we calculate the violation rate, which denotes the rate of occurrence of MR violations. A higher violation rate indicates a higher degree to which the system's behaviors deviate from the users' expectations. Based on the evaluation data, we further conduct comparison analysis to reveal the capabilities of QA systems under investigation. Our analysis mainly focuses on three aspects: both TQA and VQA's capabilities of understanding and answering questions, TQA's capabilities of understanding and processing passages, and VQA's capabilities of understanding and processing images. We also demonstrate how our analysis results can help the users to select appropriate QA systems according to their specific needs.

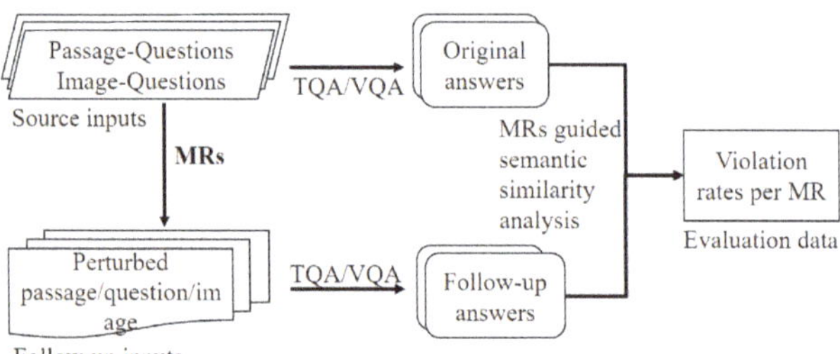

Figure 2. Overview of how metamorphic testing (MT) is applied to evaluate QA systems.

The key task of applying MT to QA systems lies in the identification of MRs by considering the characteristics of QA systems as well as the users' expectations on these systems. Moreover, upon the identification of MRs, the whole evaluation procedure can be automated.

4. Metamorphic Relations of Question Answering Systems

In order to evaluate QA systems by MT, we defined a series of MRs. These MRs consider the users' expected characteristics of QA systems, and thus the satisfaction and violation of these MRs can help users to better understand the capability and limitations of QA systems. In total, 17 MRs are identified, each of which focuses on some aspects of QA. This section presents the details of these MRs, and also gives illustrative examples for some MRs.

4.1. Output Relationships

Let t_s and t_f be a group of source and follow-up inputs of a QA system with respect to an MR, and let A_s and A_f be the corresponding source and follow-up outputs. In this study, we consider the following relationships between A_s and A_f.

- *Equivalent:* A_s and A_f are regarded to be equivalent if they have similar semantics.
- *Different:* A_s and A_f are regarded to be different if they have distinct semantics.

In order to determine whether two answers A_s and A_f have similar semantics, we first transform them into vector representations. This is done by employing the bert-as-service API [28], which encodes a sentence with a fixed length vector by using the BERT model [9]. BERT is a pre-trained transformer network built upon the attention mechanism [29]. The model has multiple layers, each of which consists of an attention sub-layer and a feed-forward network sub-layer. The former helps the model to gain a broad range of information from the input. For an input, the attention sub-layer extracts three vectors, namely, the query vector, key vector and value vector, and packs them together into matrices Q, K, and V, respectively. Based on this, it conducts the self-attention calculation as below [29].

$$Attention(Q,K,V) = softmax(\frac{QK^T}{\sqrt{d_k}})V, \tag{1}$$

where d_k represents the dimension of keys of the input, and *softmax* is a learned normalized exponential function. Specifically, BERT adopts a multi-head attention mechanism, which concats multiple attention calculations of linearly transformed queries, keys and values. The output of the attention sub-layer is provided for another sub-layer that contains a feed-forward network, which is responsible for conducting linear transformations as below [29].

$$FFN(x) = max(0, xW_1 + b_1)W_2 + b_2. \tag{2}$$

Based on the digital vectors yielded by BERT for A_s and A_f, we further apply the cosine similarity analysis [30] to decide whether or not they are semantically equivalent. Suppose that the size of the resulting vectors is n, let $vs = [vs_1, ..., vs_n]$ and $vf = [vf_1, ..., vf_n]$ be the vectors representing A_s and A_f, respectively. The semantic similarity of A_s and A_f is measured by

$$sim(A_s, A_f) = \frac{\sum_{i=1}^{n} vs_i \times vf_i}{\sqrt{\sum_{i=1}^{n} vs_i^2}\sqrt{\sum_{i=1}^{n} vf_i^2}}. \quad (3)$$

As a result, a similarity score that is higher than a threshold value indicates the equivalence of A_s and A_f in terms of their semantics.

4.2. MRs for QA Systems

The input of TQA consists of a passage and a question, and the input of VQA contains an image and a question. As such, we use P_s (or I_s) and Q_s to denotes the passage (or image) and question in t_s, and use P_f (I_f) and Q_f to denote the corresponding information in t_f. That is, t_s = (P_s, Q_s) and t_f = (P_f, Q_f) for TQA, while t_s = (I_s, Q_s) and t_f = (I_f, Q_f) for VQA. Different MRs may operate on different input parameters of t_s to construct t_f, leading to discrepancies between t_s and t_f. According to this, we classify all MRs into three categories, which are summarized in Table 1 and are explained as below.

- MR1.x has t_s = (P, Q_s) and t_f = (P, Q_f) or t_s = (I, Q_s) and t_f = (I, Q_f). That is, t_s and t_f of MR1.x have the same P (or I), but different questions Q_s and Q_f. This category of MRs operates on Q_s to construct Q_f. Hence, they focus on QA's capability of understanding and answering questions
- MR2.x has t_s = (P_s, Q) and t_f = (P_f, Q). That is, t_s and t_f of MR2.x have the same Q, but different passages P_s and P_f. This category of MRs operate on P_s to construct P_f, and they focus on the TQA's capability of processing and understanding the input passage.
- MR3.x has t_s = (I_s, Q) and t_f = (I_f, Q). That is, t_s and t_f of MR3.x have the same Q, but different images I_s and I_f. This category of MRs operate on I_s to construct I_f, and they concentrate on the VQA's capability of processing and understanding the input image.

Table 1. Summary of metamorphic relations (MRs).

	Source and Follow-Up Inputs	Number of MRs
MR1.x	t_s = (P, Q_s), t_f = (P, Q_f) t_s = (I, Q_s), t_f = (I, Q_f)	4 (MR1.1–MR1.4)
MR2.x	t_s = (P_s, Q), t_f = (P_f, Q)	5 (MR2.1–MR2.5)
MR3.x	t_s = (I_s, Q), t_f = (I_f, Q)	8 (MR3.1–MR3.8)

4.2.1. MR1.x

This category of MRs are designed to investigate the QA's capability of understanding and answering questions. For each MR, t_s and t_f use the same input passage or image but different questions, that is, (P, Q_s) and (P, Q_f) for TQA, while (I, Q_s) and (I, Q_f) for VQA. Different MRs alter Q_s in different ways to construct Q_f and also encode the relationship that is expected to be satisfied by A_s and A_f. We identify four MRs, which are described as follows.

MR1.1 (Capitalization): Q_f is constructed by replacing lowercase letters of Q_s with the corresponding uppercase letters. As a result, A_f is expected to be equivalent to A_s.

MR1.2 (Rephrasing comparative question): Suppose that Q_s contains comparative phrases. Q_f is constructed by rephrasing Q_s without changing the meaning of Q_s. As a result, A_f is expected to be equivalent to A_s.

MR1.3 (Replacing the comparative word with its antonym): Suppose that Q_s contains comparative words. Q_f is constructed by replacing a comparative word in Q_s with its antonym such that Q_f expresses a different meaning from Q_s. As a result, A_f is expected to be different from A_s.

MR1.4 (Changing the subject of a question): Q_f is constructed by changing the subject of Q_s with another noun. This change leads to different meanings of these two questions. As a result, A_f is expected to be different from A_s.

Table 2 shows some illustrative examples of Q_s and Q_f of MR1.1–MR1.4, where Q_f is highlighted with underlines. For each MR, the interpretation of MR violations is also presented.

Table 2. Interpretations and illustrations of MR1.x.

MRs	Interpretation of MR Violation	Examples of Pairs of (Q_s,Q_f)
MR1.1	QA is sensitive to the letter case of a question.	What song won Best R&B Performance? WHAT SONG WON BEST R&B PERFORMANCE?
MR1.2	QA is sensitive to questions using different comparative descriptions.	In how many years will A remain higher than B in population? In how many years will B remain lower than A in population?
MR1.3	QA cannot properly understand the questions expressed via different comparative words.	What type of residents tend to be more fluent than rural ones? What type of residents tend to be less fluent than rural ones?
MR1.4	QA cannot properly understand the questions involving different subjects.	What is the name of the final studio album from Destiny's Child? What is the name of the final studio album from Bob's Child?

4.2.2. MR2.x

This category of MRs are identified to study the TQA's capability of processing and understanding the input passage. For each MR, the source input is t_s = (P_s, Q), and the corresponding follow-up input is t_f = (P_f, Q). Every MR proposes a way of altering P_s to construct P_f and also predicts the relationships between the corresponding A_s and A_f. Table 3 summarizes this category of MRs, the details of which are presented as below.

Table 3. Summary of MR2.x.

MRs	Interpretation of MR Violations	Operation Used for Constructing P_f
MR2.1	TQA is sensitive to the letter case of a passage.	Capitalization
MR2.2	TQA is sensitive to the order of sentences in a passage.	Order reversing
MR2.3	TQA is sensitive to the added sentences that are irrelevant to the question.	Addition
MR2.4	TQA is sensitive to the deleted sentences that are irrelevant to the question.	Removal
MR2.5	TQA is incapable of properly understanding and processing the question related texts.	Replacement

MR2.1 (Capitalization): P_f is constructed by replacing lowercase letters of P_s with the corresponding uppercase letters. As a result, A_f is expected to be equivalent to A_s.

MR2.2 (Reversing the order of sentences): P_f is constructed by reversing the order of sentences of P_s. As a result, A_f is expected to be equivalent to A_s.

MR2.3 (Addition of irrelevant sentences): P_f is constructed by adding some sentences that are irrelevant to the question into P_s. As a result, A_f is expected to be equivalent to A_s.

MR2.4 (Removal of irrelevant sentences): P_f is constructed by removing sentences that are irrelevant to the question from P_s. As a result, A_f is expected to be equivalent to A_s.

MR2.5 (Replacing the answer-related words): Suppose that A_s is a numeric value, which is an answer to questions of types of how many, how old, how long, or when. P_f is constructed by replacing A_s in P_s with $A_s + n$, where n is a randomly selected numeric constant, which makes $A_s + n$ a numeric value that is different from A_s and is also unique in P_s. As a result, A_f is expected to be different from A_s but is equal to $A_s + n$.

MR2.5 is designed by considering a special case where TQA returns a numeric value as an answer to a given question. In this study, we consider four types of questions, namely, how many, how old, how long, and when. An illustrative example of MR2.5 is presented in Table 4, which demonstrates the way of constructing P_f based on both P_s and A_s. Obviously, MR2.5 can only be applied to source inputs that contain the aforementioned four types of questions.

Table 4. Example P_s and P_f of MR2.5 (n is set to be 3).

P_s:	After graduating from high school, West received a scholarship to attend Chicago's American Academy of Art in 1997 and began taking painting classes, but shortly after transferred to Chicago State University to study English. He soon realized that his busy class schedule was detrimental to his musical work, and at 20 he dropped out of college to pursue his musical dreams.This action greatly displeased his mother, who was also a professor at the university.
Q_s:	How old was Kanye when he dropped out of college?
A_s:	20
P_f:	After graduating from high school, West received a scholarship to attend Chicago's American Academy of Art in 1997 and began taking painting classes, but shortly after transferred to Chicago State University to study English. He soon realized that his busy class schedule was detrimental to his musical work, and at 23 he dropped out of college to pursue his musical dreams.This action greatly displeased his mother, who was also a professor at the university.

4.2.3. MR3.x

This category of MRs are identified for evaluating the VQA's capability of processing and understanding the input image. For each MR, the source input is $t_s = (I_s, Q)$, and the corresponding follow-up input is $t_f = (I_f, Q)$. Accordingly, each MR designs a way of altering I_s to construct I_f and also predicts the relationships between source and follow-up outputs. Researchers have proposed a series of operations, such as image scaling and image rotation, to perturb images for evaluating deep neural network based models [31]. In this study, we consider 2D input images, and identify MRs by adopting some of the operations.

We first consider the rotation operation. To rotate an image with a given angle, a rotation matrix is constructed and applied on the image (https://github.com/jrosebr1/imutils, accessed on 8 October 2020). Suppose that c is the center of the rotation, θ is the rotation angle, and x denotes the scale factor. The rotation matrix is as follows:

$$\begin{bmatrix} \alpha & \beta & (1-\alpha)*c.x - \beta*c.y \\ -\beta & \alpha & \beta*c.x + (1-\alpha)*c.y \end{bmatrix}, \tag{4}$$

where $\alpha = x * cos\theta$ and $\beta = x * sin\theta$. Three MRs, namely, MR3.1–MR3.3, are identified by adopting varying rotation angles.

MR3.1: I_f is constructed by rotating I_s by 90 degrees. As a result, A_f is expected to be equivalent to A_s.

MR3.2: I_f is constructed by rotating I_s by 180 degrees. As a result, A_f is expected to be equivalent to A_s.

MR3.3: I_f is constructed by rotating I_s by 270 degrees. As a result, A_f is expected to be equivalent to A_s.

We next consider the changing of RGB images into grayscale images. This can be implemented by using the ITU-R 601-2 (Luma transform https://github.com/python-pillow/Pillow, accessed on 8 October 2020), where each pixel of an image is expressed as 8-bits, and is transformed as below.

$$L = R * 299/1000 + G * 587/1000 + B * 114/1000, \tag{5}$$

where R, G, and B are the RGB values in range of 0–255, and L is the resulting single channel output. Based on this, MR3.4 is identified.

MR3.4: Suppose that I_s is a RGB image. I_f is constructed by converting I_s to its corresponding grayscale image. As a result, A_f is expected to be equivalent to A_s.

We further consider another two types of images operations, image flipping and resizing. Flipping an image utilizes a similar method as for rotating images but with different parameter configurations, while resizing an image can be implemented by adopting scale factors along the horizontal and vertical axes. Based on these two types of operations, the following four MRs are identified.

MR3.5: I_f is constructed by flipping I_s horizontally. As a result, A_f is expected to be equivalent to A_s.

MR3.6: I_f is constructed by flipping I_s vertically. As a result, A_f is expected to be equivalent to A_s.

MR3.7: I_f is constructed by magnifying the size of I_s by 1.5 times. As a result, A_f is expected to be equivalent to A_s.

MR3.8: I_f is constructed by reducing the size of I_s by 1.5 times. As a result, A_f is expected to be equivalent to A_s.

5. Experimental Setup

A series of experiments were conducted to evaluate four QA systems by using all of the 17 MRs. This section presents our experimental setup, including the implementation of MRs, our subject QA systems, the datasets used in the experiments, and the source inputs of MRs.

5.1. MRs Implementation

All of the identified MRs were implemented in order to automatically evaluate QA systems by MT. Some specific MR implementations are presented as below.

MR1.3: MR1.3 replaces the comparative word in Q_s with its antonym for constructing Q_f. To this end, we applied nltk (http://www.nltk.org/, accessed on 23 October 2020) for part-of-speech tagging, which can identify comparative form of an adjective or adverb in Q_s. We further searched the antonym of the given word by using PyDictionary (https://github.com/geekpradd/PyDictionary, accessed on 23 October 2020).

MR1.4: MR1.4 changes the subject of Q_s to construct Q_f. In this study, we treated a word of Q_s representing the entity of PERSON as the subject of Q_s. To identify and change the subject of Q_s, we applied the Named Entity Recognizer StanfordNERTagger (https://nlp.stanford.edu/software/CRF-NER.html, accessed on 2 November 2020). Given a Q_s, StanfordNERTagger was first applied to extract the PERSON entity from Q_s. If the the PERSON entity was successfully identified, we further replaced it with another PERSON entity that was not included in the passage.

MR3.1–MR3.3: These MRs rotate I_s to construct I_f. To automate this procedure, we utilized a package called *imutils* (https://github.com/jrosebr1/imutils, accessed on 2 November 2020), which provides a function *rotate_bound* for rotating images by given degrees.

MR3.4–MR3.8: MR3.4 changes a RGB image to a grayscale image, MR3.5 and MR3.6 flip I_s to construct I_f, while MR3.7 and MR3.8 enlarge (shrink) I_s to construct I_f. To implement these MRs, we used two libraries PIL (https://github.com/python-pillow/Pillow, accessed on 8 October 2020) and OpenCV (https://opencv.org/, accessed on 8 October 2020).

To automatically check the relationship of A_s and A_f, we employed the bert-as-service API [28], which determines the degree to which the given two sentences have similar semantics. This API represented a sentence as a fixed length vector according to BERT [9], based on which we calculated the cosine similarity of vectors of A_s and A_f to determined whether they are equivalent or different.

5.2. Subject QA Systems

In the experiments, two TQA APIs and two VQA APIs were employed as our subject systems, which are listed as below:

- AllenNLP-TQA (https://demo.allennlp.org/reading-comprehension, accessed on 10 November 2020), which is a TQA API at the AllenNLP platform [24]. AllenNLP-TQA is an implementation of the BiDAF model [8] with ELMo embeddings.
- Transformers-TQA (https://github.com/huggingface/transformers, accessed on 10 November 2020), which is a TQA API at the Transformers platform [25]. This API is built upon the the DistilBERT model [32].
- AllenNLP-VQA (https://demo.allennlp.org/visual-question-answering, accessed on 10 November 2020), which is a VQA API at the AllenNLP platform [24]. This API is built upon the ViLBERT model [11].
- CloudCV-VQA (http://vqa.cloudcv.org/, accessed on 10 November 2020), which is an API provided by the CloudCV. This API utilizes the Pythia model [33].

5.3. Datasets and Source Inputs of MRs

The SQuAD 2.0 dataset [34] was used for preparing source inputs of TQA. SQuAD2.0 contains over 150,000 questions. For VQA, we utilized the DAQUAR dataset [35], which contains 1449 images and 12,468 questions. A source input obtained from the SQuAD 2.0 dataset was a passage-question pair, while a source input extracted from the DAQUAR dataset was an image-question pair.

Nine MRs, namely, MR1.1–MR1.4 and MR2.1–MR2.5, were used to evaluate TQA systems, while 12 MRs, namely, MR1.1–MR1.4 and MR3.1–MR3.8, were used to evaluated TQA systems. Each MR was applied to individual source inputs in order to generate the relevant follow-up inputs. Note that MRs may not be applicable to some source inputs due to its preconditions and the operations used for constructing follow-up inputs. For example, MR1.3 operates on comparative words, and thus it cannot be applied to source inputs whose questions contain no comparative word. As a result, different MRs may have varying numbers of groups of source and follow-up inputs. In total, over 50,000 groups of source and follow-up inputs are used for evaluating TQA systems, and over 80,000 groups of source and follow-up inputs are used for evaluating VQA systems.

6. Results and Analysis

In this section, the MT results of evaluating the four subject QA systems are presented. Then, the capabilities of our subject QA systems are analyzed and discussed with respect to relevant MRs.

6.1. MT Results for QA Systems

To evaluate QA systems, the violation rate (VR) was used as the evaluation metric. Given an MR and a QA system, let y be the total number of groups of source and follow-up inputs of the MR that were applied to test the QA system, and x be the number of groups of source and follow-up inputs with which the system violated the MR. The VR of this QA system with respect to this MR is $\frac{y}{x}$. Obviously, a lower VR value indicated a higher degree to which the QA system conformed to the relevant MR, revealing a higher satisfaction of the users' needs. Oppositely, a higher VR value denoted that the QA system was more sensitive to the MR operations, and thus was more likely to produce unexpected answers for the given question. Particularly, a violation rate of 0 means that no violation of the relevant MR was revealed in our experiments, suggesting that the system was likely to be robust with respect to the MR and all of its source and follow-up inputs.

Table 5 summarizes the VR values of four QA systems with respect to all identified MRs. It is observed that all of the QA systems violated some MRs with different degrees, showing VR values ranging from 0.61% to 92.98%. Consider, for example, the VR value (65.10%) of AllenNLP-TQA with respect to MR1.1. This VR value indicated that among all groups of source and follow-up inputs of MR1.1 that were applied to test AllenNLP-TQA, 65.10% revealed MR violations. It can also be found from Table 5 that every QA system violated different MRs with varying VR values and that different QA systems also violated the same MR with varying VR values. This results further suggest that the proposed MRs were capable of reflecting the QA systems' capability from different aspects.

Table 5. Violation rates of question answering (QA) systems ('*' denotes that the number of source input is 0, while '-' means that the relevant MR is not applicable to the system).

	AllenNLP-TQA	Transformers-TQA	AllenNLP-VQA	CloudCV-VQA
MR1.1	65.10%	91.11%	10.34%	20.14%
MR1.2	42.86%	7.14%	*	*
MR1.3	92.98%	3.51%	*	*
MR1.4	86.97%	68.45%	*	*
MR2.1	67.37%	86.86%	-	-
MR2.2	8.12%	6.14%	-	-
MR2.3	2.05%	0.61%	-	-
MR2.4	3.73%	1.18%	-	-
MR2.5	33.18%	23.99%	-	-
MR3.1	-	-	81.10%	66.14%
MR3.2	-	-	80.79 %	62.71%
MR3.3	-	-	33.58%	66.42%
MR3.4	-	-	55.25 %	47.68%
MR3.5	-	-	48.10%	20.86%
MR3.6	-	-	79.54%	62.74%
MR3.7	-	-	32.06%	29.73%
MR3.8	-	-	32.68%	31.51%
Average	44.71%	32.11%	56.68%	48.47 %

6.2. *Further Analysis*

Based on the MT results reported in Table 5, an in-depth analysis was conducted to reveal the capabilities of the four QA systems from different perspectives. Each VR value reported in Table 5 represents the extent to which a system deviated from the properties specified by the relevant MR. Furthermore, as described and explained in Section 4, different MRs handled varying input parameters and also referred to different capabilities of QA. More importantly, a system may have performed well in some aspects but may have had bad performance in some other aspects, while different users may have had concern with varying QA capabilities due to their distinct application scenarios. It was therefore important for the users to know the strength and weakness of different systems such that appropriate systems could be selected to satisfy their needs. Because of this, we compared subject QA systems by inspecting VR values of MRs pertaining to specific QA capabilities in order to reveal the strength and weakness of individual systems from different aspects.

6.2.1. QA's Capability of Understanding and Answering Questions

Both TQA and VQA have to understand the question and then to give an appropriate answer to the question. When using these systems, the users may want to know which QA system has a better capability of processing questions. Four of the proposed MRs, namely, MR1.1–MR1.4, focus on this aspect by describing the relationships among source and follow-up inputs that differ exactly in the input questions.

Figure 3 compares different TQA systems and VQA systems based on MR1.1–MR1.4. As shown in Figure 3a, Transformers-TQA had lower VR values than AllenNLP-TQA for three out of four MRs. It can be further observed from Table 5 that the average VR value of Transformers-TQA on these four MRs was also much lower than that of AllenNLP-TQA. Therefore, as compared with AllenNLP-TQA, Transformers-TQA exhibited better capabilities of understanding and answering questions. Similarly, as shown in Figure 3b, the two VQA systems also had varying violation rates for MR1.1 (the other three MRs had 0 source input for VQA and thus no data was collected). As compared with CloudCV-VQA,

AllenNLP-TQA had a relatively lower violation rate with respect to MR1.1, suggesting that AllenNLP-VQA was more robust to the letter case of input questions. Moreover, Figure 3b also showed that the two VQA systems under investigation were of better capability of handling questions with lowercase or uppercase letters than the two TQA systems, because the former two had much lower VR values (namely, 10.34% and 20.14%) than the latter (namely, 65.10% and 91.11%) with respect to MR1.1.

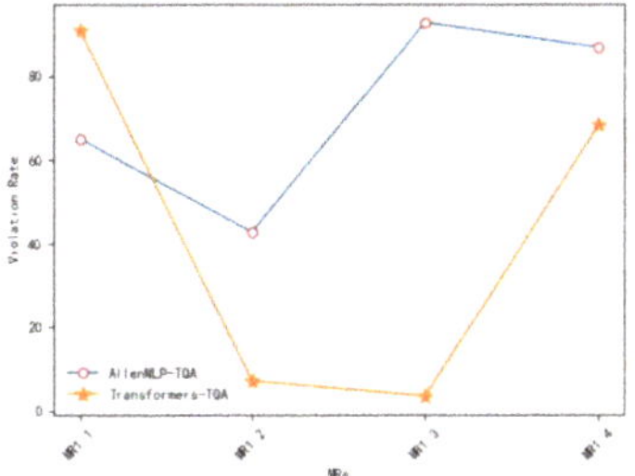

(**a**) Violation rates of TQA with respect to MR1.x.

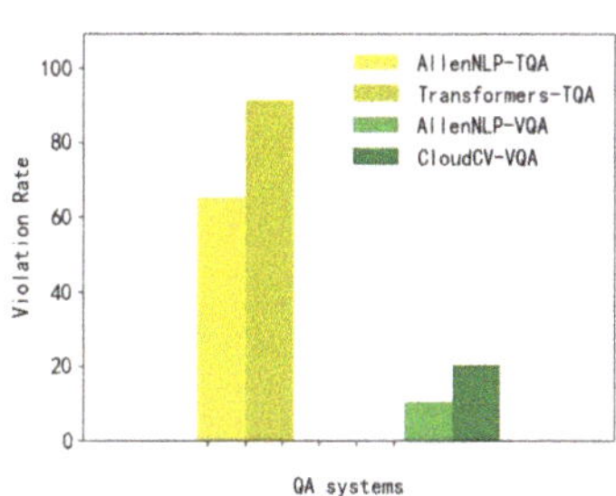

(**b**) Violation rates of QA with respect to MR1.1.

Figure 3. Violation rates of QA with respect to MR1.1–MR1.4.

6.2.2. TQA's Capability of Understanding and Processing Passages

TQA answers a given question based on a passage, it thus needs to understand and process the passage for exacting information related to the given question. We defined five MRs, MR2.1–MR2.5, for investigating TQA's capability of understanding and processing input passages.

Figure 4 compares the violation rates of the two TQA systems (AllenNLP-TQA and Transformers-TQA) with respect to MR2.1–MR2.5. Firstly, both TQA systems had much lower violation rates for MR2.2–MR2.5 (VR values are lower than 35%) as compared with those for MR2.1 (VR values are higher than 65%). These results reveal that the two TQA systems were much more robust to the adding, removing or replacing some contents of the input passage, but were less robust to the conversion of lowercase letters to uppercase letters of the input passage. Secondly, Transformers-TQA had similar violation rates as AllenNLP-TQA for MR2.2–MR2.4 (the discrepancies between the VR values of the two systems with respect to individual MRs were about 2%), but had a very different violation rates from AllenNLP-TQA for the other two MRs (the VR value of the former was about 20% higher than that of the latter with respect to MR2.1, while the VR value of the former was about 10% lower than that of the latter with respect to MR2.5). In other words, the two TQA systems had equivalent capability of dealing with passages containing sentences of different orders as well as containing more or less irrelevant sentences. Nevertheless, AllenNLP-TQA did better for handling passages containing lowercase or uppercase letters, while Transformers-TQA performed better when dealing with passages containing minor replaced contents.

6.2.3. VQA's Capability of Understanding and Processing Images

While TQA understands and processes the input passage for answering a question, VQA relies on the input image for giving an answer to a question. We identified eight MRs, MR3.1–MR3.8, for investigating the VQA's capability of understanding and processing images.

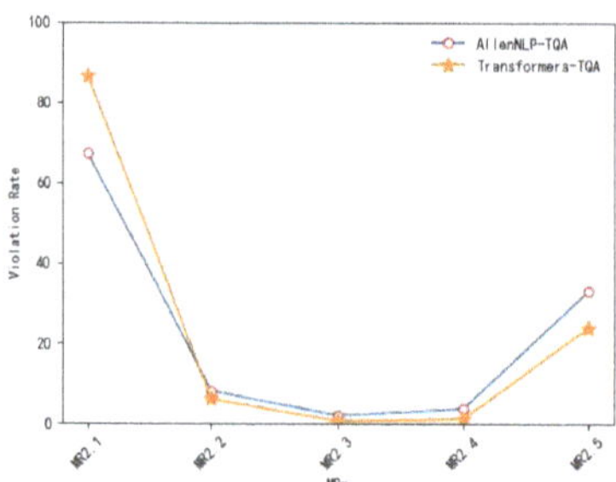

Figure 4. TQA's violation rates with respect to MR2.1–MR2.5.

Figure 5 compares AllenNLP-VQA and CloudCV-VQA with respect to MR3.1–MR3.8. It was observed that except for MR3.3, CloudCV-VQA always had lower violation rates than AllenNLP-VQA, indicating that CloudCV-VQA performed better in terms of MR3.1, MR3.2, MR3.4–MR3.8. On the other hand, both VQA systems had different violation rates for MRs involving the same image perturbation operation, such as rotation and flipping. For example, consider MR3.1–MR3.3, which rotated a source image to construct a follow-up image (but each MR rotated the image by a specific angle, such as 90 degrees, 180 degrees, and 270 degrees). For these three MRs, AllenNLP-VQA had VR values of of 81.10%, 80.79% and 33.58%, and CloudCV-VQA had VR values of 66.14%, 62.71% and 66.42%. A similar observation can also be obtained when inspecting these two VQA systems with respect to MR3.4 and MR3.5 that both flipped the source image to construct the follow-up image (but with different flipping directions).

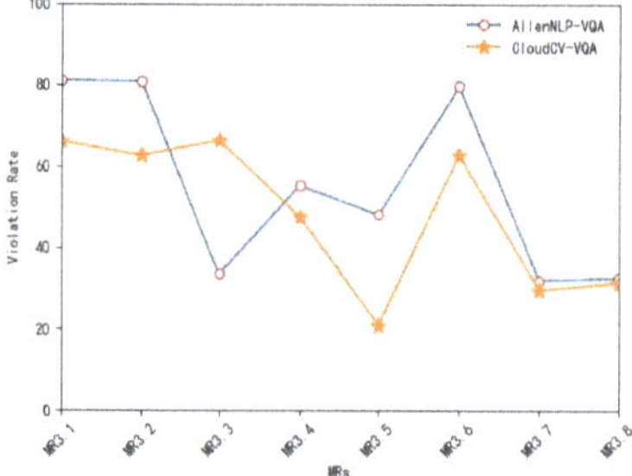

Figure 5. VQA's violation rates with respect to MR3.1–MR3.8.

6.2.4. Further Analysis and Discussion

TQA and VQA had the commonality that they both needed to understand and process the given question. Figure 3b compares our four subject systems with respect to MR1.1, showing that the two VQA systems had relatively better capabilities than the two TQA systems in terms of processing questions containing lowercase or uppercase letters. However, TQA and VQA differed in that the former relied on the passage of text while the latter relied on the image. Concerning these aspects, we respectively used MR2.x and MR3.x for evaluating TQA and VQA. It can still be found from Table 5 that the two TQA systems generally had lower violation rates for MR2.x (which focused on TQA's capability of understanding and processing passages) as compared with the VQA's violation rates for MR3.x (that concentrated on VQA's capability of understanding and processing images). These results indicated that compared with the image processing capability of the two VQA systems, the two TQA systems had better capability of processing passages. Furthermore, Table 5 presents the average violation rates across all applied MRs for individual subject QA systems (as shown in the last row of Table 5). Base on the average VR values, it was found that the two TQA systems generally performed better than the two VQA system, because the former two had average VR values of 44.71% and 32.11% while the latter two had average VR values of 56.68% and 48.47%.

In summary, the proposed 17 MRs encoded some characteristics of QA system, based on which MT results revealed the capabilities of our subject TQA and VQA systems from different perspectives. On one hand, the MT results reported the VR values for every subject system with respect to individual MRs, which could help the users to gain a better understanding about the capability and limitations of the relevant systems. For example, by inspecting the VR values of AllenNLP-TQA, the users could find that this system was good at extracting the question-related information from the passage either with or without some irrelevant sentences (as suggested by the VR value of 2.05% of MR2.3), but it was very incapable of properly understanding questions containing comparative words (as indicated by the VR value of 92.98% of MR1.3). On the other hand, the MT results supported the comparison of different QA systems by considering different aspects, which thus provided guidance for the user to select appropriate QA systems for their specific needs. For example, if the users wanted to use VQA systems without concerning the use of lowercase or uppercase letters in the question description, they could check the VQA systems' VR values with respect to MR1.1. The reason for this is that MR1.1 encoded the relationship between source and follow-up inputs to reflect to which degree a QA system was sensitive to the letter case of a question. In our experiments, AllenNLP-VQA had a VR value of 10.34%, while CloudCV-VQA had a VR value of 20.14%, with respect to MR1.1. Based on this result, it was natural that the users would utilize AllenNLP-VQA rather than CloudCV-VQA. Note that different users may have had varying needs and expectations on the QA systems, and thus MT results of different MRs should be referred in different application scenarios.

7. Related Work

A large body of studies focus on assessing the QA systems' robustness. In order to construct input data, various strategies have been proposed, such as adversarially inserting sentences into the input passages of TQA [12], perturbing questions with respect to high attribute terms [14], rephrasing questions by applying linguistic variations [36], introducing noises into questions [15,37], and applying universal adversarial triggers [38]. Another line of work focuses on improving or explaining QA systems' robustness. Chen et al. [17] proposed a model for TQA through sub-part alignment, which was able to filter out bad prediction results and thus was of higher robustness, while Patro et al. [16] proposed a collaborative correlated network for providing visual and textual explanations of the VQA's answers. Although robustness is important for evaluation, these studies are orthogonal to our focus on assessing to what degree QA systems satisfy the users' specific expectations. On the other hand, most of existing studies focused on either of TQA or VQA, and proposed strategies for changing only parts of an input (namely, question or passage). Nevertheless, our study proposed a list of MRs, which involve various operations that can be applied to both the input questions and the input passages (input images) of TQA (VQA).

Apart from focusing on the QA systems' robustness, Ribeiro et al. [39] evaluated the logic consistency of QA systems. They transformed a question and also implied the corresponding answer by considering the positive and negative implications caused by the given question with respect to the context. While useful, this method still did not take the other parts of the input (i.e., passages or images) into account, and thus the evaluation was still restricted to parts of the QA's capabilities.

Ribeiro et al. [15] introduced MT to one of the QA systems, namely, TQA, and proposed to use MT for evaluatig the TQA's robustness. However, in their work, only one MR was identified, which introduced a specific type of noises (namely, typos) into the input passage or the input question to generate follow-up inputs. In contrast, our study proposed applying MT as a comprehensive evaluation method for both TQA and VQA in a user-oriented way. We have identified a large number of MRs for QA, including MRs that reflect systems' robustness (such as the MRs adopting the capitalization operation on the inputs), and also MRs that focus on particular system functionalities (such as the MRs adopting words replacement). Moreover, these MRs are able to construct diverse test data with changes on both the input passages (images) and questions of TQA (VQA).

8. Conclusions

In recent years, question answering (QA) has emerged as a popular and powerful tool in various domains, due to its capability of enabling the machine to understand and answer question posted in natural language. Unfortunately, recent studies have adopted various techniques to evaluate QA systems, revealing a series of problems concerning different aspects. In this paper, we focused on the evaluation of two typical categories of QA systems, namely, the textual QA (TQA) and visual QA (VQA). We applied the technique of metamorphic testing (MT) to QA, and identified 17 metamorphic relations (MRs) by considering the users' varying expectations on QA systems. In the experiments, we evaluated two TQA systems and two VQA systems by using all of the MRs, and our experimental results reveal their capabilities from different perspectives. These results further suggest that the proposed MRs are capable of encoding the expected characteristics of QA and MT can be an effective evaluation method for QA.

Author Contributions: Conceptualization, M.J. and Z.D.; methodology, M.J.; software, K.T.; data curation, K.T.; writing—original draft preparation, M.J.; writing—review and editing, M.J., Z.D. and K.T.; visualization, M.J. All authors have read and agreed to the published version of the manuscript.

Funding: This research was supported by National Nature Science Foundation of China(Grant Nos. 61751210 and 61802349), and the Zhejiang Provincial Natural Science Foundation of China(Grant No. LY20F020021).

Institutional Review Board Statement: Not applicable.

Informed Consent Statement: Not applicable.

Data Availability Statement: Not applicable.

Conflicts of Interest: The authors declare no conflict of interest.

References

1. Bouziane, A.; Bouchiha, D.; Doumi, N.; Malki, M. Question Answering Systems: Survey and Trends. *Procedia Comput. Sci.* **2015**, *73*, 366–375. [CrossRef]
2. Zeng, C.; Li, S.; Li, Q.; Hu, J.; Hu, J. A Survey on Machine Reading Comprehension: Tasks, Evaluation Metrics, and Benchmark Datasets. *Appl. Sci.* **2020**, *10*, 7640. [CrossRef]
3. Liu, S.; Zhang, X.; Zhang, S.; Wang, H.; Zhang, W. Neural Machine Reading Comprehension: Methods and Trends. *Appl. Sci.* **2019**, *9*, 3698. [CrossRef]
4. Antol, S.; Agrawal, A.; Lu, J.; Mitchell, M.; Parikh, D. VQA: Visual Question Answering. *Int. J. Comput. Vis.* **2015**, *123*, 4–31.
5. Reddy, S.; Chen, D.; Manning, C.D. CoQA: A Conversational Question Answering Challenge. *Trans. Assoc. Comput. Linguist.* **2019**, *7*, 249–266. [CrossRef]
6. Cui, L.; Huang, S.; Wei, F.; Tan, C.; Zhou, M. SuperAgent: A Customer Service Chatbot for E-commerce Websites. In Proceedings of the ACL 2017, System Demonstrations, Vancouver, BC, Canada, 30 July–4 August 2017; pp. 97–102.
7. Li, H.; Wang, P.; Shen, C.; Hengel, A.V.D. Visual Question Answering as Reading Comprehension. In Proceedings of the IEEE/CVF Conference on Computer Vision and Pattern Recognition, Salt Lake City, UT, USA, 18–22 June 2018; pp. 6319–6328.
8. Seo, M.; Kembhavi, A.; Farhadi, A.; Hajishirzi, H. Bidirectional Attention Flow for Machine Comprehension. *arXiv* **2018**, arXiv:1611.01603.
9. Devlin, J.; Chang, M.W.; Lee, K.; Toutanova, K. BERT: Pre-training of Deep Bidirectional Transformers for Language Understanding. *arXiv* **2019**, arXiv:1810.04805.
10. Liu, Y.; Ott, M.; Goyal, N.; Du, J.; Joshi, M.; Chen, D.; Levy, O.; Lewis, M.; Zettlemoyer, L.; Stoyanov, V. RoBERTa: A Robustly Optimized BERT Pretraining Approach. *arXiv* **2019**, arXiv:1907.11692.
11. Lu, J.; Batra, D.; Parikh, D.; Lee, S. ViLBERT: Pretraining Task-Agnostic Visiolinguistic Representations for Vision-and-Language Tasks. *arXiv* **2019**, arXiv:1908.02265.
12. Jia, R.; Liang, P. Adversarial Examples for Evaluating Reading Comprehension Systems. In Proceedings of the 2017 Conference on Empirical Methods in Natural Language Processing, Copenhagen, Denmark, 9–11 September 2017; pp. 2021–2031.

13. Kaushik, D.; Lipton, Z.C. How Much Reading Does Reading Comprehension Require? A Critical Investigation of Popular Benchmarks. In Proceedings of the 2018 Conference on Empirical Methods in Natural Language Processing, Brussels, Belgium, 31 October–4 November 2018; pp. 5010–5015.
14. Mudrakarta, P.K.; Taly, A.; Sundararajan, M.; Dhamdhere, K. Did the Model Understand the Question? In Proceedings of the 56th Annual Meeting of the Association for Computational Linguistics, Melbourne, Australia, 15–20 July 2018; pp. 1896–1906.
15. Ribeiro, M.T.; Wu, T.; Guestrin, C.; Singh, S. Beyond Accuracy: Behavioral Testing of NLP Models with CheckList. In Proceedings of the 58th Annual Meeting of the Association for Computational Linguistics, Online, 5–10 July 2020; pp. 4902–4912.
16. Patro, B.N.; Patel, S.; Namboodiri, V.P. Robust Explanations for Visual Question Answering. In Proceedings of the 2020 IEEE Winter Conference on Applications of Computer Vision (WACV), Village, CO, USA, 1–5 March 2020; pp. 1566–1575.
17. Chen, J.; Durrett, G. Robust Question Answering Through Sub-part Alignment. *arXiv* **2020**, arXiv:2004.14648.
18. Zhou, Z.Q.; Sun, L. Metamorphic testing of driverless cars. *Commun. ACM* **2019**, *62*, 61–67. [CrossRef]
19. Xie, X.; Wong, W.E.; Chen, T.Y.; Xu, B.W. Metamorphic slice: An application in spectrum-based fault localization. *Inf. Softw. Technol.* **2013**, *55*, 866–879. [CrossRef]
20. Jiang, M.; Chen, T.Y.; Kuo, F.C.; Towey, D.; Ding, Z. A metamorphic testing approach for supporting program repair without the need for a test oracle. *J. Syst. Softw.* **2017**, *126*, 127–140. [CrossRef]
21. Jiang, M.; Chen, T.Y.; Zhou, Z.Q.; Ding, Z. Input Test Suites for Program Repair: A Novel Construction Method Based on Metamorphic Relations. *IEEE Trans. Reliab.* **2020**. [CrossRef]
22. Zhou, Z.Q.; Xiang, S.; Chen, T.Y. Metamorphic testing for software quality assessment: A study of search engines. *IEEE Trans. Softw. Eng.* **2016**, *42*, 264–284. [CrossRef]
23. Zhou, Z.Q.; Sun, L.; Chen, T.Y.; Towey, D. Metamorphic Relations for Enhancing System Understanding and Use. *IEEE Trans. Softw. Eng.* **2020**, *46*, 1120–1154. [CrossRef]
24. Gardner, M.; Grus, J.; Neumann, M.; Tafjord, O.; Dasigi, P.; Liu, N.F.; Peters, M.; Schmitz, M.; Zettlemoyer, L. AllenNLP: A Deep Semantic Natural Language Processing Platform. In Proceedings of the Workshop for NLP Open Source Software (NLP-OSS), Melbourne, Australia, 20 July 2018; pp. 1–6.
25. Wolf, T.; Debut, L.; Sanh, V.; Chaumond, J.; Delangue, C.; Moi, A.; Cistac, P.; Rault, T.; Louf, R.; Funtowicz, M.; et al. Transformers: State-of-the-Art Natural Language Processing. In Proceedings of the 2020 Conference on Empirical Methods in Natural Language Processing: System Demonstrations, Virtual Conference, 16–20 November 2020; pp. 38–45.
26. Segura, S.; Fraser, G.; Sanchez, A.B.; Ruiz-Cortés, A. A survey on metamorphic testing. *IEEE Trans. Softw. Eng.* **2016**, *42*, 805–824. [CrossRef]
27. Chen, T.Y.; Kuo, F.C.; Liu, H.; Poon, P.L.; Towey, D.; Tse, T.H.; Zhou, Z.Q. Metamorphic Testing: A Review of Challenges and Opportunities. *ACM Comput. Surv.* **2018**, *51*, 4:1–4:27. [CrossRef]
28. Xiao, H. bert-As-Service. 2018. Available online: https://github.com/hanxiao/bert-as-service (accessed on 8 November 2020).
29. Vaswani, A.; Shazeer, N.; Parmar, N.; Uszkoreit, J.; Jones, L.; Gomez, A.N.; Kaiser, L.; Polosukhin, I. Attention Is All You Need. *arXiv* **2017**, arXiv:1706.03762.
30. Reimers, N.; Gurevych, I. Sentence-BERT: Sentence Embeddings using Siamese BERT-Networks. In Proceedings of the 2019 Conference on Empirical Methods in Natural Language Processing, Hong Kong, China, 3–7 November 2019.
31. Tian, Y.; Pei, K.; Jana, S.; Ray, B. DeepTest: Automated Testing of Deep-Neural-Network-Driven Autonomous Cars. In Proceedings of the 40th International Conference on Software Engineering, Gothenburg, Sweden, 27 May–3 June 2018; pp. 303–314.
32. Sanh, V.; Debut, L.; Chaumond, J.; Wolf, T. DistilBERT, a distilled version of BERT: Smaller, faster, cheaper and lighter. *arXiv* **2020**, arXiv:1910.01108.
33. Singh, A.; Goswami, V.; Natarajan, V.; Jiang, Y.; Chen, X.; Shah, M.; Rohrbach, M.; Batra, D.; Parikh, D. MMF: A Multimodal Framework for Vision and Language Research. 2020. Available online: https://github.com/facebookresearch/mmf (accessed on 12 September 2020).
34. Rajpurkar, P.; Jia, R.; Liang, P. Know What You Don't Know: Unanswerable Questions for SQuAD. In Proceedings of the 56th Annual Meeting of the Association for Computational Linguistics (Volume 2: Short Papers), Melbourne, Australia, 15–20 July 2018; pp. 784–789.
35. Malinowski, M.; Fritz, M. A Multi-World Approach to Question Answering about Real-World Scenes based on Uncertain Input. In Proceedings of the 27th International Conference on Neural Information Processing Systems, Montreal, QC, Canada, 8–13 December 2014; pp. 1682–1690.
36. Shah, M.; Chen, X.; Rohrbach, M.; Parikh, D. Cycle-Consistency for Robust Visual Question Answering. In Proceedings of the 2019 IEEE/CVF Conference on Computer Vision and Pattern Recognition (CVPR), Long Beach, CA, USA, 15–20 June 2019; pp. 6642–6651.
37. Huang, J.H.; Dao, C.D.; Alfadly, M.; Ghanem, B. A Novel Framework for Robustness Analysis of Visual QA Models. In Proceedings of the AAAI Conference on Artificial Intelligence, New York, NY, USA, 7–12 February 2019; Volume 33, pp. 8449–8456.
38. Wallace, E.; Feng, S.; Kandpal, N.; Gardner, M.; Singh, S. Universal Adversarial Triggers for Attacking and Analyzing NLP. In Proceedings of the 2019 Conference on Empirical Methods in Natural Language Processing and the 9th International Joint Conference on Natural Language Processing (EMNLP-IJCNLP), Hong Kong, China, 3–7 November 2019; pp. 2153–2162.
39. Ribeiro, M.T.; Guestrin, C.; Singh, S. Are Red Roses Red? Evaluating Consistency of Question-Answering Models. In Proceedings of the 57th Annual Meeting of the Association for Computational Linguistics, Florence, Italy, 28 July–2 August 2019; pp. 6174–6184.

mathematics

MDPI

Article

Availability Analysis of Software Systems with Rejuvenation and Checkpointing

Junjun Zheng [1,*], Hiroyuki Okamura [2] and Tadashi Dohi [2]

[1] Department of Information Science and Engineering, Ritsumeikan University, 1-1-1 Nojihigashi, Kusatsu 5258577, Japan

[2] Graduate School of Advanced Science and Engineering, Hiroshima University, 1-4-1 Kagamiyama, Higashihiroshima 7398527, Japan; okamu@hiroshima-u.ac.jp (H.O.); dohi@hiroshima-u.ac.jp (T.D.)

* Correspondence: jzheng@asl.cs.ritsumei.ac.jp

Abstract: In software reliability engineering, software-rejuvenation and -checkpointing techniques are widely used for enhancing system reliability and strengthening data protection. In this paper, a stochastic framework composed of a composite stochastic Petri reward net and its resulting non-Markovian availability model is presented to capture the dynamic behavior of an operational software system in which time-based software rejuvenation and checkpointing are both aperiodically conducted. In particular, apart from the software-aging problem that may cause the system to fail, human-error factors (i.e., a system operator's misoperations) during checkpointing are also considered. To solve the stationary solution of the non-Markovian availability model, which is derived on the basis of the reachability graph of stochastic Petri reward nets and is actually not one of the trivial stochastic models such as the semi-Markov process and the Markov regenerative process, the phase-expansion approach is considered. In numerical experiments, we illustrate steady-state system availability and find optimal software-rejuvenation policies that maximize steady-state system availability. The effects of human-error factors on both steady-state system availability and the optimal software-rejuvenation trigger timing are also evaluated. Numerical results showed that human errors during checkpointing both decreased system availability and brought a significant effect on the optimal rejuvenation-trigger timing, so that it should not be overlooked during system modeling.

Keywords: software rejuvenation; checkpointing; optimal rejuvenation-trigger timing; steady-state system availability; phase expansion; human-error factors

Citation: Zheng, J.; Okamura, H.; Dohi, T. Availability Analysis of Software Systems with Rejuvenation and Checkpointing. *Mathematics* **2021**, *9*, 846. https://doi.org/10.3390/math9080846

Academic Editor: Vassilis C. Gerogiannis

Received: 15 March 2021
Accepted: 9 April 2021
Published: 13 April 2021

1. Introduction

In software reliability engineering, various software fault-tolerance techniques such as software rejuvenation and checkpointing are widely used for enhancing system reliability and strengthening data protection. Software rejuvenation is a countermeasure against software aging, which refers to the phenomenon that the performance or dependability of software systems degrades with time, caused by aging-related bugs [1,2], eventually resulting in system failures. In 1995, Huang et al. [3] first reported the aging phenomenon in real telecommunication billing applications where the application experienced a crash or a hang failure over time. The software-aging phenomenon exists in the real world and is inevitable, but can nevertheless be controlled or even reversed [1,2,4]. Software rejuvenation plays a central role in counteracting aging issues by refreshing the system's internal states. However, as pointed out by Alonso et al. [5], the software rejuvenation can address aging issues well, but typically involves an overhead since the system becomes unavailable during rejuvenation. That is to say, it is necessary and important to determine an optimal rejuvenation schedule for achieving the best trade-off between target performance or dependability and the associated overhead. To date, there are a number of works devoted to solving such optimization problems [6–10]. For example, Vaidyanathan and

Trivedi [6] presented a semi-Markov reward model for a UNIX operating system, and used this model to derive optimal software-rejuvenation schedules in terms of system availability or downtime cost. Dohi et al. [9] considered two basic software-rejuvenation models described by Markov regenerative processes (MRGPs), and provided transient solutions using Laplace–Stieltjes transform (LST) and their numerical inversion. In [9], an optimal software-rejuvenation policy that maximized interval system reliability was numerically determined. Wang and Liu [10] recently offered a real-time decision method for optimal software-rejuvenation timing through simulating and modeling the state-transition process of software aging and constructing the rejuvenation decision function using an analytic hierarchy process.

In the context of data protection, a typical technique is checkpointing, which is an efficient method for saving re-execution time in the presence of faults [11] through saving current data in the main memory to secondary storage. Checkpointing is easy to conduct and has been widely studied for decades [12–16]. For example, Fukumoto et al. [12], and Dohi et al. [13] introduced different checkpointing schemes for database systems, and Ranganathan and Upadhyaya [14] considered the temporal behavior related to database system states from a macroscopic viewpoint. Some of the literature also considered software rejuvenation and checkpointing together [17–20]. Okamura and Dohi [17] focused on two kinds of maintenance policies for a software system, and adopted a dynamic programming approach to comprehensively evaluate aperiodic checkpointing and rejuvenation schemes in the system. In [19], the authors introduced a stochastic reward Petri net (SRN) [21] to model a software system of which the state moves to the execution process immediately after a rollback recovery. In particular, according to SRN analysis, a non-Markovian state-transition diagram was derived. More recently, a similar to but somewhat different system from [19] was considered in [20], in which the system executes checkpointing immediately after a rollback recovery in order to update the starting point of the recovery operation from the past to the current time. In these previous works, the systems underwent both aperiodic checkpointing and software rejuvenation, and their transition diagrams are not one of the trivial stochastic models such as semi-Markov process (SMP) and MRGP. That means that common approaches such as the LST and embedded Markov chain techniques cannot be directly applied. To solve these complex non-Markovian transition diagrams, the phase (PH) expansion approach [22,23], which is an approximation technique by using phase-type (PH) distribution, was utilized and worked well in different contents. Moreover, in [19,20], it was assumed that system failures are caused by only aging problems, but in fact, human error is inescapable [24], and the system operator's misoperations during checkpointing cannot be ignored [25].

In this paper, we consider the different software systems from [19,20], where both aperiodic checkpointing and software rejuvenation were executed, and system failure occurred due to both software aging and human errors in checkpointing. A stochastic framework composed of a composite SRN and its resulting non-Markovian availability model is presented to capture the dynamics of the system from a macroscopic point of view. More specifically, the non-Markovian availability model was derived from the reachability graph of the composite SRN model. On the basis of the non-Markovian availability model, which is also a nontrivial model including multiple competitive events as in [19,20], we formulated the steady-state availability of the system by means of PH expansion, and then determined the optimal software-rejuvenation schedule that maximized steady-state system availability. The effects of human-error factors on both steady-state system availability and optimal software-rejuvenation schedule are investigated. The main differences between this work and previous ones [19,20] are that we (i) consider both aging-related and human-error-related system failures, of which the latter was overlooked in previous works; and (ii) investigate the effect of human-error factors on system availability and software rejuvenation. For brevity, the main contributions of this paper are summarized as twofold:

- stochastic modeling of software systems that undergo both software rejuvenation and checkpointing, and may fail due to both the aging problem and human errors in checkpointing;
- investigation of the effects of human-error factors on both steady-state system availability and optimal software-rejuvenation trigger timing by the comparison of cases where human-error-related system failures are considered or not.

The remainder of this paper is organized as follows. In Section 2, a stochastic framework composed of a composite SRN and its corresponding non-Markovian state-transition diagram for an operational software system with software rejuvenation and checkpointing are introduced. In particular, a reachability graph was generated from the composite SRN, and on its basis, a non-Markovian state-transition diagram was obtained. Section 3 first defines continuous PH distribution and presents an approach to formulate the steady-state system availability of the non-Markovian model by using the underlying approximate CTMC of the non-Markovian model, which was derived by replacing all general distributions with their corresponding PH distributions. In Section 4, we describe conducted numerical experiments that evaluated system availability, determined the optimal software-rejuvenation trigger timing, and quantified the effects of human-error factors. Lastly, in Section 5, we conclude this paper with some remarks.

2. Macroscopic System Model

In this section, we first introduce the system assumptions and then present a stochastic framework consisting of a composite SRN and its resulting non-Markovian transition diagram to model operational software systems from a macroscopic point of view. More specifically, the non-Markovian transition diagram was derived on the basis of a reachability graph, which was generated from analysis of the composite SRN.

2.1. System Assumptions

Consider an operational software system that aperiodically executes checkpointing for saving current data in the main memory in secondary storage. Without loss of generality, it was assumed that the system suffers from software aging, so that it may fail due to aging-related bugs, such as a memory leak and the accumulation of round-off errors. On the other hand, system failure might also be caused by incorrect operation by the operator during the execution of checkpointing. Once system failure occurred, a series of recovery operations that include checkpointed data loading and rollback recovery were conducted to recover the system. In addition, software rejuvenation was adopted to counteract the aging problem. A few other assumptions:

- the checkpointing operation just saves the current data and does not refresh system aging;
- the clock of the rejuvenation trigger is not reset and continuously accumulates even when the system executes the checkpointing;
- when a rejuvenation point is reached while the system is under checkpointing, the rejuvenation waits until the checkpointing is completed;
- the system is regarded as good as new after either rollback recovery or rejuvenation.

2.2. Stochastic Reward Nets

On the basis of the above assumptions, the dynamics of the system are described by a composite SRN as in Figures 1 and 2. Concretely, the composite SRN contains three submodels: clock model for system aging (Figure 1a), clock model for software rejuvenation (Figure 1b), and SRN model for system behavior (Figure 2). In these SRNs, transitions are divided into three types: (i) immediate (IMM) transition (represented by a thin black bar), which means the zero firing time transition; (ii) exponential (EXP) transition (represented by a white rectangle), which refers to the exponentially distributed firing time transition; and (iii) general (GEN) transition (represented by a thick black bar), which is generally distributed firing time transition. The places are defined as follows:

- P_{fclock}: software aging accumulates as time passes.
- $P_{fsignal}$: it is time for an aging-related system failure to occur.
- P_{rclock}: time is accumulated to trigger a rejuvenation.
- $P_{rsignal}$: a rejuvenation point was reached.
- P_{normal}: the system waits for checkpointing and rejuvenation in the normal execution process.
- $P_{checkpointing}$: the system is under checkpointing.
- $P_{rejuvenation}$: the system is under rejuvenation.
- $P_{failure}$: the system fails due to either aging-related bugs or human-error factors, and checkpointed data are loaded for rollback recovery.
- $P_{recovery}$: rollback recovery is executed to recover the failed system.
- $P_{completed}$: the system becomes as good as new after the completion of either rejuvenation or rollback recovery.

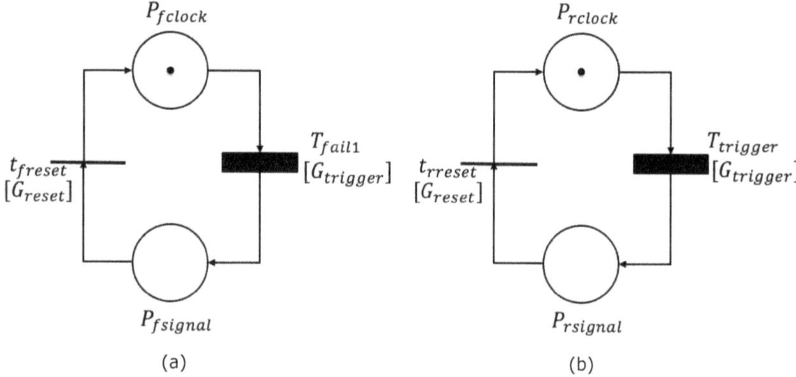

Figure 1. Clock models for (**a**) system aging and (**b**) software rejuvenation.

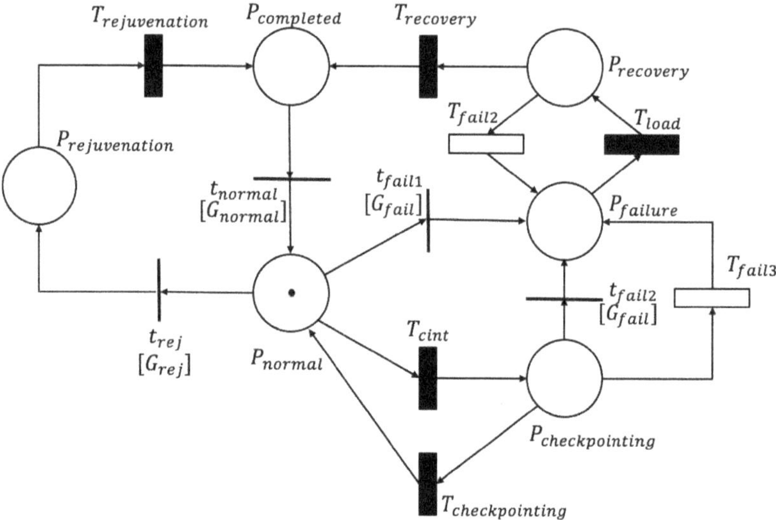

Figure 2. Stochastic (Petri) reward net (SRN) model for system behavior.

On the other hand, transitions T_{cint}, $T_{trigger}$, and T_{fail1} correspond to the trigger intervals of checkpointing and rejuvenation, and system lifetime, respectively. Transitions

$T_{checkpointing}$, $T_{rejuvenation}$, T_{load}, and $T_{recovery}$ separately represent the operations of checkpointing, rejuvenation, loading of checkpointed data, and rollback recovery. Transitions T_{fail2} and T_{fail3} are both EXP transitions, representing failures caused by incorrect operations by the operators. Once IMM transition t_{rej} fires with satisfied guard function G_{rej}, the system is immediately rejuvenated. If a token appears in place $P_{fsignal}$, either transition t_{fail1} or transition t_{fail2} fires due to the exhausted lifetime. Transitions t_{freset} and t_{rreset} represent the reset of the clocks, and t_{normal} means that the system becomes normal again at the same time as when clock reset. The details of guard functions are shown in Table 1.

Table 1. Guard functions.

Guard	Guard Function
G_{normal}	$\#(P_{fclock}) = 1$ && $\#(P_{rclock}) = 1$
G_{fail}	$\#(P_{fsignal}) = 1$
G_{rej}	$\#(P_{rsignal}) = 1$ && $\#(P_{fsignal}) = 0$
$G_{trigger}$	$\#(P_{normal}) = 1$ && $\#(P_{checkpointing}) = 1$
G_{reset}	$\#(P_{completed}) = 1$

2.3. Reachability Graph

A Petri net's reachability graph is also a directed graph composed of nodes and edges, each of which representing a reachable marking and a transition between two reachable markings, respectively. According to analysis of the composite SRN described in Section 2.2, a reachability graph, starting with the initial marking $\{P_{normal} : 1, P_{fclock} : 1, P_{rclock} : 1\}$ (here no token places are not shown for brevity), is generated and depicted as in Figure 3. The description of nodes in the graph are summarized in Table 2. For example, node GEN ($T_{cint} \rightarrow$ enable $T_{fail1} \rightarrow$ enable $T_{trigger} \rightarrow$ enable) is the initial marking and represents the normal execution state of the system in which all transitions T_{cint}, T_{fail1}, and $T_{trigger}$ are enable. Both nodes GEN ($T_{checkpointing} \rightarrow$ enable $T_{fail1} \rightarrow$ enable $T_{trigger} \rightarrow$ enable) and GEN ($T_{checkpointing} \rightarrow$ enable $T_{fail1} \rightarrow$ enable) correspond to the checkpointing execution states, and the difference between them is whether a rejuvenation point was reached. Node GEN ($T_{load} \rightarrow$ enable) means that the system failed, and the loading of checkpointed data is being executed. This graph shows that there exist two edges from either node GEN ($T_{checkpointing} \rightarrow$ enable $T_{fail1} \rightarrow$ enable $T_{trigger} \rightarrow$ enable) or node GEN ($T_{checkpointing} \rightarrow$ enable $T_{fail1} \rightarrow$ enable) to node GEN ($T_{load} \rightarrow$ enable). This is explained by the fact that, during checkpointing, the system may fail due to aging-rated bugs or human-error factors, that is, among two edges, one represents the GEN transition T_{fail1} and another corresponds to the EXP transition T_{fail3}.

Table 2. Nodes in reachability graph.

Node	Description
GEN ($T_{cint} \rightarrow$ enable $T_{fail1} \rightarrow$ enable $T_{trigger} \rightarrow$ enable)	Initial marking representing the normal execution state
GEN ($T_{checkpointing} \rightarrow$ enable $T_{fail1} \rightarrow$ enable $T_{trigger} \rightarrow$ enable)	Marking representing checkpointing-execution state with disabled rejuvenation
GEN ($T_{checkpointing} \rightarrow$ enable $T_{fail1} \rightarrow$ enable)	Marking representing checkpointing-execution state with enabled rejuvenation
GEN ($T_{load} \rightarrow$ enable)	Marking representing system-failure state
GEN ($T_{recovery} \rightarrow$ enable)	Marking representing rollback-recovery state
GEN ($T_{rejuvenation} \rightarrow$ enable)	Marking representing rejuvenation-execution state

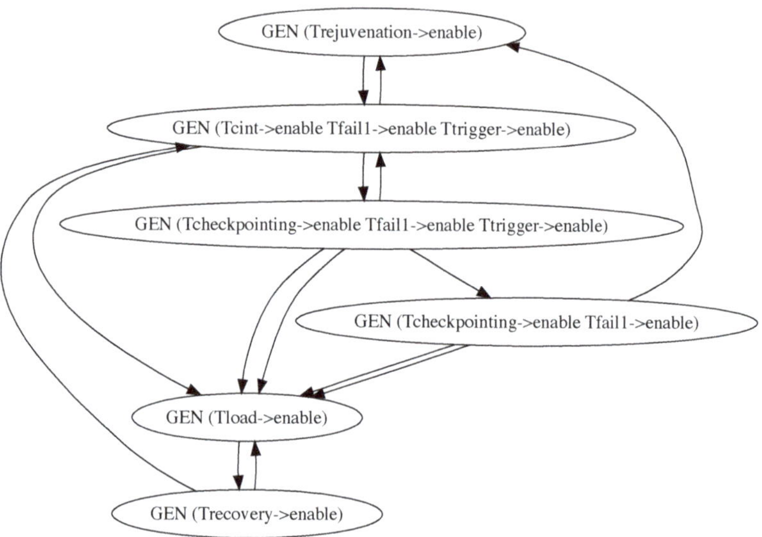

Figure 3. Reachability graph.

2.4. Non-Markovian State-Transition Diagram

From the reachability graph in Section 2.3, a non-Markovian state-transition diagram was derived as shown in Figure 4. This model consisted of seven states: *Normal*, *Checkpointing*, *Checkpointing′*, *Rejuvenation*, *Failure*1, *Recovery*, and *Failure*2. State *Normal* is the initial state and represents that the system is in the normal execution process in the main memory and waits for the checkpointing and rejuvenation. Once a checkpoint is reached prior to the rejuvenation point, the system state becomes *Checkpointing*, in which data on the main memory are saved in secondary storage. Since the checkpointing operation does not reset the clock of the rejuvenation trigger, a rejuvenation point may be reached during checkpointing. In such a case, the system enters state *Checkpointing′*, which represents the checkpoint execution with enabled rejuvenation. After the completion of checkpointing, the system transitions from state *Checkpointing′* to state *Rejuvenation*. If a rejuvenation point is reached prior to the checkpoint, the system immediately executes rejuvenation and enters state *Rejuvenation* from state *Normal*. As mentioned in Section 2.1, system failure may occur due to aging-related bugs and human-error factors. Thus, two failure states, *Failure*1 and *Failure*2, were defined to distinguish two kinds of system failures. When the system fails, a series of recovery operations, including checkpointed data loading and the rollback recovery, are conducted to recover the system from failure. Lastly, the system becomes *Normal* again from state *Recovery*. Of course, the system may fail before both checkpointing and rejuvenation. The details of state notation are given in Table 3.

Table 4 summarizes the cumulative distribution functions (CDFs) of the corresponding transitions in the state-transition diagram. In this table, GEN represents general distribution, and EXP means exponential distribution. The reasons for making such assumptions of probability distributions can be found in [20]. The checkpoint interval was assumed to follow general distribution $G_{intv}(t)$, and the CDF of the time needed for checkpointing is given by $G_{cp}(t)$. The time for an aging-related failure to occur follows a general distribution $G_{fail}(t)$ with increasing failure rate (IFR), while the time distributions for failures occurring during both rollback recovery and checkpointing due to incorrect operations by operators are given by $F_{fail1}(t)$ and F_{fail2} with constant failure rates (CFRs) λ_{fail1} and

λ_{fail2}, respectively. Similarly, the rejuvenation-trigger interval distribution is described by $G_{trig}(t)$, and its relevant overhead distribution is represented by $G_{rej}(t)$. The probability distribution of loading time of checkpointed data and the time needed for rollback recovery are given by $G_{load}(t)$ and $G_{rc}(t)$, respectively.

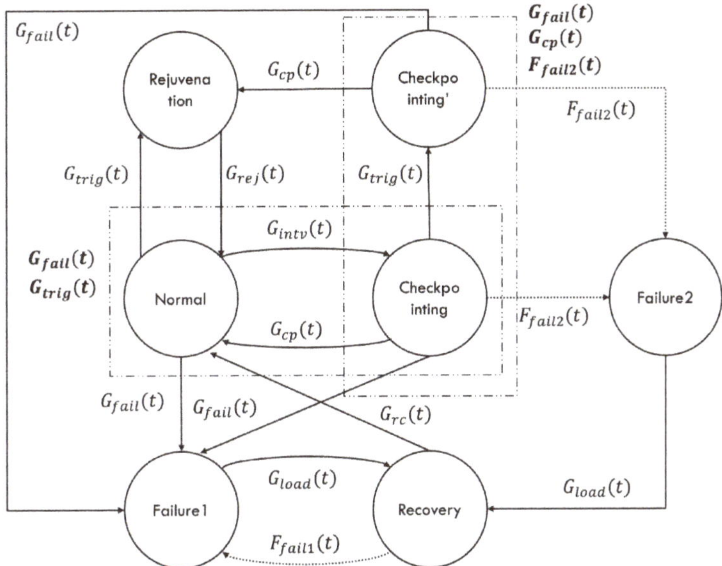

Figure 4. Non-Markovian state-transition diagram.

Table 3. State notation in non-Markovian state-transition diagram.

State	Description
Normal	Normal execution process in the main memory
Checkpointing	Checkpointing execution with a disabled rejuvenation
Checkpointing′	Checkpointing execution with an enabled rejuvenation
Failure1	Aging-related system failure
Failure2	Human-error-related system failure
Recovery	Rollback recovery to recover from system failure
Rejuvenation	Software-rejuvenation execution to refresh system's internal states

Table 4. Cumulative distribution functions (CDFs) of transitions in state-transition diagram.

CDF	Description	Type
$G_{intv}(t)$	CDF of checkpoint interval.	GEN
$G_{fail}(t)$	CDF of time for an aging-related failure to occur.	GEN
$G_{cp}(t)$	CDF of time needed for checkpointing.	GEN
$G_{load}(t)$	CDF of loading time of checkpointed data.	GEN
$G_{rc}(t)$	CDF of time needed for rollback recovery.	GEN
$G_{trig}(t)$	CDF of time required to trigger a rejuvenation.	GEN
$G_{rej}(t)$	CDF of rejuvenation overhead.	GEN
$F_{fail1}(t)$	CDF of time for failure to occur during rollback recovery.	EXP
$F_{fail2}(t)$	CDF of time for a human-error-related failure to occur during checkpointing execution.	EXP

Figure 4 shows states *Normal* and *Checkpointing*, highlighted by a dashed rectangle with $G_{fail}(t)$ and $G_{trig}(t)$, indicating that these GEN transitions regarding $G_{fail}(t)$ and

$G_{trig}(t)$ are enabled and could fire under either the *Normal* or the *Checkpointing* state. In the same way, the dashed rectangle for *Checkpointing* and *Checkpointing'* means the possible firings of GEN and EXP transitions regarding $G_{fail}(t)$, $G_{cp}(t)$, and $F_{fail2}(t)$. This implies that the non-Markovian state-transition diagram under consideration is neither the SMP nor the MRGP, resulting in difficult numerical analysis. To cope with this issue, in this paper we consider the PH expansion approach [22], which proved to be efficient for solving such kind of non-Markovian state-transition models [19,20,26].

3. System Availability Analysis

This section first introduces the well-known continuous PH distribution [22] and then derives the underlying approximate CTMC for the non-Markovian state-transition diagram in Figure 4 via PH expansion approach, of which the essential idea is to replace general distribution with its corresponding PH distribution at a high accuracy level. Lastly, the stationary solution for the model in Figure 4 through CTMC analysis is presented. The measure of interest is steady-state system availability, which is defined as the probability that the system is operational in the steady state.

3.1. Continuous PH Distribution

Continuous PH distribution is defined as the probability distribution of absorbing time in a finite CTMC with absorbing states, and it is widely applied in various fields, such as reliability assessment [26], queueing systems [27], and random telegraph noise analysis [28]. Without loss of generality, we define $\boldsymbol{Q}$ as an infinitesimal generator matrix of a CTMC that has m transient states and one absorbing state, and then partition $\boldsymbol{Q}$ into four parts as below:

$$\boldsymbol{Q} = \left(\begin{array}{c|c} \boldsymbol{T} & \boldsymbol{\xi} \\ \hline \boldsymbol{0} & 0 \end{array} \right). \tag{1}$$

In the above, $\boldsymbol{T}$ and $\boldsymbol{\xi}$ represent transition rates among transient states and exit rates from transient states to the absorbing state, respectively. Defining $\boldsymbol{\alpha}$ as an initial probability vector over the transient states, we have the CDF and probability density function (PDF) for the continuous PH distribution:

$$F_{PH}(t) = 1 - \boldsymbol{\alpha} \exp(\boldsymbol{T}t)\boldsymbol{1}, \quad f_{PH}(t) = \boldsymbol{\alpha} \exp^{\boldsymbol{T}t} \boldsymbol{\xi}, \tag{2}$$

where $\boldsymbol{1}$ is a column vector of ones. Exit vector $\boldsymbol{\xi}$ is given by $\boldsymbol{\xi} = -\boldsymbol{T}\boldsymbol{1}$. Transient states are called phases in general.

Continuous PH distribution can be categorized into several subclasses according to the structure of $\boldsymbol{T}$ [29]. When phase transition is acyclic, the corresponding PH distribution is called acyclic PH distribution (APH). The APH is the widest class among mathematically tractable PH distributions, and it can be converted into the canonical form (CF), which is the minimal representation of APH with the smallest number of free parameters [30]. The APH and its CF are important from the viewpoint of practical applications because it covers some well-known probability distributions, such as exponential distribution, Erlang distribution, and their mixtures. In particular, canonical form 1 (CF1) is usually considered and defined by

$$\boldsymbol{\alpha} = \begin{pmatrix} \alpha_1 & \alpha_2 & \cdots & \alpha_m \end{pmatrix}, \tag{3}$$

$$\boldsymbol{T} = \begin{pmatrix} -\beta_1 & \beta_1 & & & \boldsymbol{O} \\ & -\beta_2 & \beta_2 & & \\ & & \ddots & \ddots & \\ & & & -\beta_{m-1} & \beta_{m-1} \\ \boldsymbol{O} & & & & -\beta_m \end{pmatrix}, \tag{4}$$

$$\boldsymbol{\xi} = \begin{pmatrix} 0 \\ 0 \\ \vdots \\ 0 \\ \beta_m \end{pmatrix}, \tag{5}$$

where $\alpha_i \geq 0$, $\sum_i \alpha_i = 1$ and $0 < \beta_1 \leq \cdots \leq \beta_m$ for m phases.

In this paper, continuous PH distribution was applied to approximate all general distributions in the non-Markovian state-transition diagram, that is, to determine PH distribution with parameters $(\boldsymbol{\alpha}, \boldsymbol{T}, \boldsymbol{\xi})$, which can fit the target distribution well by means of maximum likelihood estimation (MLE) approach [22].

3.2. PH-Expanded CTMC

According to the definition of PH distribution in Section 3.1, we define the general distributions in Table 4 by PH distributions with appropriate phases as follows:

$$F_{intv}^{PH}(t) = 1 - \boldsymbol{\alpha}_{intv} \exp(\boldsymbol{T}_{intv} t) \mathbf{1}_{intv}, \quad f_{intv}^{PH}(t) = \boldsymbol{\alpha}_{intv} \exp(\boldsymbol{T}_{intv} t) \boldsymbol{\xi}_{intv}, \tag{6}$$

$$F_{fail}^{PH}(t) = 1 - \boldsymbol{\alpha}_{fail} \exp(\boldsymbol{T}_{fail} t) \mathbf{1}_{fail}, \quad f_{fail}^{PH}(t) = \boldsymbol{\alpha}_{fail} \exp(\boldsymbol{T}_{fail} t) \boldsymbol{\xi}_{fail}, \tag{7}$$

$$F_{cp}^{PH}(t) = 1 - \boldsymbol{\alpha}_{cp} \exp(\boldsymbol{T}_{cp} t) \mathbf{1}_{cp}, \quad f_{cp}^{PH}(t) = \boldsymbol{\alpha}_{cp} \exp(\boldsymbol{T}_{cp} t) \boldsymbol{\xi}_{cp}, \tag{8}$$

$$F_{load}^{PH}(t) = 1 - \boldsymbol{\alpha}_{load} \exp(\boldsymbol{T}_{load} t) \mathbf{1}_{load}, \quad f_{load}^{PH}(t) = \boldsymbol{\alpha}_{load} \exp(\boldsymbol{T}_{load} t) \boldsymbol{\xi}_{load}, \tag{9}$$

$$F_{rc}^{PH}(t) = 1 - \boldsymbol{\alpha}_{rc} \exp(\boldsymbol{T}_{rc} t) \mathbf{1}_{rc}, \quad f_{rc}^{PH}(t) = \boldsymbol{\alpha}_{rc} \exp(\boldsymbol{T}_{rc} t) \boldsymbol{\xi}_{rc}, \tag{10}$$

$$F_{trig}^{PH}(t) = 1 - \boldsymbol{\alpha}_{trig} \exp(\boldsymbol{T}_{trig} t) \mathbf{1}_{trig}, \quad f_{trig}^{PH}(t) = \boldsymbol{\alpha}_{trig} \exp(\boldsymbol{T}_{trig} t) \boldsymbol{\xi}_{trig}, \tag{11}$$

$$F_{rej}^{PH}(t) = 1 - \boldsymbol{\alpha}_{rej} \exp(\boldsymbol{T}_{rej} t) \mathbf{1}_{rej}, \quad f_{rej}^{PH}(t) = \boldsymbol{\alpha}_{rej} \exp(\boldsymbol{T}_{rej} t) \boldsymbol{\xi}_{rej}. \tag{12}$$

Here, PH parameters $(\boldsymbol{\alpha}_x, \boldsymbol{T}_x, \boldsymbol{\xi}_x)$, $x \in \{intv, fail, cp, load, rc, trig, rej\}$ were estimated on the basis of MLE using an expectation–maximization (EM) algorithm [22,31]. Using the above-estimated PH distributions to replace general distributions, the non-Markovian transition diagram was expanded into an approximate CTMC, alternatively called PH-expanded CTMC, of which the infinitesimal generator matrix is given by

$$Q = \begin{pmatrix}
\boldsymbol{T}_{intv} \oplus \boldsymbol{T}_{fail} \oplus \boldsymbol{T}_{trig} & (\boldsymbol{\xi}_{intv} \boldsymbol{\alpha}_{cp}) \otimes \boldsymbol{I} \otimes \boldsymbol{I} & & (\mathbf{1}_{intv} \otimes \mathbf{1}_{fail} \otimes \boldsymbol{\xi}_{trig}) \boldsymbol{\alpha}_{rej} & (\mathbf{1}_{intv} \otimes \boldsymbol{\xi}_{fail} \otimes \mathbf{1}_{trig}) \boldsymbol{\alpha}_{load} & & \\
(\boldsymbol{\xi}_{cp} \boldsymbol{\alpha}_{intv}) \otimes \boldsymbol{I} \otimes \boldsymbol{I} & \boldsymbol{T}_{cp} \oplus \boldsymbol{T}_{fail} \oplus \boldsymbol{T}_{trig} \oplus (-\lambda_{fail2}) & \boldsymbol{I} \otimes \boldsymbol{I} \otimes \boldsymbol{\xi}_{trig} & & (\mathbf{1}_{cp} \otimes \mathbf{1}_{trig} \otimes \boldsymbol{\xi}_{fail}) \boldsymbol{\alpha}_{load} & & (\mathbf{1}_{cp} \otimes \mathbf{1}_{trig} \otimes \mathbf{1}_{fail} \otimes \lambda_{fail2}) \boldsymbol{\alpha}_{load} \\
& & \boldsymbol{T}_{fail} \oplus \boldsymbol{T}_{cp} \oplus (-\lambda_{fail2}) & (\mathbf{1}_{fail} \otimes \boldsymbol{\xi}_{cp}) \boldsymbol{\alpha}_{rej} & (\boldsymbol{\xi}_{fail} \otimes \mathbf{1}_{cp}) \boldsymbol{\alpha}_{load} & & (\mathbf{1}_{fail} \otimes \mathbf{1}_{cp} \otimes \lambda_{fail2}) \boldsymbol{\alpha}_{load} \\
\boldsymbol{\xi}_{rej} (\boldsymbol{\alpha}_{intv} \otimes \boldsymbol{\alpha}_{fail} \otimes \boldsymbol{\alpha}_{trig}) & & & \boldsymbol{T}_{rej} & & & \\
& & & & \boldsymbol{T}_{load} & \boldsymbol{\xi}_{load} \boldsymbol{\alpha}_{rc} & \\
\boldsymbol{\xi}_{rc} (\boldsymbol{\alpha}_{intv} \otimes \boldsymbol{\alpha}_{fail} \otimes \boldsymbol{\alpha}_{trig}) & & & & (\lambda_{fail1} \otimes \mathbf{1}_{rc}) \boldsymbol{\alpha}_{load} & (-\lambda_{fail1}) \oplus \boldsymbol{T}_{rc} & \\
& & & & & \boldsymbol{\xi}_{load} \boldsymbol{\alpha}_{rc} & \boldsymbol{T}_{load}
\end{pmatrix}. \tag{13}$$

The infinitesimal generator matrix is derived on the basis of the Kronecker representation [23], and the order of states is {*Normal, Checkpointing, Checkpointing′, Rejuvenation, Failure1, Recovery, Failure2*}. In Equation (13), $\oplus$ and $\otimes$ are the Kronecker product and sum [32], $\boldsymbol{I}$ is an identity matrix, and $1/\lambda_{fail1}$ and $1/\lambda_{fail2}$ are the mean values of EXP distributions $F_{fail1}(t)$ and $F_{fail2}(t)$, say the mean times to failure during rollback recovery and checkpointing, respectively.

Entry $(\boldsymbol{\xi}_{intv} \boldsymbol{\alpha}_{cp} \otimes \boldsymbol{I} \otimes \boldsymbol{I})$ shows that the clock of the rejuvenation trigger is not reset and continuously accumulates, even when the system executes the checkpointing. Since the checkpointing operation just saves the current data and does not refresh system aging, entry $(\boldsymbol{\xi}_{cp} \boldsymbol{\alpha}_{intv}) \otimes \boldsymbol{I} \otimes \boldsymbol{I}$ indicates that only the clock of checkpointing trigger is reset. When a rejuvenation point is reached while the system is under checkpointing, rejuvenation waits until checkpointing is completed; in such a case, the system transits from *Checkpointing* to *Checkpointing′* with entry $\boldsymbol{I} \otimes \boldsymbol{I} \otimes \boldsymbol{\xi}_{trig}$. Entries $(\mathbf{1}_{intv} \otimes \boldsymbol{\xi}_{fail} \mathbf{1}_{trig}) \boldsymbol{\alpha}_{load}$, $(\mathbf{1}_{cp} \otimes \mathbf{1}_{trig} \otimes \boldsymbol{\xi}_{fail}) \boldsymbol{\alpha}_{load}$, and $(\boldsymbol{\xi}_{fail} \otimes \mathbf{1}_{cp}) \boldsymbol{\alpha}_{load}$ indicate aging-related failures in both normal and checkpointing states, while entries $(\mathbf{1}_{cp} \otimes \mathbf{1}_{trig} \otimes \mathbf{1}_{fail} \otimes \lambda_{fail2}) \boldsymbol{\alpha}_{load}$ and $(\mathbf{1}_{fail} \otimes \mathbf{1}_{cp} \otimes \lambda_{fail2}) \boldsymbol{\alpha}_{load}$ represent human-error-related failures during checkpointing. In addition, the system is regarded to be as good as new after either rollback recovery or rejuvenation, so the corresponding transitions are represented by entries $\xi_{rej}(\boldsymbol{\alpha}_{intv} \otimes \boldsymbol{\alpha}_{fail} \otimes \boldsymbol{\alpha}_{trig})$, and $\xi_{rc}(\boldsymbol{\alpha}_{intv} \otimes \boldsymbol{\alpha}_{fail} \otimes \boldsymbol{\alpha}_{trig})$, where $(\boldsymbol{\alpha}_{intv} \otimes \boldsymbol{\alpha}_{fail} \otimes \boldsymbol{\alpha}_{trig})$ implies that the clocks of checkpointing trigger, system aging, and rejuvenation trigger are refreshed at the same time.

3.3. *Steady-State System Availability*

Steady-state system availability gives the probability that the system is operational in the steady state, so that it provides a significant insight into the long-term performance of a repairable system. Let A_{ss} define the steady-state system availability. Then, we can obtain it by

$$A_{ss} = \boldsymbol{\pi}_{ss}\boldsymbol{r}, \tag{14}$$

where $\boldsymbol{\pi}_{ss}$ is the steady-state probability vector of the PH-expanded CTMC, $\boldsymbol{Q}$, and can be computed by solving the following linear equation [33]:

$$\boldsymbol{\pi}_{ss}\boldsymbol{Q} = \mathbf{1}, \quad \boldsymbol{\pi}_{ss}\mathbf{1} = 1, \tag{15}$$

and $\boldsymbol{r}$ is the reward (column) vector of the PH-expanded CTMC and given by

$$\boldsymbol{r} = \begin{pmatrix} 1 \otimes \mathbf{1}_{intv} \otimes \mathbf{1}_{fail} \otimes \mathbf{1}_{trig} \\ 0 \otimes \mathbf{1}_{cp} \otimes \mathbf{1}_{fail} \otimes \mathbf{1}_{trig} \\ 0 \otimes \mathbf{1}_{fail} \otimes \mathbf{1}_{cp} \\ 0 \otimes \mathbf{1}_{rej} \\ 0 \otimes \mathbf{1}_{load} \\ 0 \otimes \mathbf{1}_{rc} \\ 0 \otimes \mathbf{1}_{load} \end{pmatrix}. \tag{16}$$

It is clear that the system is only available in the normal execution process state. In this paper, one problem of interest is to determine optimal software-rejuvenation timing that maximizes steady-state system availability.

4. Numerical Illustration

This section is devoted to the numerical illustration of the presented model in Figure 4 by means of phase expansion. Model parameters are summarized in Table 5, where all values are given according to the related literature [13,20,34]. All general distributions were accurately approximated by PH distributions with appropriate phases, that is, 100 phases for $G_{intv}(t)$, $G_{cp}(t)$, $G_{load}(t)$, $G_{rc}(t)$, $G_{trig}(t)$, and $G_{rej}(t)$ and 10 phases for $G_{fail}(t)$ (see [20] for more details); eventually, we obtained a large approximate CTMC consisting of 201,400 PH-expanded states. Similar to [20], in order to evaluate the effects of the checkpoint interval and the rejuvenation-trigger interval on system availability, the mean checkpoint interval (MCI) was varied from 1 to 10 h, and the mean rejuvenation-trigger interval (MRTI) was changed from 5 to 35 h. In addition, human-error-related system failures both were and were not considered, aiming at quantifying the effects of human-error factors on both system availability and optimal software-rejuvenation timing.

Table 5. Model parameters.

CDF	Distribution	Mean (h)	CV
$G_{intv}(t)$	Lognormal	1–10	0.2
$G_{fail}(t)$	Weilbull	10	0.5
$G_{cp}(t)$	Lognormal	0.05	0.2
$G_{load}(t)$	Lognormal	0.5	0.2
$G_{rc}(t)$	Lognormal	0.5	0.2
$G_{trig}(t)$	Lognormal	5–35	0.1
$G_{rej}(t)$	Lognormal	0.5	0.2
$F_{fail1}(t)$	Exponential	16.67	1
$F_{fail2}(t)$	Exponential	1.5	1

4.1. Steady-State System Availability

Here, we show the steady-state availabilities of a system that may fail due to human error in checkpointing under different cases of MRTI and MCI. The corresponding results are given in Table 6, which shows that steady-state system availability increased as the value of MCI increased under each MRTI case. This means that too-frequent checkpointing decreases system availability because the system becomes unavailable during checkpointing. The effect of MRTI on system availability is now examined. For each MCI, steady-state system availability increases at the beginning and subsequently decreases with increasing MRTI, implying that an optimal MRTI might exist for maximizing steady-state system availability.

Table 6. Steady-state system availability (with human-error-related system failures). Note: MCI, mean checkpoint interval; MRTI, mean rejuvenation-trigger interval.

MCI (h)	MRTI = 5 h	MRTI = 7 h	MRTI = 10 h	MRTI = 13 h	MRTI = 15 h
1	0.83333	0.84600	0.85168	0.85226	0.85194
2	0.86380	0.87684	0.88245	0.88259	0.88192
3	0.87494	0.88747	0.89309	0.89305	0.89227
4	0.87897	0.89335	0.89846	0.89836	0.89752
5	0.88327	0.89598	0.90182	0.90155	0.90069
6	0.88679	0.89801	0.90404	0.90369	0.90278
7	0.88849	0.90022	0.90531	0.90529	0.90430
8	0.88908	0.90204	0.90635	0.90637	0.90546
9	0.88925	0.90318	0.90740	0.90714	0.90630
10	0.88929	0.90377	0.90838	0.90779	0.90694

Moreover, by comparing results in Tables 6 and 7, the latter of which gives the steady-state system availability without considering human-error-related system failures, it is reasonable to say that human-error factors significantly decreased system availability, especially in the case where the value of MCI was small. In other words, although frequent checkpointing can save data in a timely manner, it also brings a higher risk of system failure, caused by incorrect operations. Therefore, it is crucial to determine a suitable frequency of executing checkpointing to satisfy target system availability. For example, given a target steady-state system availability of 0.9 and an MRTI of 10 h, an MCI equal to or larger than 5 h is a good choice.

Table 7. Steady-state system availability (without human-error-related system failures).

MCI (h)	MRTI = 5 h	MRTI = 7 h	MRTI = 10 h	MRTI = 13 h	MRTI = 15 h
1	0.84850	0.86206	0.86796	0.86821	0.86758
2	0.87067	0.88438	0.89024	0.89025	0.88942
3	0.87876	0.89200	0.89788	0.89779	0.89692
4	0.88154	0.89626	0.90174	0.90162	0.90073
5	0.88469	0.89810	0.90415	0.90393	0.90303
6	0.88735	0.89954	0.90576	0.90548	0.90456
7	0.88867	0.90117	0.90666	0.90665	0.90567
8	0.88913	0.90254	0.90741	0.90744	0.90652
9	0.88926	0.90341	0.90818	0.90800	0.90714
10	0.88929	0.90387	0.90892	0.90849	0.90761

4.2. Optimal Rejuvenation-Trigger Timing

This subsection discusses optimal software-rejuvenation timing maximizing steady-state system availability. Figure 5 illustrates the sensitivity of steady-state system availability with respect to the mean rejuvenation-trigger interval in the cases of MCI $= 2, 4, 6, 8$ and 10. The figure plots unimodal curves of the steady-state system availabilities, which reveals the existence of optimal rejuvenation-trigger timing maximizing steady-state system availability in each case. Specifically, the overhead incurred by frequent rejuvenation (i.e., short MRTI) largely affects system availability. Conversely, downtime due to system failures caused by a less frequent execution of rejuvenation smoothly decreases system availability.

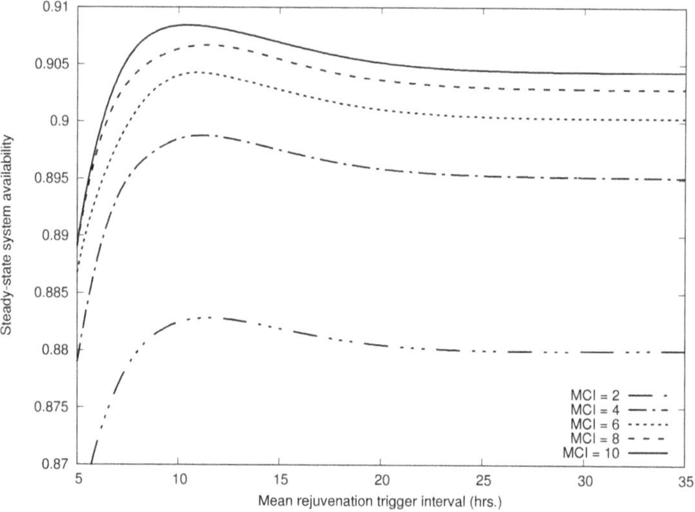

Figure 5. Sensitivity of steady-state system availability with respect to mean rejuvenation-trigger timing.

Optimal rejuvenation-trigger timings and their corresponding maximal steady-state system availabilities in all cases are presented in Table 8. We present all optimal rejuvenation timings for the system regardless of considering human-error-related system failures. Optimal MRTIs for all cases of MCI were very similar, which means that the optimal rejuvenation-trigger timing is not very sensitive to checkpoint interval. Optimal MRTIs in the case where human-error-related system failures were not considered were slightly smaller than those in the case with human-error-related failure when the value of MCI was small, and vice versa when the MCI had a large value, for example, MCI $= 9, 10$.

Table 8. Optimal rejuvenation-trigger timings.

MCI (h)	with Human-Error-Related Failures		without Human-Error-Related Failures	
	MRTI (h)	A_{ss}	MRTI (h)	A_{ss}
1	12.3	0.85230	11.6	0.86841
2	11.5	0.88283	11.3	0.89059
3	11.3	0.89339	11.2	0.89819
4	11.2	0.89878	11.2	0.90206
5	11.0	0.90196	11.1	0.90435
6	10.9	0.90428	11.0	0.90603
7	11.3	0.90572	11.3	0.90708
8	11.4	0.90668	11.4	0.90777
9	11.0	0.90753	11.1	0.90838
10	10.5	0.90842	10.7	0.90902

5. Conclusions

In this paper, we presented a composite stochastic Petri reward net and its resulting non-Markovian availability model for operational software systems where both checkpointing and software rejuvenation are adopted to protect data and to enhance the system availability, and the system may fail due to both the aging problem and human errors during checkpointing. More specifically, the non-Markovian availability model was derived on the basis of a reachability graph that was generated from the original SRNs. In particular, the PH expansion approach was applied to solve the stationary solution of the non-Markovian availability model since the model was not one of the trivial stochastic models such as SMP and MRGP, so that common approaches such as LST and embedded Markov chain techniques do not work. Numerical results showed that human-error factors both decreased steady-state system availability and brought a significant effect on optimal rejuvenation-trigger timing, which means that human-error factors during system modeling should not be overlooked.

The model presented in this paper was based on a macroscopic view, providing a fundamental idea of how to model such a software system that undergoes both checkpointing and software rejuvenation, and in which the system behaves with multiple competitive events. The system's actual behavior is very complex, and more possible events need to be considered, for example, software environment upgrades and time-scope limitations of used versions of libraries. Although this improvement may vastly increase difficulty in numerical analysis, it is significant to take a microscopic look at system behavior, which will be one of our future directions. This paper only considered both aperiodic checkpointing and software rejuvenation, but to the best of our knowledge, there exist various kinds of checkpointing [35] and rejuvenation techniques [8]. In the future, we aim to extend this work to solve more complicated software systems considering different rejuvenation and checkpointing schemes.

Author Contributions: Conceptualization, J.Z., H.O. and T.D.; methodology, J.Z., H.O. and T.D.; formal analysis, J.Z.; investigation, J.Z.; writing—original draft preparation, J.Z.; writing—review and editing, H.O. and T.D.; supervision, H.O. and T.D. All authors have read and agreed to the published version of the manuscript.

Funding: This research received no external funding.

Institutional Review Board Statement: Not applicable.

Informed Consent Statement: Not applicable.

Data Availability Statement: Not applicable.

Conflicts of Interest: The authors declare no conflict of interest.

Abbreviations

The following abbreviations are used in this manuscript:

MRGP	Markov regenerative process
LST	Laplace–Stieltjes transform
SRN	Stochastic (Petri) reward net
PH	Phase or phase-type
CTMC	Continuous-time Markov chain
IMM	Immediate
EXP	Exponential
GEN	General
APH	Acyclic PH distribution
CF	Canonical form
MLE	Maximum-likelihood estimation
MCI	Mean checkpoint interval
MRTI	Mean rejuvenation-trigger interval

References

1. Grottke, M.; Trivedi, K.S. Fighting bugs: remove, retry, replicate, and rejuvenate. *IEEE Comput.* **2007**, *40*, 107–109. [CrossRef]
2. Dohi, T.; Trivedi, K.S.; Avritzer, A. *Handbook of Software Aging and Rejuvenation: Fundamentals, Methods, Applications, and Future Directions*; World Scientific: Singapore, 2020.
3. Huang, Y.; Kintala, C.; Kolettis, N.; Funton, N.D. Software rejuvenation: Analysis, module and applications. In Proceedings of the 25th IEEE International Symposium on Fault Tolerant Computing (FTC'95), Pasadena, CA, USA, 27–30 June 1995; pp. 381–390.
4. Trivedi, K.S.; Vaidyanathan, K. Software aging and rejuvenation. In *Wiley Encyclopedia of Computer Science and Engineering*; John Wiley and Sons: Hoboken, NJ, USA, 2007; pp. 1–8.
5. Alonso, J.; Matias, R.; Vicente, E.; Maria, A.; Trivedi, K.S. A comparative experimental study of software rejuvenation overhead. *Perform. Eval.* **2013**, *70*, 231–250. [CrossRef]
6. Vaidyanathan, K.; Trivedi, K.S. A comprehensive model for software rejuvenation. *IEEE Trans. Depend. Secur. Comput.* **2005**, *2*, 124–137. [CrossRef]
7. Ning, G.; Zhao, J.; Lou, Y.; Alonso, J.; Matias, R.; Trivedi, K.S.; Yin, B.B.; Cai, K.Y. Optimization of two-granularity software rejuvenation policy based on the Markov regenerative process. *IEEE Trans. Reliab.* **2016**, *65*, 1630–1646. [CrossRef]
8. Zheng, J.; Okamura, H.; Li, L.; Dohi, T. A comprehensive evaluation of software rejuvenation policies for transaction systems with Markovian arrivals. *IEEE Trans. Reliab.* **2017**, *66*, 1157–1177. [CrossRef]
9. Dohi, T.; Zheng, J.; Okamura, H.; Trivedi, K.S. Optimal periodic software rejuvenation policies based on interval reliability criteria. *Reliab. Eng. Syst. Saf.* **2018**, *180*, 463–475. [CrossRef]
10. Wang, S.; Liu, J. HARRD: Real-time software rejuvenation decision based on hierarchical analysis under weibull distribution. In Proceedings of the 20th IEEE International Conference on Software Quality, Reliability and Security (QRS'20), Macau, China, 11–14 December 2020; pp. 83–90.
11. Zhang, Y.; Chakrabarty, K. Fault recovery based on checkpointing for hard real-time embedded systems. In Proceedings of the 18th IEEE Symposium on Defect and Fault Tolerance in VLSI Systems (DFT'03), Boston, MA, USA, 5 November 2003; pp. 320–327.
12. Fukumoto, S.; Kaio, N.; Osaki, S. Optimal checkpointing policies using the checkpointing density. *J. Inf. Process.* **1992**, *15*, 87–92.
13. Dohi, T.; Osajima, S.; Kaio, N.; Osaki, S. On the effects of checkpoint institution methods for a macroscopic database model. *Electron. Commun. Jpn. Part III Fundam. Electron. Sci.* **2000**, *83*, 23–33. [CrossRef]
14. Ranganathan, A.; Upadhyaya, S.J. Performance evaluation of rollback-recovery techniques in computer programs. *IEEE Trans. Reliab.* **1993**, *42*, 220–226. [CrossRef]
15. Bajunaid, N.; Menascé, D.A. Efficient modeling and optimizing of checkpointing in concurrent component-based software systems. *J. Syst. Softw.* **2018**, *139*, 1–13. [CrossRef]
16. Sigdel, P.; Tzeng, N.F. Coalescing and deduplicating incremental checkpoint files for restore-express multi-level checkpointing. *IEEE Trans. Parallel Distrib. Syst.* **2018**, *29*, 2713–2727. [CrossRef]
17. Okamura, H.; Dohi, T. Comprehensive evaluation of aperiodic checkpointing and rejuvenation schemes in operational software system. *J. Syst. Softw.* **2010**, *83*, 1591–1604. [CrossRef]
18. Levitin, G.; Xing, L.; Luo, L. Joint optimal checkpointing and rejuvenation policy for real-time computing tasks. *Reliab. Eng. Syst. Saf.* **2019**, *182*, 63–72. [CrossRef]
19. Zheng, J.; Okamura, H.; Dohi, T. A phase expansion for non-Markovian availability models with time-based aperiodic rejuvenation and checkpointing. *Commun. Stat-Theory Methods* **2020**, *49*, 3712–3729. [CrossRef]
20. Zheng, J.; Okamura, H.; Dohi, T. Optimal rejuvenation policies for non-Markovian availability models with aperiodic checkpointing. *IEICE Trans. Inf. Syst.* **2020**, *E103-D*, 2133–2142. [CrossRef]
21. Bolch, G.; Greiner, S.; De Meer, H.; Trivedi, K.S. *Queueing Networks and Markov Chains: Modeling and Performance Evaluation with Computer Science Applications*, 2nd ed.; John Wiley and Sons: New York, NY, USA, 2006.

22. Okamura, H.; Dohi, T. Fitting phase-type distributions and Markovian arrival processes: Algorithms and tools. In *Principles of Performance and Reliability Modeling and Evaluation*; Lance, F., Antonio, P., Eds.; Springer: Berlin/Heidelberg, Germany, 2016; pp. 49–75.
23. Trivedi, K.S.; Bobbio, A. *Reliability and Availability Engineering: Modeling, Analysis, and Applications*; Cambridge University Press: Cambridge, UK, 2017.
24. Brown, A. An Overview of Human Error. *CS294-4 ROC Semin.* **1990**, *54*. Available online: http://roc.cs.berkeley.edu/294fall01/slides/human-error.pdf (accessed on 10 December 2020).
25. Yanagihara, M.; Odagiri, M.; Osaki, S.; Kaio, N. Optimal checkpointing procedures taking into account system failure caused by checkpointing. *Electron. Commun. Jpn. Part III Fundam. Electron. Sci.* **1995**, *78*, 69–79. [CrossRef]
26. Zheng, J.; Okamura, H.; Dohi, T. A transient interval reliability analysis for software rejuvenation models with phase expansion. *Softw. Qual. J.* **2020**, *28*, 173–194. [CrossRef]
27. Yang, X.; Alfa, A.S. A class of multi-server queueing system with server failures. *Comput. Ind. Eng.* **2009**, *56*, 33–43. [CrossRef]
28. Ruiz-Castro, J.E.; Acal, C.; Aguilera, A.M.; Roldán, J.B. A complex model via phase-type distributions to study random telegraph noise in resistive memories. *Mathematics* **2021**, *9*, 390. [CrossRef]
29. Kemper, P.; Müller, D.; Thümmler, A. Combining response surface methodology with numerical methods for optimization of Markovian models. *IEEE Trans. Depend. Secur. Comput.* **2006**, *3*, 259–269. [CrossRef]
30. Cumani, A. On the canonical representation of homogeneous Markov processes modelling failure-time distributions. *Microelectron. Reliab.* **1982**, *22*, 583–602. [CrossRef]
31. Okamura, H.; Dohi, T.; Trivedi, K.S. Improvement of EM algorithm for phase-type distributions with grouped and truncated data. *Appl. Stoch. Model. Bus. Ind.* **2013**, *29*, 141–156. [CrossRef]
32. Dayar, T. *Analyzing Markov Chains Using Kronecker Products: Theory and Applications*; Springer Science and Business Media: New York, NY, USA, 2012.
33. Trivedi, K.S. *Probability and Statistics with Reliability, Queuing, and Computer Science Applications*, 2nd ed.; John Wiley and Sons: Hoboken, NJ, USA, 2001.
34. Leung, C.H.C.; Currie, E. The effect of failures on the performance of long-duration database transactions. *Comput. J.* **1995**, *38*, 471–478. [CrossRef]
35. Tantawi, A.N.; Ruschitzka, M. Performance analysis of checkpointing strategies. *ACM Trans. Comput. Syst.* **1984**, *2*, 123–144. [CrossRef]

Article

DICER 2.0: A New Model Checker for Data-Flow Errors of Concurrent Software Systems

Dongming Xiang [1,*], Fang Zhao [2] and Yaping Liu [3]

1 The School of Information Science and Technology, Zhejiang Sci-Tech University, Hangzhou 310018, China
2 The MoE Key Lab of Embedded System & Service Computing, Tongji University, Shanghai 201804, China; 1610471@tongji.edu.cn
3 The School of Transportation Management, Zhejiang Institute of Communications, Hangzhou 310018, China; liuyp061054@zjvtit.edu.cn
* Correspondence: dmxiang@zstu.edu.cn

Abstract: Petri nets are widely used to model concurrent software systems. Currently, there are many different kinds of Petri net tools that can analyze system properties such as deadlocks, reachability and liveness. However, most tools are not suitable to analyze data-flow errors of concurrent systems because they do not formalize data information and lack efficient computing methods for analyzing data-flows. Especially when a concurrent system has so many concurrent data operations, these Petri net tools easily suffer from the state–space explosion problem and pseudo-states. To alleviate these problems, we develop a new model checker DICER 2.0. By using this tool, we can model the control-flows and data-flows of concurrent software systems. Moreover, the errors of data inconsistency can be detected based on the unfolding techniques, and some model-checking can be done via the guard-driven reachability graph (GRG). Furthermore, some case studies and experiments are done to show the effectiveness and advantage of our tool.

Keywords: petri net; concurrent software systems; model-checking; data-flows

Citation: Xiang, D.; Zhao, F.; Liu, Y. DICER 2.0: A New Model Checker for Data-Flow Errors of Concurrent Software Systems. *Mathematics* **2021**, *9*, 966. https://doi.org/10.3390/math9090966

Academic Editors: Tadashi Dohi and Vassilis C. Gerogiannis

Received: 11 March 2021
Accepted: 22 April 2021
Published: 25 April 2021

1. Introduction

Presently, concurrent software systems are widely used in our daily life. In particular, they are successfully applied in so many safety-critical scenarios, e.g., health-care, intelligent traffic, and stock exchange. Thus, how to guarantee the correctness of concurrent systems has become a bone of contention for people's lives and properties. In reality, the correctness of concurrent systems is closely related with control-flows and data-flows. However, the most existing studies mainly focus on the error detections of control-flows such as deadlocks, livelocks and compatibility [1–3]. In fact, concurrent systems are vulnerable to data-flow errors, e.g., missing data, lost data and data inconsistency [4–6]. Although the testing-based methods can detect these errors, they need to design a series of test cases to cover as many execution paths as possible. Due to the difficulty in the completeness of test cases, it is hard for these methods to guarantee a concurrent system error-free.

The Petri net-based model-checking is a prominent method/technique for analyzing data-flows of concurrent software systems. This is because Petri nets [7–10] have a great capability of explicitly specifying parallelism, concurrency and synchronization [11,12]. Thus, many different kinds of Petri nets are used to check data-flow errors, such as algebraic Petri net (or extended concurrent algebraic nets, ECANets), predicate/transitions net (PrTNet), and colored Petri nets (CPN), etc. Kheldoun et al. [13] transformed BPMN (Business Process Model and Notation) models of complex business processes into to Recursive ECATNets (RECATNets), which combine the expressive power of abstract data types with recursive Petri nets. Furthermore, they used rewriting logics to check proper terminations and LTL properties. Buchs et al. [14] proposed Concurrent Object-Oriented Petri

Nets (CO-OPN/2) to ensure the specifications of control/data-flows in a large distributed system. Barkaoui et al. [15] provided an approach for detecting data consistency with respect to a multilevel security policy based on ECATNets. He et al. [16] modeled smart contracts by predicate/transition nets, and then checked their correctness of pre/post-conditions. Wu et al. [17] developed a model-based method for quantitative safety analysis of safety-critical systems by Timed Colored Petri Nets (TCPNs). Yu et al. [18] proposed an E-commerce Business Process Net (EBPN) to verify the rationality and transaction consistency between trading parties. All these methods place emphasis on the formalizations of data structures and abstract data types. Thus, they are suitable to check data-flow errors caused by these aspects.

By comparison, some checking methods based on Petri nets focus on the modeling of conceptual data operations, e.g., *read*, *write* and *delete*. Dual Flow Nets (DFNs) [19] were proposed to model control- and data-flows of embedded systems. Awad et al. [20] mapped BPMN models into Petri nets, and then detected and repaired errors based on the work in [21]. Contextual net (C-net) [22,23] was proposed to model a concurrent read operation. Furthermore, its unfolding technique was developed to generate a minimal test suite for multi-threaded programs [24]. Referring to contextual nets, Petri Net with Data Operations (PN-DO) [5] was given to detect data-flow errors of concurrent software systems. However, these explicit formalizations of read/write arcs and data places easily increase the scales and complexity of Petri net models. Fortunately, *WFD-net* (WorkFlow net with Data) [4,25,26], as a high-level Petri net [8], is extended with conceptual labeling data operations and guards. Thus, on the one hand, a WFD-net can greatly model control-flows and data-flows of concurrent systems. On the other hand, the model scales of WFD-nets are much smaller than other Petri nets with data-flow arcs (e.g., read arcs, write arcs and delete arcs), such as C-net and PN-DO. Furthermore, WFD-net has been widely used to do model-checking, e.g., soundness [25], completion requirements [27] and data consistencies [28], although it is an easy way to model software systems. In general, these verification/analysis methods are based on the classical reachability graphs (CRG) [25] of WFD-nets. However, they easily suffer from the problems of state–space explosion and illegal states (or pseudo-states). This is because a state may have an exponential number of successor states since they are produced based on the possible values of all guards. Moreover, the exclusive logical relations (e.g., multiple choice conditions) between guards easily lead to pseudo-states. In order to alleviate these problems, we proposed a guard-driven reachability graph (GRG) of WFD-nets in our previous work [29].

Although a GRG of WFD-nets can describe all running information of concurrent systems and save their state–space compared with CRG, it still likely suffers from the state–space explosion problem. As shown in Figure 1, it easily leads to a rapid increase of state–space with the increase of concurrent operations of WFD-nets. This is because the interleaving semantics of GRG is based on the partial orders of fired transitions, and it describes the behaviors of concurrent systems only by global states. Thus, a GRG-based analysis method needs to find out all precedence relations between activities, and generates their successor states. Compared with the interleaving-semantics-based methods, some studies are conducted on a concurrency analysis of Petri nets [30,31]. In particular, the unfolding technique [32] can both alleviate the state space explosion problem and characterize the concurrency relations due to its true concurrency semantics [33]. Currently, this technique has been successfully applied in different kinds of model-checking, e.g., fault diagnosis [34], concurrent planning [35], test case generations [36], deadlock detection [37], and verifying soundness [38], reachability and coverability [39]. Thus, in view of these advantages, we proposed an unfolding-based method [5] to check errors of data inconsistency. Specifically, we use an acyclic net to represent all behaviors of a Petri net with data (PD-net [5]). On the one hand, all concurrent operations can be directly recorded in this acyclic net. On the other hand, this formal model can store all states and save much more space especially when a system has so many concurrent activities.

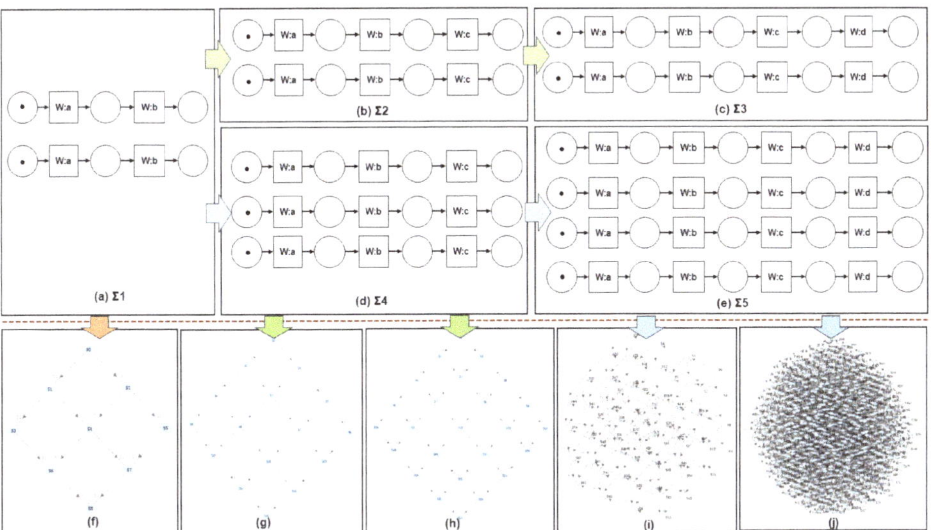

Figure 1. The state–space (reachability graphs) of WFD-nets and state–space explosion problems. (**f**) is the reachability graph of Σ_1 in (**a**); (**g**) is the reachability graph of Σ_2 in (**b**); (**h**) is the reachability graph of Σ_3 in (**c**); (**i**) is the reachability graph of Σ_4 in (**d**); and (**j**) is the reachability graph of Σ_5 in (**e**).

To support and improve the above previous work [5,29], we develop a new model checker DICER 2.0. Currently, there are many Petri net tools [40–42] such as PIPE, Snoopy, CPN Tools, Protos, and ProM. These tools can support different kinds of Petri net modeling, e.g., Place/Transition nets [7], Timed Petri nets [9], Stochastic Petri nets [10] and High-level Petri net [8]. Furthermore, they can be used to do structural analysis, generate condensed state spaces, construct reachability graphs, and analyze place/transition invariants. However, most of these tools fail in unfolding a Petri net. Although Mole, ERVunfold and Punf can do this work and conduct some model-checking (e.g., deadlocks, reachability and coverability), they cannot support the modeling and checking of data-flows that have been considered in some abstracted models, such as WFD-net [43] and PD-net [5]. Therefore, the most existing Petri net tools are not suitable to analyze data-flow errors of concurrent systems especially based on the unfolding techniques. The specified comparisons between some Petri net tools are summarized as Table 1.

In this paper, we develop DICER 2.0 to analyze data-flows of concurrent systems. Specifically, we can use this tool to model concurrent systems by general Petri nets, WFD-nets and PD-nets. Meanwhile, we can draw, edit, import and export these models in DICER 2.0. Moreover, the errors of data inconsistency can be detected based on the unfolding technique of PD-nets, and some GRG-based model-checking can be done in our tool.

This paper is organized as follows. Section 2 presents some basic notations. Section 3 introduces our model checker DICER 2.0. Section 4 gives two case studies on concurrent systems. Section 5 conducts a group of experiments to show the advantages of our tool. The last section concludes this paper.

Table 1. The comparison between some Petri net tools

Tools	Petri Nets	Functions	Branching Process	The Unfolding Techniques within Data-Flows	Data-Flow Error Detection
Snoopy CPN Tools ProM PIPE2 PROTOS	P/T net Timed Petri net High-level Petri net	Graphical editor Reachability graph Condensed state spaces P/T invariants Structural analysis	×	×	×
Maude Acceleo+Maude PIPE+	ECATNet RECATNet PrTNet	Rewriting logic LTL model-checking Transform RECATNets into rewriting logics Modeling & simulating	×	×	✓
ERVunfold Tours	P/T net	Deadlocks Test-case generation	✓	×	×
PUNF MOLE	Safe C-net	Reachability Coverability	✓	×	×
DICER 2.0	WFD-net PD-net P/T net WF-net	Detecting data inconsistency Deadlocks Reachability	✓	✓	✓

2. Basic Notations

A *net* is a triple $N = (P, T, F)$, where P and T are two finite and disjoint sets, and they are called *place* and *transition*, respectively. $F \subseteq (P \times T) \cup (T \times P)$ denotes a *flow* relation. A *marking* of a net is a mapping function m: $P \to \mathbb{N}$, where $\mathbb{N}$ is a set of non-negative integers. In details, we use a multi-set to represent a marking. A net N with an initial marking m_0 is called a *Petri net* Σ [7] , i.e., $\Sigma = (N, m_0)$. Given a node $x \in P \cup T$, its *preset* and *postset* are respectively denoted by $^\bullet x$ and $x^\bullet$, where $^\bullet x = \{y \mid (y, x) \in F\}$ and $x^\bullet = \{y \mid (x, y) \in F\}$.

As a particular class of Petri net, *workflow net* (WF-net) is widely used to model control-flows of concurrent systems.

Definition 1. *A net $N = (P, T, F)$ is a WF-net (workflow net) [43,44] if*

(1) *there exists only one source place i and one sink place o satisfying $^\bullet i = \emptyset$ and $o^\bullet = \emptyset$; and*

(2) *each node $x \in P \cup T$ is on a path from i to o.*

Besides modeling control-flows of concurrent systems, we can use a net with some data information to formalize data-flows.

Definition 2. *A 7-tuple $N = (P, T, F, D, Read, Write, Delete)$ is a net with data (D-net) [5], if*

(1) *(P, T, F) is a net;*

(2) *D is a finite set of data elements;*

(3) *Read: $T \to 2^D$ is a labeling function of reading data;*

(4) *Write: $T \to 2^D$ is a labeling function of writing data; and*

(5) *Delete: $T \to 2^D$ is a labeling function of deleting data.*

Given two nodes $x, y \in P \cup T$ in an acyclic D-net $N = (P, T, F, D, Read, Write, Delete)$,

(1) x and y are in *causality* relation if the net N contains a path from x to y, which is denoted by $x \preceq y$. In particular, $x \prec y$ if $x \neq y$;

(2) x and y are in *conflict* relation if $\exists\, t_1, t_2 \in T$: $t_1 \preceq x$, $t_2 \preceq y$ and $^\bullet t_1 \cap {^\bullet t_2} \neq \emptyset$, which is denoted by $x \# y$;

(3) x and y are in *backward-conflict* relation if $x^\bullet \cap y^\bullet \neq \emptyset$, which is denoted by $x \widetilde{\#} y$; or

(4) x and y are in *concurrency* relation if $\neg(x \prec y \vee y \prec x \vee x\#y)$, which is denoted by x *co* y, i.e., x and y are neither in causality relation nor in conflict relation.

OD-net (Occurrence Net with Data) is a simple acyclic net, which can be used in the unfolding technique of PD-nets [5].

Definition 3. *A D-net* $N = (P, T, F, D, Read, Write, Delete)$ *is an OD-net (Occurrence net with Data) [5] if*
(1) $\forall p \in P$: $|{}^{\bullet}p| \leq 1$;
(2) $\forall x, y \in P \cup T$: $x \prec y \Rightarrow y \nprec x$; *and*
(3) *no transition is in self-conflict relation, i.e.,* $\forall t \in T{:}\neg(t\#t)$.

In an OD-net, places and transitions are called *conditions* and *events*, respectively. In general, we use $O = (B, E, G, D, Rd, Wr, De)$ to denote an occurrence net with data for convenience. With respect to this formalization, B, E and G are conditions, events and arcs, respectively. Rd, Wr and De are labeling functions of data operations (read, write and delete), respectively.

3. DICER 2.0

DICER 2.0 is developed to model and analyze the control-/data-flows of concurrent systems. It is the derivative version of our model checker for detecting data inconsistency [45]. Currently, we can use it to do many more model checking.

3.1. The Modeling of Concurrent Systems Based on the Petri Net with Data Information

As is well known, we usually use read/write arcs, data places, labeling functions of data operations and guards to formalize data-flows of concurrent systems [4,19,46]. In these formalizations, Petri nets such as DFN [19], PN-DO [47] and Awad method [20] mainly use data places and flow relations to model data operations, e.g., read, write and delete. Although these methods are suitable to accurately model the control structures of data-flows, it lacks formal semantic descriptions about shared reading and overwriting. Contextual net [46] can describe the concurrent (shared) reading operation by read arcs, but it needs extra data places and flow relations to formalize data-flows, and thus may be much more complex [48].

Compared with the above modeling methods, WFD-net [4,49] has a prominent advantage. It combines the traditional workflow nets with conceptual data operations, and uses labeling functions and guards to describe data operations and routing conditions, respectively. Thus, it is not only greatly suitable to model the control-flows and data-flows of a concurrent system but also much smaller than other Petri nets with data-operation arcs (e.g., contextual net and PN-DO) in the scales of nodes and arcs [48]. Now, this modeling method has been widely applied to various model-checking, e.g., detecting data-flow errors [4] and data inconsistency in the migrations of service cases [28], checking data inaccuracy [50] and completed requirements [27], and verifying may/must soundness of workflow systems [25].

Definition 4. *A workflow net with data (WFD-net) is a 9-tuple* $N = (P, T, F, D, GD, Read, Write, Delete, Guard)$ *[25], if*
(1) (P, T, F) *is a WF-net;*
(2) D *is a finite set of data elements;*
(3) $Read$: $T \rightarrow 2^D$ *is a labeling function of reading data;*
(4) $Write$: $T \rightarrow 2^D$ *is a labeling function of writing data;*
(5) $Delete$: $T \rightarrow 2^D$ *is a labeling function of deleting data;*
(6) GD *is a finite set of guards that are related with data elements in* D*; and*
(7) $Guard$: $T \rightarrow GD$ *is a labeling function of assigning guards to transitions.*

Referring to the labeling functions of data operations in WFD-nets, a *Petri net with data* (PD-net) [5] is proposed, i.e., a PD-net Σ is a D-net N with an initial marking m_0, i.e., $\Sigma = (N, m_0)$. Although this modeling method neglects the formalization of guards, it is much suitable for generating the unfolding of Peri nets with data information due to its simple structural semantics. For example, Σ is a WFD-net in Figure 2a, while Σ' is a PD-net in Figure 2c,d is its unfolding.

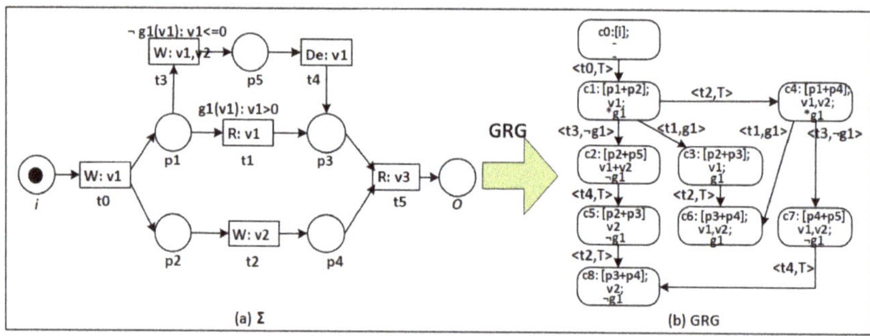

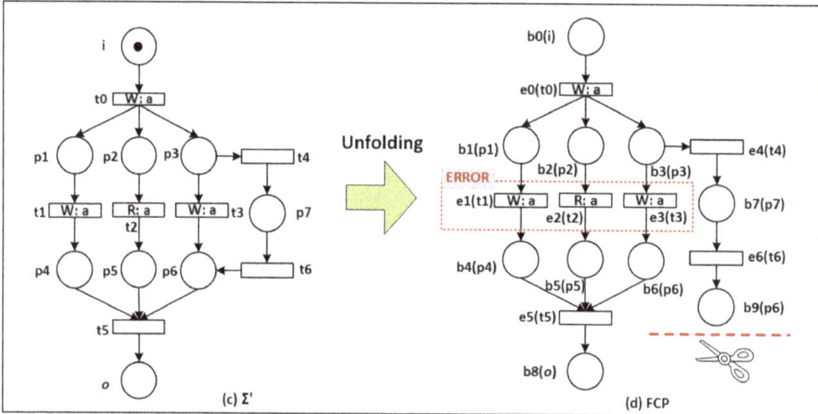

Figure 2. (**a**) A WFD-net Σ; (**b**) the guard-driven reachability graph (GRG) of Σ; (**c**) a PD-net Σ'; (**d**) the unfolding FCP of Σ'.

DICER 2.0 supports the modeling of WFD-nets and PD-nets. By this tool, we can formalize the control-/data-flows of concurrent systems. Furthermore, it provides a series of model-checking based on the guard-driven methods and unfolding techniques.

3.2. The Model-Checking Based on the GRG of WFD-Nets

The classical reachability graph [25] is a fundamental method for analyzing a WFD-net. However, this method easily suffers from the problems of state–space explosion and pseudo-states (or illegal states) due to its guard evaluations and their exclusive relations. Hence, we proposed a *Guard-driven Reachability Graph* (GRG) in our previous work [29], and now achieve this function in DICER 2.0.

To construct a GRG of WFD-nets, we define a state as a *weak configuration* in DICER 2.0, which includes a marking and some evaluations of data and guards.

Definition 5. *(Weak configuration) Given a WFD-net $N = (P, T, F, D, GD, Read, Write, Delete, Guard)$, $c = \langle m, \sigma, \eta \rangle$ is a weak configuration, if*

(1) (P, T, F) is a WF-net, and m is its marking;

(2) a mapping function $\sigma : D \rightarrow \{\top, \bot\}$ assigns a defined value ($\top$) or an undefined value ($\bot$) to each data element; and

(3) *a mapping function* $\eta : GD \rightarrow \{TRUE, FALSE, \bot, \top\}$ *assigns the values of TRUE , FALSE,* $\bot$ *or* $\top$ *to each guard.*

In DICER 2.0, we also define the basic enabling/firing rules of WFD-nets based on weak configurations.

Definition 6. *(Enabling/firing rules) Given a WFD-net N =(P, T, F, D, GD, Read, Write, Delete, Guard) and its weak configuration* $c = \langle m, \sigma, \eta \rangle$, *a transition t is enabled at c and denoted by* $c[t\rangle$, *if*

(1) $m[t\rangle$;

(2) $\forall v \in Read(t) : \sigma(v) = \top$; *and*

(3) $\forall v \in Varb(Guard(t))$: $\sigma(v) \neq \bot \wedge \eta(Guard(t)) \in \{TRUE, \top\}$, *where the function Varb is to obtain all variables in a guard.*

After firing a transition t at the weak configuration c, a new weak configuration $c' = \langle m', \sigma', \eta' \rangle$ *can be generated, i.e.,* $c[t\rangle c'$, *where*

(1) $m[t\rangle m'$;

(2) $\forall v \in Write(t) \setminus De(t) : \sigma'(v) = \top$;

(3) $\forall v \in Delete(t) : \sigma'(v) = \bot$;

(4) $\forall v \in D \setminus (Write(t) \cup Delete(t)) : \sigma'(v) = \sigma(v)$;

(5) $\exists g \in Guard(t) : Write(t) \cap Var(g) = \emptyset \Rightarrow \eta'(g) = TRUE$; *and*

(6) $\forall g \in GD, \forall v \in Varb(g) : (\sigma'(v) = \top \Rightarrow \eta'(g) = \top) \wedge ((Write(t) \cap Varb(g) = \emptyset \wedge g \notin Guard(t)) \Rightarrow \eta'(g) = \eta(g))$.

Let c_1 and c_2 be two weak configurations of a WFD-net. c_2 is *may-reachable* from c_1, denoted as $c_1 \rightarrow^*_{may} c_2$, if there exist some weak configurations $c^{(1)}, c^{(2)}, \cdots, c^{(n)}$ such that $c_1[t_1\rangle c^{(1)}[t_2\rangle c^{(2)}[t_3\rangle \cdots c^{(n)}[t_3\rangle c_2$. Furthermore, a set of may-reachable weak configurations from c_1 is denoted by $R(c_1)$. Based on may-reachable sets and enabling/firing rules, we can formalize a GRG in DICER 2.0 as follows.

Definition 7. *Given a WFD-net N =(P, T, F, D, GD, Read, Write, Delete, Guard) and its initial weak configuration* c_0, $GRG(N) = (V^+, E^+, \ell^+)$ *is a guard-driven reachability graph (GRG), where*

(1) $V^+ = R(c_0)$, $E^+ = \{(c, c') \mid \exists c, c' \in R(c_0), \exists t \in T : c[t\rangle c'\}$, *and*

(2) ℓ^+: $E^+ \rightarrow T \times GD$ *such that* $(c, c') \in E^+ \wedge c[t\rangle c' \wedge \ell^+(c, c') = \langle t, Guard(t) \rangle$.

For example, Figure 2b shows a guard-driven reachability graph of Figure 2a, where g_1 and $\neg g_1$ are two exclusive guards, $c_0 = \langle [i], -, - \rangle$ and $c_1 = \langle [p_1 + p_2], \{v_1\}, \{*g_1\} \rangle$ are two weak configurations such that $c_0[t_0\rangle c_1$.

Since a GRG of a WFD-net contains all execution information of a concurrent system, we can traverse its reachable weak configurations by DICER 2.0 to do some model-checking such as deadlocks [51] and proper completeness [27], i.e., given a WFD-net N and its guard-driven reachability graph $GRG(N)$, o is its sink place and $c = \langle m, \sigma, \eta^+ \rangle$ is a weak configuration such that $c \in R(c_0)$.

- If $m(o) = 0$ and no transition is enabled at the weak configuration c, then c is a *deadlock*. Thus, we can check deadlocks in N according to this formal specification. For example, the WFD-net in Figure 2a have a deadlock at the weak configuration $c_8 : \langle [p_3 + p_4], \{v_2\}, \{\neg g_1\} \rangle$ because t_5 cannot read the data v_3 and no transition is enabled at this time.
- If $\forall c \in R(c_0) : m(o) > 0 \Rightarrow m = \{o\}$, then N is *properly completed*. For example, the WFD-net in Figure 2a is not properly completed since the final weak configuration is not reachable from the initial weak configuration and the sink place o has no token at this time.

3.3. The Model-Checking Based on the Unfolding Techniques of PD-Nets

Besides the model-checking based on GRGs of WFD-nets, DICER 2.0 can be used to detect errors of data inconsistency based on the unfolding techniques of PD-nets. We first define branching processes in DICER 2.0.

Definition 8. *Given a PD-net* $\Sigma = (N, m_0) = (P, T, F, m_0, D, Read, Write, Delete)$ *and an OD-net* $O = (B, E, G, D, Rd, Wr, De)$, *the mapping* $h : B \cup E \to P \cup T$ *is a homomorphism between* Σ *and* O. (O, h) *is a branching process if satisfying:*

(1) $h(E) \subseteq T$ *and* $h(B) \subseteq P$;

(2) *for each event e belonging to E, the restriction of h onto* ${}^\bullet e$ *(resp.,* $e^\bullet$*) is a bijection between* ${}^\bullet e$ *and* ${}^\bullet h(e)$ *(resp., between* $e^\bullet$ *and* $h(e)^\bullet$*);*

(3) *the restriction of h onto* $Min(O)$ *is a bijection between* $Min(O)$ *and* m_0;

(4) $\forall e_1, e_2 \in E$: $({}^\bullet e_1 = {}^\bullet e_2) \wedge (h(e_1) = h(e_2)) \Rightarrow e_1 = e_2$; *and*

(5) $\forall e \in E : Rd(e) = Read(h(e)) \wedge Wr(e) = Write(h(e)) \wedge De(e) = Delete(h(e))$.

Given two branching processes $(O_i, h_i) = (B_i, E_i, G_i, D, Rd_i, Wr_i, De_i, h_i)$ and $i \in \{1, 2\}$, (O_1, h_1) is a prefix of (O_2, h_2) if $B_1 \subseteq B_2 \wedge E_1 \subseteq E_2$. All branching processes of a PD-net Σ forms a partial order set *w.r.t* the binary relation of *prefix*, and its greatest element is *Unfolding* [46], which is denoted by $Unf(\Sigma)$. Please note that the unfolding of a PD-net is also an occurrence net with data. Although the unfolding of a PD-net records its running information, it may be infinite if there exists an infinite execution path. Therefore, it needs to be truncated so as to get a *finite complete prefix* (FCP) [52]. In DICER 2.0, we refer to the ERV method [52] to cut off the unfolding of PD-nets, and then generate its FCP.

As a matter of fact, ERV method does not consider the Petri net modeling with data information. Moreover, it does not specify a highly efficient calculations on configurations, cuts and cut-off events. This is mainly caused by the following two facts. On the one hand, the most computing methods of configurations and cuts need a lot of repetitive calculations. On the other hand, once some new events are added into a given finite prefix, these methods usually match up them with all existing events and determine whether they are cut-off events or not. In order to solve these problems, DICER 2.0 uses recursion formulas and contextual information of events to compute configurations, concurrent conditions and cuts. Meanwhile, it uses backward conflicts to guide the calculations of cut-off events.

After generating an FCP of a PD-net Σ in DCIER 2.0, we can use its matrix manipulations to detect data inconsistencies since it contains the same behavioral information as the reachability graph of Σ (i.e., the completeness property [5] of FCP). In details, we first get an incidence matrix of this FCP, and then use Warshell algorithm to calculate its causality matrix $J^{\#}_{unf(\Sigma)}$. Afterwards, we obtain a conflict matrix $J^{\#}_{unf(\Sigma)}$ according to the mathematical definition of conflicts. Then, a concurrency matrix $J^{co}_{unf(\Sigma)}$ is calculated by $J^{<}_{unf(\Sigma)}$ and $J^{\#}_{unf(\Sigma)}$, *i.e.*, two events are in concurrency relation if they are neither in causality relation nor in conflict relation, i.e., $J^{\#}_{unf(\Sigma)} = [a_{(i,j)}]_{n\times n}$, $J^{co}_{unf(\Sigma)} = [a'_{(i,j)}]_{n\times n}$ and $J^{co}_{unf(\Sigma)} = [a''_{(i,j)}]_{n\times n}$, where $e_i, e_j \in E$ $(i, j \in \mathbb{N})$, and

$$a_{(i,j)} = \begin{cases} 1 & if\, e_i \# e_j \\ 0 & otherwise \end{cases} \quad a'_{(i,j)} = \begin{cases} 1 & if\, e_i \# e_j \\ 0 & otherwise \end{cases} \quad a''_{(i,j)} = \begin{cases} 1 & if\, e_i\, co\, e_j \\ 0 & otherwise \end{cases}$$

Based on the concurrency matrix $J^{co}_{unf(\Sigma)}$, we can check the errors of data inconsistency in Σ, i.e., there exists an error of data inconsistency if two concurrent events e_1 and e_2 have some data operations on a share data element, *i.e.*,

$$(Read(e_1) \cup Write(e_1) \cup Delete(e_1)) \cap (Write(e_2) \cup Delete(e_2)) \neq \emptyset.$$

For example, Figure 2d is an FCP of the PD-net in Figure 2c. Its related matrix calculations are conducted as shown in Figure 3. From this concurrency matrix, we can find that e_1, e_2 and e_3 are three concurrent events. Furthermore, they suffer from the errors of data inconsistency because $Write(e_1) \cap Read(e_2) \cap Write(e_3) \neq \emptyset$.

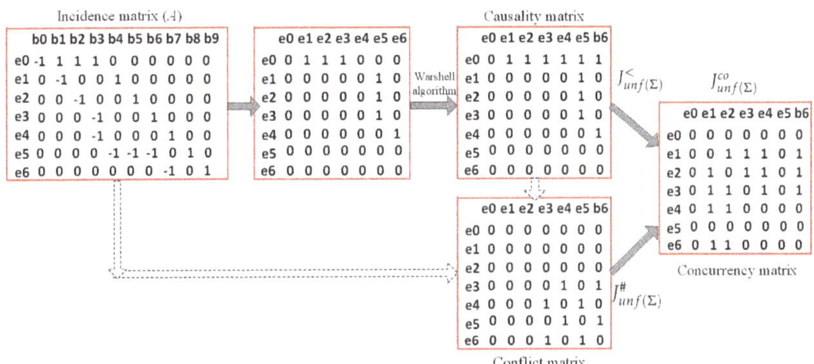

Figure 3. Some matrix manipulations on the FCP in Figure 2b.

3.4. The Implementations of DICER 2.0

Corresponding to the specified modeling and checking methods, we now introduce the basic framework and implementations of DICER 2.0.

Figures 4 and 5 show the user interface (UI) and basic functions of DICER 2.0, respectively. Its framework is made up of two modules: graphical user interface (GUI) and model checker (MC), as shown in Figure 6. These two modules respectively correspond to the menus of drawing and model-checking in Figure 4.

- In the module of graphical user interface, Place/Transition nets, WFD-nets and PD-nets can be imported, exported, drawn and edited. The labeling functions of data operations (e.g., read, write and delete) can be added, deleted and modified in DICER 2.0. Moreover, different kinds of Petri nets are imported and exported in the format of an extended Petri Net Markup Language [53] (ePNML). In fact, ePNML provides a common interchange format for all types of Petri nets based on XML, and defines specifications of data operations and guard functions. As shown in Figure 7, the label $\langle isData \rangle$ formalizes data-flows of concurrent systems, including labeling functions of *read, write, delete* and *guards*. Since ePNML is an XML-based document, we can create or parse these Petri nets according to some configuration files, e.g., *GenerateObjectList*.xsl and *GeneratePNML*.xsl.
- In the module of model checker, Place/Transition nets and PD-nets can be unfolded, and then we can get their FCPs. As for the FCPs of PD-nets, we can use their matrix calculations (e.g., causality matrix, conflict matrix and concurrency matrix) to find out all concurrent events and then check errors of data inconsistency. Additionally, both classical reachability graphs and guard-driven reachability graphs of WFD-nets can be constructed in DICER 2.0. Furthermore, they are used to analyze some data-flow properties of concurrent systems, e.g., deadlocks, data inconsistency and soundness [29].

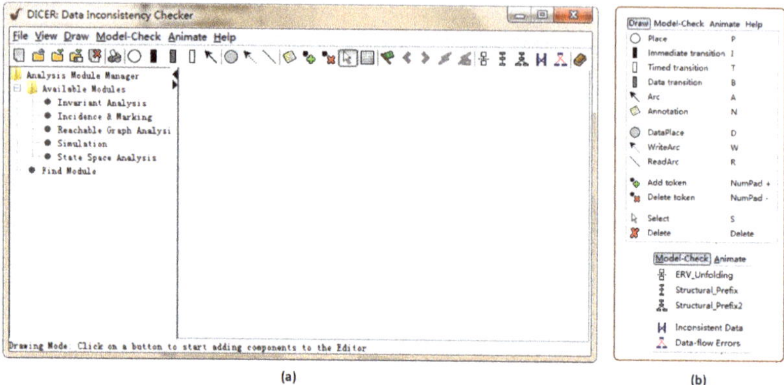

Figure 4. DICER 2.0 [45]. (**a**) Software interface; (**b**) the drawing menu and the model-checking menu.

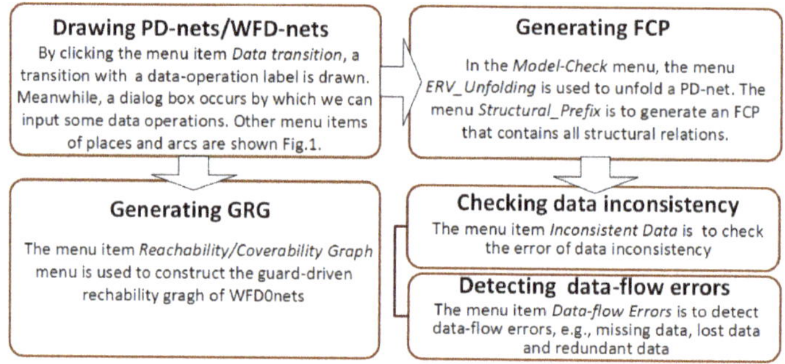

Figure 5. The basic functions of DICER 2.0.

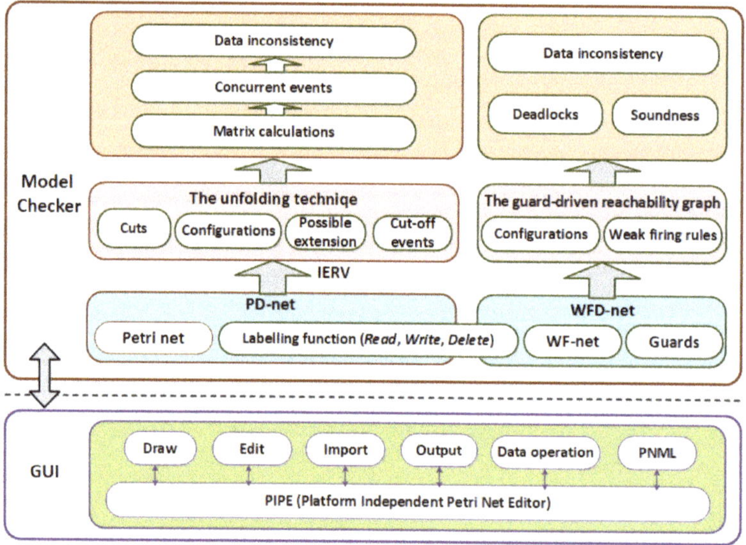

Figure 6. The basic framework of DICER 2.0.

```
<?xml version="1.0" encoding="ISO-8859-1"?>
- <pnml>
  - <net type="P/T net" id="Net-One">
      <tokenclass id="Default" blue="0" green="0"
        red="0" enabled="true"/>
    + <place id="P0">
    + <place id="P1">
    + <place id="P2">
    - <transition id="T0">
      + <graphics>
      + <name>
      + <orientation>
      + <rate>
      + <timed>
      - <isData>
          <value>true</value>
          <read/>
          <write>a</write>
          <delete/>
          <guards>isHigh(a)</guards>
        </isData>
      + <infiniteServer>
      + <priority>
      </transition>
    + <transition id="T1">
    + <arc id="P0 to T0" target="T0" source="P0">
    + <arc id="P0 to T1" target="T1" source="P0">
    + <arc id="T0 to P1" target="P1" source="T0">
    + <arc id="T1 to P2" target="P2" source="T1">
    </net>
  </pnml>
```

Figure 7. An extended PNML [53] (ePNML) document of Petri nets with data operations and guards.

DICER 2.0 is developed-based on Platform Independent Petri Net Editor (PIPE) [40], which is an open source and graphical tool for drawing and analyzing Petri nets. In details, it is made up of a series of Java classes. Figure 8 shows the main hierarchy of these classes, which includes some flow information, inheritance relations, interfaces and methods.

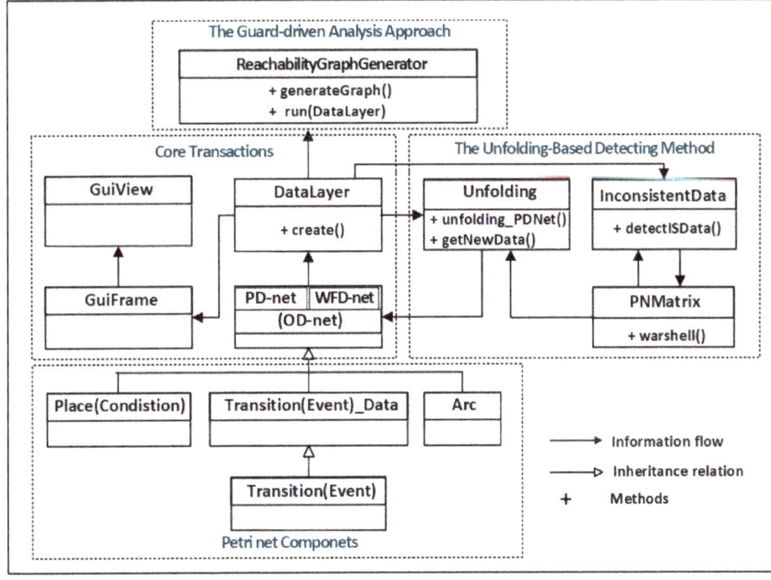

Figure 8. Main class hierarchy.

- The class *DataLayer* acts on the Petri net modeling of concurrent systems. It can be used to create, edit (e.g., add, move, or modify), import and export a PD-net or a

WFD-net. In this class, the method *getNewData*() is to obtain some information about the Petri net components of FCPs such as events, conditions and arcs.

- The class *Unfolding* is developed to unfold a PD-net or a Place/Transition net. Their FCPs can be generated by the method of *unfolding_PDNet(visual, "ERV", null)*. In this Java method, the parameter *visual* indicates whether an FCP needs to be displayed in the software interface, and the parameter *ERV* means a selected unfolding method, such as ERV, merged process, and directed unfolding.
- The class *ReachabilityGraphGenerator* is used to construct a guard-driven reachability graph of WFD-nets, and the methods *generateGraph*() and *run(DataLayer)* correspond to this function.
- The class *InconsistentData* is developed to check errors of data inconsistencies based on the unfolding of PD-nets, and the method *detectISData*() achieves this work in details.
- The classes *GuiView* and *GuiFrame* are used to create the front end, and display the software interface of DICER 2.0.
- A homomorphism from conditions to places (or from events to transitions) is represented by a hashmap. Its keys and values are in the form of $\langle Place, Place \rangle$ or $\langle Transition, Transition \rangle$, where *Place* and *Transition* are Java classes of Petri net components. Additionally, in order to improve the unfolding efficiency of PD-nets, we use some linked hash tables to store the contextual information of events and concurrent conditions, e.g., local configurations, pre/post-sets and cuts.

4. Case Study

To show the application scenarios of DICER 2.0, we give the following case studies.

4.1. Case _1: Intelligent Traffic Light System (ITIC)

Our first case study is conducted on an intelligent traffic light controller (ITIC) [54,55] for a North–South and East–West intersection. In this case study, the North–South (NS) is a main road, and the East–West (EW) is a rarely used country road. The North–South traffic light is always GREEN if the sensor of East–West Road is not activated. Otherwise, the North–South light will change from GREEN to YELLOW so as to give way to the East–West traffic. Additionally, some emergency vehicles can activate an emergency sensor. At this time, both the North–South and the East–West traffic lights need to turn RED.

In this case, of ITIC, we first use a WFD-net to model its business process, as shown in Figure 9. Table 2 shows all places and their meanings. The Boolean functions *select* $(EmgSensor, EWSensor)$ and *select*$(EmgSensor, EWSensor)$ are two exclusive guards on t_2 and t_3, respectively. By using DICER 2.0, we can draw and edit this WFD-net. Then, a guard-driven reachability graph is constructed, as shown in Figure 10. Based on this GRG, some properties can be verified by traversing each weak configuration (or state). For example, there is no deadlock in this ITIC system because there always exist enabled transitions at any weak configurations. Moreover, there is no error of data inconsistency since all concurrent transitions do not access a shared data element.

Table 2. Places and their meanings.

Place ID	Meanings
p_2	The yellow light of NS Road
p_3	The red light of NS Road
p_4	The green light of NS Road
p_6	The pre-green light of EW Road
p_7	The green light of EW Road
p_8	The yellow light of EW Road
p_{10}	The red light of EW Road
p_0, p_1, p_5, p_9	(*Control places*)

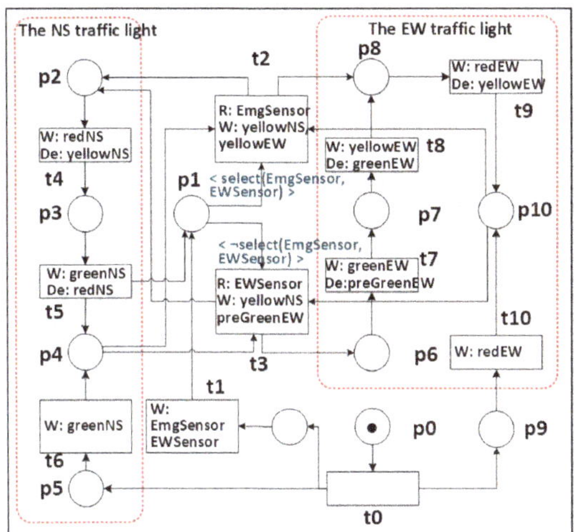

Figure 9. A WFD-net that models an intelligent traffic light system.

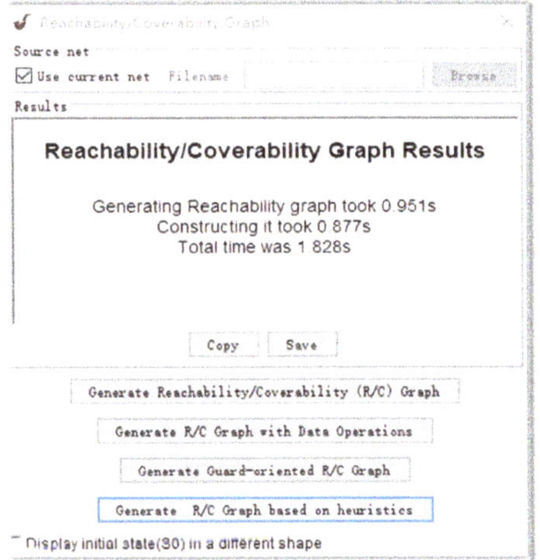

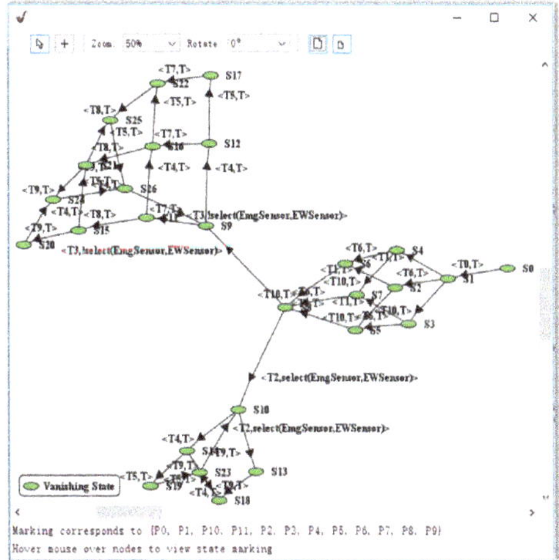

Figure 10. A guard-driven reachability graph (GRG) of Figure 9. (**a**) A user interface for generating a GRG; (**b**) the visualization of a GRG.

4.2. Case _2: Health-Care Cyber-Physical System (HCPS)

The health-care cyber-physical system (HCPS) [56] consists of a series of devices such as e-health sensors, ambulance drones and ambulance vehicles. When an e-health sensor detects a cardiac arrest from patients, they will transmit this information to a controller, and then some warnings are sent to an emergency center. This center can also directly receive an emergency call from patients. After receiving these emergency messages, both drones and ambulances are ordered and sent to the emergency scene according to specific locations of patients.

In this case, of HCPS, we first use a PD-net to model its business process, as shown in Figure 11. Table 3 lists all transitions and their meanings. By using DICER 2.0, we can draw and edit this PD-net. Then, an FCP is generated, and some errors of data inconsistency are detected, which are respectively shown in Figure 12a,b. From Figure 12b, we can easily find that 12 concurrent events suffer from the errors of data inconsistency.

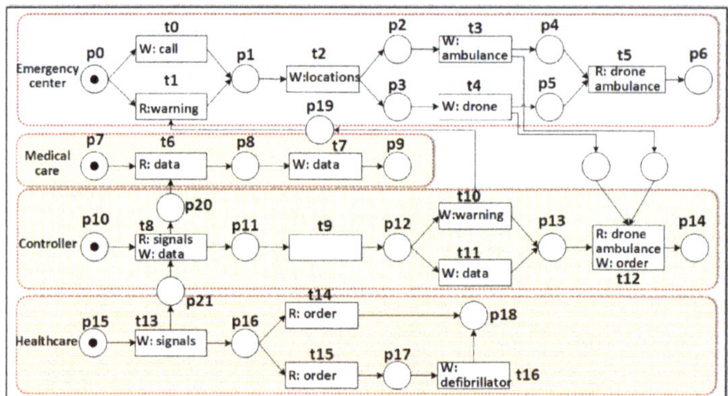

Figure 11. A PD-net that models a health-care cyber-physical system.

Table 3. Transitions and their meanings.

Transition ID	Meanings	Transition ID	Meanings
t_0	Receive emergency call	t_9	Control activity
t_1	Receive warning	t_{10}	Send warming
t_2	Find location	t_{11}	Store data
t_3	Send ambulance	t_{12}	Receive order
t_4	Send drone	t_{13}	Measure vital signals (E-health)
t_5	Supervise Drone	t_{14}	Movement of the ambulance
t_6	Receive data	t_{15}	Movement of the drone
t_7	Storage task	t_{16}	Install defibrillator
t_8	Send data		

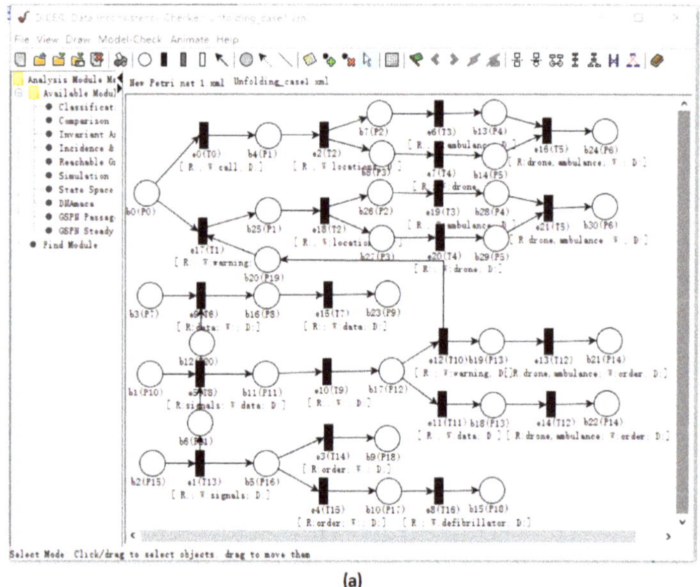

Figure 12. *Cont.*

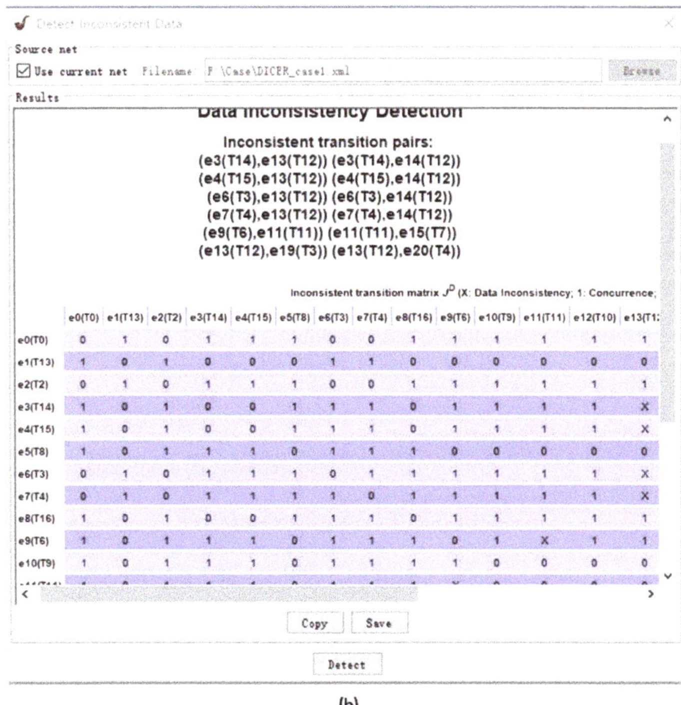

(b)

Figure 12. Detecting errors of data inconsistency based on the unfolding techniques of PD-nets; (**a**) an FCP of the PD-net in Figure 11; (**b**) the detection results.

5. Experiments

5.1. Benchmarks

A group of experiments are done based on the following benchmarks to show the advantages of DICER 2.0. Please note that all of these experiments are implemented on a PC with 4.0G memory and Intel Core i5-2400 CPU.

- The *Index* program [57] is widely used for the experimental evaluation of multi-threads.
- The *Prime* benchmark (http://docs.oracle.com/cd/E19205-01/820-0619/gdvwv/index.html, accessed on 16 April 2021) is a tutorial program for detecting data race.
- The *Child_benefit* benchmark [58] is an example of transactional payment processes for child benefits.
- The *SystemC* benchmark [59] illustrates a SystemC (a modeling language) module.
- The *Driver* [60] benchmark describes a simplified model of bluetooth drivers.
- *AddGlobal* [61] gives an example of concurrency bugs.
- The *AppLoan* benchmark [62] describes a business process of approving property loan.
- The *Airport* benchmark [63] shows a business process of an airport check-in system.
- *Case_1* and *Case_2* are two case studies of intelligent traffic light system and health-care cyber-physical system, respectively.

5.2. Implementation and Results

(1) The experiments on the GRG of WFD-nets

The guard-driven reachability graph (GRG) of WFD-nets is an improved method for analyzing data-flows of concurrent systems. In this experiment, we use DICER 2.0 to compare it with the classical reachability graph (CRG) in terms of state–space and runtime.

We first use some WFD-nets to mode the benchmarks of *SystemC, AddGlobal, ApproveLoan, AirportCheck,* and *Driver* in DICER 2.0, and then respectively obtain their CRGs and GRGs. Table 4 shows the results of these experiments. Obviously, the scale of GRG is much smaller than RG. Meanwhile, our GRG-based method spends less time to produce a reachability graph than the CRG-based method.

Please note that although the GRGs of WFD-nets in Table 4 can save the state–space of concurrent systems compared with CRGs, they still likely suffer from the state–space explosion problem especially with the increase of concurrent (data) operations. In order to alleviate this problem, we conduct the following experiments based on the unfolding techniques.

(2) The experiments on the unfolding of PD-nets

The errors of data inconsistencies are usually detected based on reachability graphs (RGs). Thus, all states and arcs of RGs need to be traversed to do this work at worst. In this experiment, DICER 2.0 are used to detect these errors based on the unfolding techniques of PD-nets. In details, we compare their FCPs with RGs in terms of state–space, runtime and detection time.

We first use some PD-nets to model the benchmarks of *Child_benefit, Index* and *Prime* in DICER 2.0. Afterwards, their FCPs are generated, and some errors of data inconsistency are detected. Table 5 shows the scales (i.e., the numbers of nodes and arcs) of FCPs and RGs. Obviously, FCPs take up much smaller space than RGs. Meanwhile, this table also lists the time of generating FCPs and RGs. Thus, we can easily find that the former has a significant advantage over the latter.

Table 4. The experimental results of GRG and CRG in DICER 2.0.

Benchmarks	CRG			GRG		
	Nos. of States	**Nos. of Arcs**	**Time of Constructing CRGs**	**Nos. of States**	**Nos. of Arcs**	**Time of Constructing GRGs**
SystemC	33	62	76.6	25	39	62.5
AddGlob	50	101	125.1	30	37	72.8
AppLoan	51	112	149	17	22	63
Airport	15	16	320	12	13	220
Driver(2)	409	864	1987	172	283	532
Driver(4)	4117	14,696	14,863	2215	6094	6793
Driver(6)	22,921	105,988	95,333	13,754	48,346	45,461

CRG: Classical Reachability Graph; GRG: Guard-driven Reachability Graph. Time: (ms).

Table 5. The experimental results of unfolding PD-nets in DICER 2.0.

Benchmarks	FCPs					RGs		
	$\lvert E \cup B \rvert$	$\lvert G \rvert$	**Time of Unfolding**	**Time of Error Detection**	**Nos. of Errors**	**Nos. of States**	**Nos. of Arcs**	**Time of Constructing RGs**
Child_benefit	10	13	22	3	0	37	79	45
Index (5)	45	50	90	18	2	462	1680	557
Index (10)	90	100	180	44	3	7686	38,691	11,104
Index (15)	135	150	270	86	8	39,234	226,459	63,910
Index (20)	180	200	360	150	15	101,341	616,469	178,974
Prime (2)	37	39	75	13	0	82	197	102
Prime (4)	69	73	141	29	1	1369	5829	1795
Prime (6)	101	107	207	54	3	12,380	69,893	19,922
Prime (8)	133	141	273	92	7	75,538	509,004	160,541

Time: (ms).

(3) The comparison experiments between DICER 2.0 and other Petri net tools.

To further show the advantage of DICER 2.0, we make some comparisons between DICER 2.0 and other existing Petri net tools, e.g., PIPE, Tina and Punf. We select these tools based on the following considerations.

- The same or similar runtime environments.
- The same or similar functions and features.
- Available installations.

In these experiments, we first implement the benchmarks of *Case_1* and *Case_2* into different Petri net tools, and then we can get their experimental results. Tables 6 and 7 respectively show comparisons on the performance and functions of different Petri net tools. From these tables, we can find that DICER 2.0 supports the WFD-net modeling of concurrent systems, constructing GRGs, unfolding PD-nets and detecting errors of data inconsistency, while other Petri net tools do not. Please note that we must model data operations by data places and their related flows in Tina, PIPE and Punf because these tools cannot support the formalizations of labeling functions and guard functions. With respect to this modeling method, we can find that the model scales of *Case_1* and *Case_2* by these tools is much larger than WFD-net by DICER 2.0. Meanwhile, due to the lack of guard functions, these tools cannot model routing path conditions. Naturally, its reachability graph (by Tina and PIPE) is smaller than our GRG. Additionally, we cannot get an FCP of *Case_2* by Punf because it cannot support the unfolding of unsafe Petri nets.

Table 6. The comparison experiments on the performance of DICER 2.0 and other Petri net tools.

Tools	**Case_1**			**Case_2**			
	Modeling ($\vert P \cup T \cup F\vert$)	**CRG**	**GRG**	**Modeling** ($\vert P \cup T \cup F\vert$)	**RG**	**FCP** ($\vert B \cup E \cup G\vert$)	**Detecting Data Inconsistency**
DICER 2.0	31	77	68	87	608	137	1.0 (*ms*)
PunF	87	–	–	125	–	–	–
Improved PIPE	31	77	–	87	608	–	–
Tina	87	53	–	125	608	–	–
PIPE	87	53	–	125	608	–	–

CRG: Classical Reachability Graph; GRG: Guard-driven Reachability Graph; RG: Reachability Graph. Data operations are modeled by data places and their related flows in Tina, PIPE and Punf because they cannot support the formalizations of labeling functions, guard functions and data-flow arcs.

Table 7. The comparison experiments on the functions of DICER 2.0 and other Petri net tools.

	Functions \ Tools	**DICER 2.0**	**Tina**	**PIPE**	**Punf**	**Improved PIPE**
Case_1	WFD-net	■	□	□	□	■
	Reachability graph	■	■	■	■	■
	Guard-driven reachability graph	■	□	□	□	□
	Unfolding	■	□	□	■	□
	Unfolding within data-flows	□	□	□	□	□
	Checking data inconsistency	■	□	□	□	□
Case_2	WFD-net	■	□	□	□	■
	Reachability graph	■	■	■	■	■
	Guard-driven reachability graph	■	□	□	□	□
	Unfolding	■	□	□	□	□
	Unfolding within data-flows	■	□	□	□	□
	Checking data inconsistency	■	□	□	□	□

6. Conclusions

Data-flow analysis plays an important role in the correctness verification of concurrent software systems. Petri net-based model checkings are a prominent method/technique for analyzing these data-flows. Currently, many different kinds of Petri nets have been used to do this work such as algebraic Petri net, predicate/transitions net, and colored Petri nets. WFD-net, as a high-level Petri net, is extended with conceptual labeling data operations. Thus, it can greatly model control/data- flows of concurrent systems. Moreover, its model scale is much smaller than other Petri nets with data-flow arcs such as C-net and PN-DO. Furthermore, WFD-net has been widely used to do model checkings. However, concurrent data operations and guard functions easily lead to the problems of state–space explosion and pseudo-states. In order to alleviate these problems, we proposed some efficient methods to detect data-flow errors and verify some properties. In this paper, we develop a new model checker DICER 2.0. By this tool, we can do a series of model checkings, e.g., detecting data inconsistencies based on the unfolding technique of PD-nets, and checking deadlocks via the GRG of WFD-nets.

In the future work, we plan to do the following studies:

(1) The unfolding methods of WFD-nets are studied to check many more data-flow errors and concurrency bugs [64,65] of concurrent systems;

(2) DICER 2.0 is further improved to support many more efficient model checkings; and

(3) Timed concurrent systems are modeled and checked by the unfolding techniques of Petri nets.

Author Contributions: D.X. proposed the idea in this paper and prepared the software application; D.X. and F.Z. designed the experiments; D.X. performed the experiments; Y.L. analyzed the data; D.X. wrote the paper; All authors have read and agreed to the published version of the manuscript.

Funding: This work was supported in part by National Natural Science Foundation of China under Grant 62002328, Zhejiang Provincial Natural Science Foundation of China under Grant *LQ20F020002*, and in part by the Key Laboratory of Embedded System and Service Computing (Ministry of Education) under Grant *ESSCKF* 2019-02.

Institutional Review Board Statement: Not applicable.

Informed Consent Statement: Not applicable.

Data Availability Statement: Not applicable.

Conflicts of Interest: The authors declare no conflict of interest.

References

1. Liu, G.; Jiang, C.; Zhou, M. Process nets with channels. *IEEE Trans. Syst. Man Cybern. Part A Syst. Hum.* **2012**, *42*, 213–225. [CrossRef]
2. You, D.; Wang, S.G.; Seatzu, C. Verification of Fault-predictability in Labeled Petri Nets Using Predictor Graphs. *IEEE Trans. Autom. Control* **2019**, *64*, 4353–4360. [CrossRef]
3. Li, W.; Xia, Y.; Zhou, M.; Sun, X.; Zhu, Q. Fluctuation-aware and predictive workflow scheduling in cost-effective Infrastructure-as-a-Service clouds. *IEEE Access* **2018**, *6*, 61488–61502. [CrossRef]
4. Trčka, N.; Van der Aalst, W.M.; Sidorova, N. Data-flow anti-patterns: Discovering data-flow errors in workflows. In *International Conference on Advanced Information Systems Engineering*; Springer: Berlin/Heidelberg, Germany, 2009; pp. 425–439.
5. Xiang, D.; Liu, G.; Yan, C.; Jiang, C. Detecting data inconsistency based on the unfolding technique of petri nets. *IEEE Trans. Ind. Inform.* **2017**, *13*, 2995–3005. [CrossRef]
6. Liu, C.; Zeng, Q.; Duan, H.; Wang, L.; Tan, J.; Ren, C.; Yu, W. Petri net based data-flow error detection and correction strategy for business processes. *IEEE Access* **2020**, *8*, 43265–43276. [CrossRef]
7. Murata, T. Petri nets: Properties, analysis and applications. *Proc. IEEE* **1989**, *77*, 541–580. [CrossRef]
8. Gerogiannis, V.C.; Kameas, A.D.; Pintelas, P.E. Comparative study and categorization of high-level petri nets. *J. Syst. Softw.* **1998**, *43*, 133–160. [CrossRef]
9. Zuberek, W.M. Timed Petri nets definitions, properties, and applications. *Microelectron. Reliab.* **1991**, *31*, 627–644. [CrossRef]
10. Balbo, G. Introduction to generalized stochastic Petri nets. In *Proceedings of the 7th International Conference on Formal Methods for Performance Evaluation*; Springer: Berlin/Heidelberg, Germany, 2007; pp. 83–131.

11. Luan, W.; Qi, L.; Zhao, Z.; Liu, J.; Du, Y. Logic Petri Net Synthesis for Cooperative Systems. *IEEE Access* **2019**, *7*, 161937–161948. [CrossRef]
12. Moutinho, F.; Gomes, L. Asynchronous-channels within Petri net-based GALS distributed embedded systems modeling. *IEEE Trans. Ind. Inform.* **2014**, *10*, 2024–2033. [CrossRef]
13. Kheldoun, A.; Barkaoui, K.; Ioualalen, M. Formal verification of complex business processes based on high-level Petri nets. *Inf. Sci.* **2017**, *385*, 39–54. [CrossRef]
14. Buchs, D.; Guelfi, N. A formal specification framework for object-oriented distributed systems. *IEEE Trans. Softw. Eng.* **2000**, *26*, 635–652. [CrossRef]
15. Barkaoui, K.; Ayed, R.B.; Boucheneb, H.; Hicheur, A. Verification of workflow processes under multilevel security considerations. In Proceedings of the 2008 Third International Conference on Risks and Security of Internet and Systems, Tozeur, Tunisia, 28–30 October 2008; pp. 77–84.
16. He, X. Modeling and Analyzing Smart Contracts using Predicate Transition Nets. In Proceedings of the 2020 IEEE 20th International Conference on Software Quality, Reliability and Security Companion (QRS-C), Macau, China, 11–14 December 2020; pp. 108–115.
17. Wu, D.; Zheng, W. Formal model-based quantitative safety analysis using timed Coloured Petri Nets. *Reliab. Eng. Syst. Saf.* **2018**, *176*, 62–79. [CrossRef]
18. Yu, W.; Yan, C.; Ding, Z.; Jiang, C.; Zhou, M. Modeling and validating e-commerce business process based on Petri nets. *IEEE Trans. Syst. Man Cybern. Syst.* **2013**, *44*, 327–341. [CrossRef]
19. Varea, M.; Al-Hashimi, B.M.; Cortés, L.A.; Eles, P.; Peng, Z. Dual Flow Nets: Modeling the control/data-flow relation in embedded systems. *ACM Trans. Embed. Comput. Syst. (TECS)* **2006**, *5*, 54–81. [CrossRef]
20. Awad, A.; Decker, G.; Lohmann, N. Diagnosing and repairing data anomalies in process models. In *International Conference on Business Process Management*; Springer: Berlin/Heidelberg, Germany, 2009; pp. 5–16.
21. Sharma, D.; Pinjala, S.; Sen, A.K. Correction of Data-flow Errors in Workflows. In Proceedings of the 25th Australasian Conference on Information Systems (ACIS), Auckland, New Zealand, 8–10 December 2014.
22. Baldan, P.; Bruni, A.; Corradini, A.; König, B.; Rodríguez, C.; Schwoon, S. Efficient unfolding of contextual Petri nets. *Theor. Comput. Sci.* **2012**, *449*, 2–22. [CrossRef]
23. Montanari, U.; Rossi, F. Contextual nets. *Acta Inform.* **1995**, *32*, 545–596. [CrossRef]
24. Kähkönen, K.; Heljanko, K. Testing Programs with Contextual Unfoldings. *ACM Trans. Embed. Comput. Syst. (TECS)* **2017**, *17*, 1–25. [CrossRef]
25. Sidorova, N.; Stahl, C.; Trčka, N. Soundness verification for conceptual workflow nets with data: Early detection of errors with the most precision possible. *Inf. Syst.* **2011**, *36*, 1026–1043. [CrossRef]
26. Yang, B.; Liu, G.; Xiang, D.; Yan, C.; Jiang, C. A Heuristic Method of Detecting Data Inconsistency Based on Petri Nets. In Proceedings of the 2018 IEEE International Conference on Systems, Man, and Cybernetics (SMC), Miyazaki, Japan, 7–10 October 2018; pp. 202–208.
27. Trecka, N.; van der Aalst, W.; Sidorova, N. Workflow completion patterns. In Proceedings of the 2009 IEEE International Conference on Automation Science and Engineering, Bangalore, India, 22–25 August 2009; pp. 7–12.
28. Zou, J.; Liu, X.; Sun, H.; Zeng, J. Live instance migration with data consistency in composite service evolution. In Proceedings of the 2010 6th World Congress on Services, Miami, FL, USA, 5–10 July 2010; pp. 653–656.
29. Xiang, D.; Liu, G.; Yan, C.G.; Jiang, C. A Guard-driven Analysis Approach of Workflow Net With Data. *IEEE Trans. Serv. Comput.* **2018**. [CrossRef]
30. Wisniewski, R.; Karatkevich, A.; Adamski, M.; Costa, A.; Gomes, L. Prototyping of Concurrent Control Systems With Application of Petri Nets and Comparability Graphs. *IEEE Trans. Control Syst. Technol.* **2017**, *26*, 575–586. [CrossRef]
31. Wisniewski, R.; Wisniewska, M.; Jarnut, M. C-exact Hypergraphs in Concurrency and Sequentiality Analyses of Cyber-Physical Systems Specified by Safe Petri Nets. *IEEE Access* **2019**, *7*, 13510–13522. [CrossRef]
32. McMillan, K.L. Using unfoldings to avoid the state explosion problem in the verification of asynchronous circuits. In *Computer Aided Verification*; Springer: Berlin/Heidelberg, Germany; 1992; pp. 164–177.
33. Franco, A.; Baldan, P. *True Concurrency and Atomicity: A Model Checking Approach with Contextual Petri Nets*; LAP LAMBERT Academic Publishing: Saarbrucken, Germany; 2015.
34. Haar, S. Types of asynchronous diagnosability and the reveals-relation in occurrence nets. *IEEE Trans. Autom. Control* **2010**, *55*, 2310–2320. [CrossRef]
35. Hickmott, S.L.; Rintanen, J.; Thiébaux, S.; White, L.B. Planning via Petri Net Unfolding. *Int. Jt. Conf. Artif. Intell.* **2007**, *7*, 1904–1911.
36. de León, H.P.; Saarikivi, O.; Kähkönen, K.; Heljanko, K.; Esparza, J. Unfolding Based Minimal Test Suites for Testing Multithreaded Programs. In Proceedings of the 15th International Conference on Application of Concurrency to System Design, Brussels, Belgium, 21–26 June 2015; pp. 40–49.
37. Khomenko, V.; Koutny, M. LP deadlock checking using partial order dependencies. In *International Conference on Concurrency Theory*; Springer: Berlin/Heidelberg, Germany, 2000; pp. 410–425.
38. Liu, G.; Reisig, W.; Jiang, C. A Branching-process-based method to check soundness of workflow systems. *IEEE Access* **2016**, *4*, 4104–4118. [CrossRef]

39. Rodriguez, C.; Schwoon, S. Verification of Petri nets with read arcs. In *International Conference on Concurrency Theory;* Springer: Berlin/Heidelberg, Germany, 2012; pp. 471–485.
40. Dingle, N.J.; Knottenbelt, W.J.; Suto, T. PIPE2: A tool for the performance evaluation of generalised stochastic Petri Nets. *ACM SIGMETRICS Perform. Eval. Rev.* **2009**, *36*, 34–39. [CrossRef]
41. Heiner, M.; Herajy, M.; Liu, F.; Rohr, C.; Schwarick, M. Snoopy—A unifying Petri net tool. In *International Conference on Application and Theory of Petri Nets and Concurrency;* Springer: Berlin/Heidelberg, Germany, 2012; pp. 398–407.
42. Jensen, K.; Kristensen, L.M.; Wells, L. Coloured Petri Nets and CPN Tools for modelling and validation of concurrent systems. *Int. J. Softw. Tools Technol. Transf.* **2007**, *9*, 213–254. [CrossRef]
43. Aalst, W.M.P.V.D.; Hee, K.M.V.; Hofstede, A.H.M.T.; Sidorova, N.; Wynn, M.T. Soundness of workflow nets: Classification, decidability, and analysis. *Form. Asp. Comput.* **2011**, *23*, 333–363. [CrossRef]
44. Liu, C.; Zeng, Q.; Cheng, L.; Duan, H.; Zhou, M.; Cheng, J. Privacy-preserving behavioral correctness verification of cross-organizational workflow with task synchronization patterns. *IEEE Trans. Autom. Sci. Eng.* **2020**. [CrossRef]
45. Xiang, D.; Liu, G.; Yan, C.; Jiang, C. DICER: Data Inconsistency CheckER based on the unfolding technique of Petri net. In Proceedings of the 2017 IEEE 14th International Conference on Networking, Sensing and Control (ICNSC), Calabria, Italy, 16–18 May 2017; pp. 115–120.
46. Saarikivi, O.; Ponce-De-León, H.; Kähkönen, K.; Heljanko, K.; Esparza, J. Minimizing test suites with unfoldings of multithreaded programs. *ACM Trans. Embed. Comput. Syst. (TECS)* **2017**, *16*, 45. [CrossRef]
47. Xiang, D.; Liu, G.; Yan, C.; Jiang, C. Detecting data-flow errors based on Petri nets with data operations. *IEEE/CAA J. Autom. Sin.* **2017**, *5*, 251–260. [CrossRef]
48. Xiang, D.; Liu, G. Checking Data-Flow Errors Based on The Guard-Driven Reachability Graph of WFD-Net. *Comput. Inform.* **2020**, *39*, 193–212. [CrossRef]
49. De Masellis, R.; Di Francescomarino, C.; Ghidini, C.; Tessaris, S. Enhancing workflow-nets with data for trace completion. In *International Conference on Business Process Management;* Springer: Berlin/Heidelberg, Germany, 2017; pp. 89–106.
50. Evron, Y.; Soffer, P.; Zamansky, A. Incorporating data inaccuracy considerations in process models. In *Enterprise, Business-Process and Information Systems Modeling;* Springer: Berlin/Heidelberg, Germany, 2017; pp. 305–318.
51. Lu, F.; Tao, R.; Du, Y.; Zeng, Q.; Bao, Y. Deadlock detection-oriented unfolding of unbounded Petri nets. *Inf. Sci.* **2019**, *497*, 1–22. [CrossRef]
52. Esparza, J.; Römer, S.; Vogler, W. An improvement of McMillan's unfolding algorithm. *Form. Methods Syst. Des.* **2002**, *20*, 285–310. [CrossRef]
53. Hillah, L.-M.; Kordon, F.; Petrucci, L.; Treves, N. Pnml framework: an extendable reference implementation of the petri net markup language. In Proceedings of the International Conference on Applications and Theory of Petri Nets, Braga, Portugal, 21–25 June 2010; pp. 318–327.
54. Aziz, M.W.; Rashid, M. Domain specific modeling language for cyber physical systems. In Proceedings of the 2016 International Conference on Information Systems Engineering (ICISE), Los Angeles, CA, USA, 20–22 April 2016; pp. 29–33.
55. Qi, L.; Zhou, M.; Luan, W. A two-level traffic light control strategy for preventing incident-based urban traffic congestion. *IEEE Trans. Intell. Transp. Syst.* **2018**, *19*, 13–24. [CrossRef]
56. Graja, I.; Kallel, S.; Guermouche, N.; Kacem, A.H. BPMN4CPS: A BPMN extension for modeling cyber-physical systems. In Proceedings of the 2016 IEEE 25th International Conference on Enabling Technologies: Infrastructure for Collaborative Enterprises (WETICE), Paris, France, 13–15 June 2016; pp. 152–157.
57. Flanagan, C.; Godefroid, P. Dynamic partial-order reduction for model checking software. *ACM Sigplan Not.* **2005**, *40*, 110–121. [CrossRef]
58. Lodde, A.; Schlechter, A.; Bauler, P.; Feltz, F. Data Consistency in Transactional Business Processes. In *International Conference on Business Informatics Research;* Springer: Berlin/Heidelberg, Germany, 2011; pp. 83–95.
59. Blanc, N.; Kroening, D. Race analysis for SystemC using model checking. *ACM Trans. Des. Autom. Electron. Syst. (TODAES)* **2010**, *15*, 1–32. [CrossRef]
60. Razavi, N.; Ivančić, F.; Kahlon, V.; Gupta, A. Concurrent test generation using concolic multi-trace analysis. In *Asian Symposium on Programming Languages and Systems;* Springer: Berlin/Heidelberg, Germany, 2012; pp. 239–255.
61. Sinha, N.; Wang, C. Staged concurrent program analysis. In Proceedings of the Eighteenth ACM SIGSOFT International Symposium on Foundations of Software Engineering, Santa Fe, NM, USA, 7–11 November 2010; pp. 47–56.
62. Sun, S.X.; Zhao, J.L.; Nunamaker, J.F.; Sheng, O.R.L. Formulating the data-flow perspective for business process management. *Inf. Syst. Res.* **2006**, *17*, 374–391. [CrossRef]
63. Xiang, D.; Tao, X.; Liu, Y. An Incremental and Backward-Conflict Guided Method for Unfolding Petri Nets. *Symmetry* **2021**, *13*, 392. [CrossRef]
64. Kim, K.H.; Yavuz-Kahveci, T.; Sanders, B.A. JRF-E: Using model checking to give advice on eliminating memory model-related bugs. *Autom. Softw. Eng.* **2012**, *19*, 491–530. [CrossRef]
65. Zhang, M.; Wu, Y.; Shan, L.U.; Qi, S.; Ren, J.; Zheng, W. A Lightweight System for Detecting and Tolerating Concurrency Bugs. *IEEE Trans. Softw. Eng.* **2016**, *42*, 899–917. [CrossRef]

 *mathematics*

Article

Application of EM Algorithm to NHPP-Based Software Reliability Assessment with Generalized Failure Count Data

Hiroyuki Okamura * and Tadashi Dohi

Graduate School of Advanced Science and Engineering, Hiroshima University, 1-4-1 Kagamiyama, Higashi-Hiroshima 7398527, Japan; dohi@hiroshima-u.ac.jp
* Correspondence: okamu@hiroshima-u.ac.jp

Abstract: Software reliability models (SRMs) are widely used for quantitative evaluation of software reliability by estimating model parameters from failure data observed in the testing phase. In particular, non-homogeneous Poisson process (NHPP)-based SRMs are the most popular because of their mathematical tractability. In this paper, we focus on the parameter estimation algorithm for NHPP-based SRMs and discuss the EM algorithm for generalized fault count data. The presented algorithm can be applied for failure time data, failure count data, and their mixture. The paper derives the EM-step formulas for basic 12 NHPP-based SRMs and demonstrate a numerical experiment to present the convergence property of our algorithms. The developed algorithms are suitable for an automatic tool for software reliability evaluation.

Keywords: software reliability model; maximum likelihood estimation; EM algorithm; non-homogeneous Poisson process; generalized failure count data

Citation: Okamura, H.; Dohi, T. Application of EM Algorithm to NHPP-Based Software Reliability Assessment with Generalized Failure Count Data. *Mathematics* **2021**, *9*, 985. https://doi.org/10.3390/math9090985

Academic Editor: Frank Werner

Received: 15 March 2021
Accepted: 26 April 2021
Published: 27 April 2021

1. Introduction

Software reliability models (SRMs) are used to assess quantitative reliability and to control the quality of software products. Since Jelinski and Moranda [1], and Goel and Okumoto [2] presented SRMs based on stochastic processes, numerous SRMs have been proposed [3–8]. In particular, non-homogeneous Poisson process (NHPP)-based SRMs have become popular in representing the dynamics of failure occurrence processes in a variety of situations [9–13]. By using an NHPP-based SRM, we predict the future behavior of software failure, i.e, the number of failures experienced in future, and estimate the quantitative measure of software reliability.

The advantage of NHPP-based SRMs is simplifying the stochastic analysis. NHPPs are generally dominated by *mean value functions*. The mean value function indicates the expected number of failures experienced at arbitrary testing time. By choosing appropriate mean value functions, NHPP-based SRMs can fit any observed failure data. The NHPP-based SRMs and the mean value functions have a one-to-one correspondence.

The Goel–Okumoto model [2]; Goel model [2]; Musa–Okumoto model [14]; Ohba [15,16]; Yamada, Ohba, and Osaki model [17]; Zhao and Xie model [18] are early NHPP-based SRMs. They were constructed with the deterministic debugging scenarios of mean value functions. Pham [19] solved a generalized differential equation by which the mean value function in the NHPP-based SRM is governed and proposed a generalized SRM with many redundant parameters.

Apart from such a deterministic modeling framework, almost all NHPP-based SRMs can be characterized as Markov processes. Shantikumar [20] discussed a modeling framework to integrate time-homogeneous Markov processes and NHPPs by using a binomial-type stochastic point process. Langberg and Singpurwalla [21] presented a unified modeling framework for almost all NHPP-based SRMs. Chen and Singpurwalla [22] also discussed the framework with a self-exciting point process. Miller [23] introduced the concept of exponential order statistics and drastically extended Langberg and Singpurwalla's idea. In fact, the realizations of NHPP-based SRM can be described by either the general order statistics or record value statistics of the underlying software fault data, where the fault-detection times are assumed to be independent and identically distributed (i.i.d.) random variables. Specifically, the general order statistics are based on the order of all the fault detection times, and the record value statistics focus on their maximum detection time.

In this paper, we focus on the parameter estimation problem of NHPP-based SRMs. In general, there are three steps to evaluating the software reliability with NHPP-based SRM. (i) Collect the failure data such as the number of detected bugs in the testing phase, (ii) estimate the model parameters of NHPP-based SRM to fit it to the collected data, and (iii) compute reliability measures from the NHPP-based SRM with the estimated parameters. Based on quantitative measures, we control the software development process. As a typical usage of NHPP-based SRM, we estimate the number of failures that will be experienced in the future and decide whether to continue testing the software or the software can be released. In other words, the parameter estimation of NHPP-based SRM is frequently executed in the software development phase. The computation cost of the estimation should be small in practice.

Therefore, many authors have concerns about the parameter estimation problem on NHPP-based SRMs. Nevertheless, in actual software reliability assessments, a few NHPP-based SRMs and familiar maximum likelihood (ML) estimation methods are still used conventionally. The main reason for this is that the practitioners wish to use intuitively simple statistical methods, which exclude empirically based tuning parameters for a few SRMs that have survived a long history of software reliability engineering. In fact, the Bayesian estimation methods are still minor in software engineering practice, although its theoretical benefit is recognized. On the other hand, ML estimation is based on the maximization of likelihood function with software failure data and possesses several rational properties such as asymptotic efficiency. Hossain and Dahiya [24] derived necessary and sufficient conditions that the maximum likelihood estimates (MLEs), which satisfy the non-linear likelihood equations, exist in Goel and Okumoto SRM [2]. Knafl and Morgan [25] presented a method to systematically solve the likelihood equations with two model parameters. Joe [26] also discussed the confidence interval of MLEs. Zhao and Xie [18] provided the MLEs for an extended Goel and Okumoto SRM. Jeske and Pham [27] discovered empirically that the MLEs in Goel and Okumoto SRM are not statistically consistent. It should be noted, however, that ML estimation is always possible for all NHPP-based SRMs. Even if the likelihood functions are strictly concave in model parameters, it is difficult to solve the likelihood equations analytically. For instance, in the cases where the likelihood functions are not concave and where there exists no solution for the likelihood equations inside the parameter space, the conventional methods to calculate the MLEs cannot be used. Usually, the Newton method and the Nelder–Mead method are used to solve the maximization problem in the existing literature. From the recent development of computational ability, it is becoming easier to handle a large-scale complex optimization problem.

On the other hand, it is known that the local convergence property of the Newton method is a weakness for practical application. The local convergence property means that the convergence radius of algorithm is limited, and thus, it may fail to obtain a result if we set unsuitable initial guesses. For example, when we develop the application to automatically obtain parameter estimates from given data, the local convergence property becomes troublesome when choosing the initial guesses. Therefore, the Newton method is not suitable for this purpose. Additionally, the Nelder–Mead method is one of the direct search methods. The convergence property of the Nelder–Mead method is improved from the Newton method. However, some design parameters should be provided appropriately for the Nelder–Mead method. Even if we use the Nelder–Mead algorithm, the convergence of algorithm is not always guaranteed for any given data.

From the early 2000s, our research group has developed an alternative parameter estimation algorithm based on the EM (expectation maximization) principle [28,29] and applied it to the software reliability assessment based on NHPP-based SRMs [30–40]. As another examples of EM algorithms in SRMs, Kimura and Yamada [41], Leadoux [42], and Okamura and Dohi [43] attempted to use EM algorithms to estimate the imperfect debugging model [44] and architecture-based SRMs [45,46]. Their models were based on the continuous-time Markov chain and are closely related to Markov-modulated Poisson processes and/or Markovian arrival processes. Additionally, Zeephongsekul et al. [47] and Nagaraju et al. [48] proposed ECM (expectaton conditional maximization) algorithms for NHPP-based SRMs to handle several specific models.

The EM algorithm is an algorithm that finds maximum likelihood estimates for a statistical model with incomplete data. The idea behind our EM algorithms is to find the incomplete data structure of NHPP-based SRMs. Concretely, in NHPP-based SRMs, we assume that the number of failures is finite due to a finite number of software bugs, but all of them cannot be observed, i.e., the number of reaming software failures can be regarded as missing data. From this insight, the EM algorithm for an individual NHPP-based SRM is developed. Although the convergence speed of EM algorithm is generally slower than other general-purpose numerical methods such as the Newton method, it has a global convergence property. This property allows us to reduce efforts in choosing good initial guesses for the model parameters and is suitable for automating the estimation procedure. In our past work [49], we summarized EM algorithms for 12 NHPP-based SRMs when the failure data were time data. The failure time data consisted of a set of exact failure times experienced. In practice, it is difficult to obtain exact failure times. Generally, we record failure count data consisting of the number of failures experienced for time intervals. For example, it is reasonable to record the number of failures per working day. From this reason, this paper presents the EM algorithms for 12 basic NHPP-based SRMs when the failure count data are given. In particular, we consider the generalized failure count data that involve both failure time and count data formats, and thus, the developed EM algorithms can be applied to either failure time data, failure count data, or their mixture.

We highlight our contributions here: (i) we derive the EM-step formula for NHPP-based SRMs with a finite number of failures under generalized fault count data, (ii) we derive concrete EM-step formulas for 12 basic NHPP-based SRMs, and (iii) we demonstrate the performance on the convergence property of the presented algorithms with real software failure data. To our best knowledge, this is the first paper that presents the EM algorithm for the generalized fault count data in 12 basic NHPP-based SRMs.

This paper is organized as follows. In Section 2, we introduce NHPP-based SRMs that are considered in this paper. In particular, NHPP-based SRMs are classified by failure time distribution and present the relationship between basic 12 NHPP-based SRMs and their failure time distributions. In Section 3, we derive the EM-step formulas for 12 basic NHPP-based SRMs. Section 4 is devoted to a numerical example to compare the convergence properties of EM algorithm, the Nelder–Mead method, and the quasi-Newton method. Finally, we conclude the paper with remarks in Section 5.

2. NHPP-Based SRMs

2.1. Model Description

Let $\{X(t), t \geq 0\}$ denote the number of software failures experienced before time t. We make the following model assumptions [21]:

- Assumption A: A software failure occurs at a random time. The probability distribution of all failure times are identical and mutually independent.
- Assumption B: The number of inherent software faults causing failures is finite.

Here, $F(t)$ and N are the cumulative distribution function of the failure time and the number of inherent faults. Then, the probability mass function of the cumulative number of failures experienced by time t is

$$P(X(t) = n) = \binom{N}{n} F(t)^n \overline{F}(t)^{N-n}, \tag{1}$$

where $\overline{F}(\cdot) = 1 - F(\cdot)$. This is often called the framework of generalized order statistics [21]. For instance, when the failure distribution is an exponential distribution, the corresponding SRM, the so-called exponential order statistics model, is the same as the Jelinski–Moranda SRM [1].

Most NHPP-based SRMs are advanced models of the generalized order statistics models. We make an additional model assumption [21]:

- Assumption C: The number of inherent faults is unknown, but prior information is given by a Poisson distribution.

When the expected number of inherent faults is ω, the cumulative number of software failures at time t has the following probability mass function:

$$P(X(t) = n) = \frac{(\omega F(t))^n}{n!} e^{-\omega F(t)}. \tag{2}$$

Equation (2) is equivalent to the probability mass function of NHPP with mean value function $\omega F(t)$. In this modeling framework, the failure time distribution $F(t)$ specifies an NHPP-based SRM.

Since the NHPP-based SRM is characterized by the failure time distribution, there have been a number of NHPP-based SRMs that change the failure time distribution. In this paper, we propose basic NHPP-based SRMs using well-known statistical distributions as the failure time distribution. Table 1 shows 11 basic NHPP-based SRMs and their failure time distributions. In the table, most of the basic NHPP-based SRMs correspond to the existing traditional NHPP-based SRMs. 'exp' is the so-called Goel and Okumoto model [2], 'gamma' is a generalized delayed S-shaped model [17,18], 'pareto' is a modified Duane model [50], 'tlogis' is an inflection S-shaped model [15], and 'lxvmin' is the Goel (Weibull) model [51].

Table 1. Basic NHPP-based SRMs.

Model	Failure Time Distribution
exp	Exponential distribution [2]
gamma	Gamma distribution [17,18]
pareto	Pareto type-II distribution [50]
tnorm	Truncated normal distribution [34]
lnorm	Log-normal distribution [34]
tlogis	Truncated logistic distribution [15]
llogis	Log-logistic distribution [52]
txvmax	Truncated extreme-value distribution (max) [35]
lxvmax	Log-extreme-value distribution (max) [35]
txvmin	Truncated extreme-value distribution (min) [35]
lxvmin	Log-extreme-value distribution (min) [35,51]

2.2. Parameter Estimation

As mentioned before, the model parameters of NHPP-based SRMs should be estimated from software failure data to predict the future tendency of a software failure. The most commonly used technique for parameter estimation is maximum likelihood (ML) estimation. In the context of ML estimation, we found model parameters that maximize the log-likelihood function (LLF). Since the LLF depends on the failure data experienced, the ML estimation of NHPP-based SRMs has been discussed for two types of data: failure time data and count data. The failure time data is a set of exact times in which a software failure occurs in the testing phase. The count data, equivalently called grouped data, consists of the number of failures experienced for time intervals. The estimation problems for these two data structures have been discussed separately.

This paper deals with a generalized data structure to express both failure time and count data. Our data structure is $\mathcal{D} := \{(t_1, x_1, u_i), \ldots, (t_k, x_k, u_k)\}$, where x_i failures that occur at the ith time interval, (x_{i-1}, x_i). In addition, if $u_i = 1$, an additional failure occurs at the end of the ith time interval, i.e, at time x_i. Otherwise, if $u_i = 0$, no failure occurs at the instant. If $u_i = 0$ for all time intervals, the data turns out the failure count data. If $x_i = 0$ and $u_i = 1$ for all i, $\mathcal{D}$ is the failure time data.

Based on the generalized data, the LLF for NHPP-based SRMs is written in the following form:

$$\begin{aligned} \mathrm{LLF}(\omega, \boldsymbol{\theta}) &= \sum_{i=1}^{k}(x_i + u_i)\log\omega + \sum_{i=1}^{k} x_i \log\{F(t_i;\boldsymbol{\theta}) - F(t_{i-1};\boldsymbol{\theta})\} \\ &+ \sum_{i=1} u_i \log f(t_i;\boldsymbol{\theta}) - \log x_i! - \omega F(t_k;\boldsymbol{\theta}). \end{aligned} \tag{3}$$

Then, the problem is to find the optimal $(\omega, \boldsymbol{\theta})$, so-called maximum likelihood estimates (MLEs), maximizing $\mathrm{LLF}(\omega, \boldsymbol{\theta})$. However, it is noted that we cannot derive the closed form solution of MLEs. That is, we need to utilize numerical optimization techniques such as the Newton method, quasi-Newton method, and Nelder–Mead method.

Although conventional methods such as the Newton method and the Nelder–Mead method may be occasionally useful in computing MLEs of the NHPP-based SRMs, it is worth noting that these aim to solve unconstrained optimization problems in ML estimation. However, in many cases, we have to cope with constrained optimization problems because almost all of the model parameters of NHPP-based SRMs are implicitly constrained, such as positive constraint.

3. EM Algorithms for NHPP-Based SRMs

This paper develops numerical procedures to compute MLEs for NHPP-based SRMs with generalized data. The proposed estimation algorithms are based on the EM principle. The EM algorithm is one of the statistical approaches to compute the MLEs for incomplete data and is numerically stable because of its global convergence property. Moreover, the proposed EM algorithms for NHPP-based SRMs are based on the closed forms of MLEs for an arbitrary fault-detection time distribution and are capable of solving constrained optimization problems. Although we have already developed EM algorithms for failure time data and failure count data for several basic NHPP-based SRMs, this paper revisits their EM algorithm when generalized data are given.

3.1. EM Algorithm

The EM algorithm is an iterative method for computing ML estimates with incomplete data [28,29]. Let $\mathcal{D}$ and $\mathcal{U}$ be observable and unobservable data vectors, respectively, and $\boldsymbol{\theta}$ be a model parameter vector $\boldsymbol{\theta}$ to be estimated from only the observable data. In the ML estimation, we find a parameter vector by maximizing the following log-likelihood function (LLF) $\mathcal{L}(\boldsymbol{\theta}; \mathcal{D})$:

$$\mathcal{L}(\boldsymbol{\theta}; \mathcal{D}) = \log p(\mathcal{D}; \boldsymbol{\theta}) = \log \int p(\mathcal{D}, \mathcal{U}; \boldsymbol{\theta}) d\mathcal{U}, \tag{4}$$

where $p(\cdot)$ is any probability density or mass function and thus $p(\mathcal{D}, \mathcal{U}; \boldsymbol{\theta})$ denotes the likelihood function for complete data $(\mathcal{D}, \mathcal{U})$.

Let $Q(\boldsymbol{\theta}|\boldsymbol{\theta}')$ denote the conditional expected LLF with respect to the complete data vector $(\mathcal{D}, \mathcal{U})$ using the posterior distribution for unobservable data vector with a given parameter vector $\boldsymbol{\theta}'$:

$$\begin{aligned} Q(\boldsymbol{\theta}|\boldsymbol{\theta}') &= \mathrm{E}[\log p(\mathcal{D}, \mathcal{U}; \boldsymbol{\theta}) | \mathcal{D}; \boldsymbol{\theta}'] \\ &= \int p(\mathcal{U}|\mathcal{D}; \boldsymbol{\theta}') \log p(\mathcal{D}, \mathcal{U}; \boldsymbol{\theta}) d\mathcal{U}. \end{aligned} \tag{5}$$

Then, the EM algorithm consists of an E-step and an M-step. The E-step computes the conditional expected LLF with respect to the complete data vector $(\mathcal{D}, \mathcal{U})$ using the posterior distribution for unobservable data vector with provisional parameter vector $\boldsymbol{\theta}'$, i.e., $Q(\boldsymbol{\theta}|\boldsymbol{\theta}')$. In the M-step, we find a new parameter vector $\boldsymbol{\theta}''$ that maximizes the expected LLF:

$$\boldsymbol{\theta}'' := \underset{\boldsymbol{\theta}}{\operatorname{argmax}}\, Q(\boldsymbol{\theta}|\boldsymbol{\theta}'), \tag{6}$$

and $\boldsymbol{\theta}''$ becomes a provisional parameter vector at the next E- and M-steps. These steps surely increase the marginal LLF. The E- and M-steps are repeatedly executed until the parameters converge.

3.2. EM Algorithm for NHPP-Based SRMs

Consider the complete data in NHPP-based SRMs, $T_1 < T_2 < \ldots < T_N$, where T_i is the ith failure time and N is the number of all the failures. It is worth noting that the number of all the failures in software is unobserved. Since N is the Poisson-distributed random variable and T_i obeys $F(\cdot;\boldsymbol{\theta})$, the complete LLF is given by

$$\mathrm{LLF}(\omega, \boldsymbol{\theta}) = N \log \omega - \omega + \sum_{i=1}^{N} \log f(T_i; \boldsymbol{\theta}). \tag{7}$$

From the standard argument of MLEs, the MLEs of ω and $\boldsymbol{\theta}$ can be derived as

$$\omega = N \tag{8}$$

and

$$\boldsymbol{\theta} = \underset{\boldsymbol{\theta}}{\operatorname{argmax}} \sum_{i=1}^{N} \log f(T_i; \boldsymbol{\theta}) \tag{9}$$

respectively. These imply that the estimation problem of NHPP-based SRMs under complete data can be decomposed into separate problems for two distribution functions: Poisson distribution and the failure time distribution.

Since the number of failures and the exact failure time in intervals are unobserved, the generalized data $\mathcal{D} := \{(t_1, x_1, u_1), \ldots, (t_k, x_k, u_k)\}$ are incomplete data. By applying the EM algorithm, we have the following EM-step formulas for NHPP-based SRMs with the generalized data:

$$\omega \leftarrow \mathrm{E}[N|\mathcal{D}; \omega', \boldsymbol{\theta}'] \tag{10}$$

and

$$\boldsymbol{\theta} \leftarrow \underset{\boldsymbol{\theta}}{\operatorname{argmax}}\, \mathrm{E}\left[\sum_{i=1}^{N} \log f(T_i; \boldsymbol{\theta}) \middle| \mathcal{D}; \omega', \boldsymbol{\theta}' \right] \tag{11}$$

Additionally, we obtain the following formula to compute the expected values. For any measurable function $h(\cdot)$, the expected value with the generalized data is expressed as

$$\begin{aligned} \mathrm{E}\left[\sum_{i=1}^{N} h(T_i) \middle| \mathcal{D}; \omega', \boldsymbol{\theta}' \right] = & \sum_{i=1}^{n} \left\{ \frac{x_i \int_{t_{i-1}}^{t_i} h(z) f(z; \boldsymbol{\theta}') dz}{\int_{t_{i-1}}^{t_i} f(z; \boldsymbol{\theta}') dz} + u_i h(t_i) \right\} \\ & + \omega' \int_{t_k}^{\infty} h(z) f(z; \boldsymbol{\theta}') dz, \end{aligned} \tag{12}$$

where $f(z;\boldsymbol{\theta})$ is a probability density function (p.d.f.) of failure time provided that the parameter vector is $\boldsymbol{\theta}$. The derivation of this formula is given in Appendix A.

exp: 'exp' is the model where the failure time distribution is an exponential distribution. This model is exactly the same as the Goel–Okumoto model [2]. Define the c.d.f. of failure time as

$$F(t; \beta) = 1 - \exp(-\beta t), \quad \beta > 0. \tag{13}$$

Since the MLE of an ordinary exponential distribution is given by a closed from, the EM-step formulas for exp are directly derived from Equations (10) and (11);

$$\omega \leftarrow \mathrm{E}[N|\mathcal{D};\omega',\beta'] \tag{14}$$

$$\beta \leftarrow \frac{\mathrm{E}[N|\mathcal{D};\omega',\beta']}{\mathrm{E}[\sum_{i=1}^{N} T_i|\mathcal{D};\omega',\beta']}. \tag{15}$$

By applying the formula for the expected value, we have

$$\omega \leftarrow \sum_{i=1}^{k}(x_i+u_i)+\omega'\exp(-\beta' t_k) \tag{16}$$

$$\beta \leftarrow \frac{\sum_{i=1}^{k}(x_i+u_i)+\omega'\exp(-\beta' t_k)}{\sum_{i=1}^{k}(x_i\tau_i+u_i t_i)+\omega'(t_k+1/\beta')\exp(-\beta' t_k)} \tag{17}$$

where

$$\tau_i = \frac{(t_{i-1}+1/\beta')\exp(-\beta' t_{i-1})-(t_i+1/\beta')\exp(-\beta' t_i)}{\exp(-\beta' t_{i-1})-\exp(-\beta' t_i)}. \tag{18}$$

gamma: The failure time distribution becomes the following gamma distribution:

$$F(t;\alpha,\beta)=\int_0^t \frac{\beta^\alpha u^{\alpha-1}\exp(-\beta u)}{\Gamma(\alpha)}du, \quad \alpha>0, \quad \beta>0, \tag{19}$$

where α and β are shape and scale parameters, respectively. When $\alpha = 2$ is fixed, the model reduces the delayed S-shaped SRM [17].

Similar to exp, the EM-step formulas are given using Equations (10) and (11):

$$\omega \leftarrow \mathrm{E}[N|\mathcal{D};\omega',\alpha',\beta'] \tag{20}$$

$$\alpha \leftarrow \left\{\alpha \,\middle|\, \log\alpha-\Psi(\alpha)=\log\left(\frac{\mathrm{E}[\sum_{i=1}^{N} T_i|\mathcal{D};\omega',\alpha',\beta']}{\mathrm{E}[N|\mathcal{D};\omega',\alpha',\beta']}\right) - \frac{\mathrm{E}[\sum_{i=1}^{N}\log T_i|\mathcal{D};\omega',\alpha',\beta']}{\mathrm{E}[N|\mathcal{D};\omega',\alpha',\beta']}\right\} \tag{21}$$

$$\beta \leftarrow \frac{\alpha\mathrm{E}[N|\mathcal{D};\omega',\alpha',\beta']}{\mathrm{E}[\sum_{i=1}^{N} T_i|\mathcal{D};\omega',\alpha',\beta']}, \tag{22}$$

where $\psi(\cdot)$ is the digamma function, i.e., $\psi(\alpha) = d\log\gamma(\alpha)/d\alpha$. Additionally, we use the updated α to compute β. Note that Equation (21) can easily be solved with the non-linear equation solver such as a bisection method. In addition, $\mathrm{E}[N|\mathcal{D};\omega',\alpha',\beta']$ and $\mathrm{E}[\sum_{i=1}^{N} T_i|\mathcal{D};\omega',\alpha',\beta']$ are obtained as follows:

$$\mathrm{E}[N|\mathcal{D};\omega',\alpha',\beta'] = \sum_{i=1}^{k}(x_i+u_i)+\omega'\overline{F}(t_k;\alpha',\beta') \tag{23}$$

$$\mathrm{E}\left[\sum_{i=1}^{N} T_i \,\middle|\, \mathcal{D};\omega',\alpha',\beta'\right] = \sum_{i=1}^{k}(x_i\tau_i+u_i t_i)+\omega'(\alpha'/\beta')\overline{F}(t_k;\alpha'+1,\beta'), \tag{24}$$

$$\tau_i = \frac{\alpha'}{\beta'}\frac{\overline{F}(t_{i-1};\alpha'+1,\beta')-\overline{F}(t_i;\alpha'+1,\beta')}{\overline{F}(t_{i-1};\alpha',\beta')-\overline{F}(t_i;\alpha',\beta')}, \tag{25}$$

where $\overline{F}(t;\alpha,\beta)$ is the complementary c.d.f. of gamma distribution with parameters α and β. On the other hand, we need the numerical integration to obtain $\mathrm{E}[\sum_{i=1}^{N}\log T_i|\mathcal{D};\omega',\alpha',\beta']$. It should be noted that, if the shape parameter α is fixed, then the computation algorithm becomes simple because we ignore solving the nonlinear equation and computing the expected value $\mathrm{E}[\sum_{i=1}^{N}\log T_i|\mathcal{D};\omega',\alpha',\beta']$.

pareto: 'pareto' is the SRM where the failure time distribution is the Pareto distribution of the second kind. The Pareto distribution of the second kind is called Lomax distribution:

$$F(t) = 1 - \frac{c^a}{(x+c)^a}, \quad a > 0, \quad c > 0. \tag{26}$$

This model was proposed as the modified Duane model [50].

Since the Pareto distribution of the second kind is a mixture of exponential distribution, the EM algorithm for 'pareto' is constructed using this property. In general, the mixture distribution is defined as a superposition of original statistical distributions with mixture ratio. Let $G(\xi;\boldsymbol{\theta})$ be the c.d.f. of mixture ratio distribution for the parameter ξ. Then, the mixture distribution is given by

$$F_M(x;\boldsymbol{\theta}) = \int F(x;\xi)dG(\xi;\boldsymbol{\theta}). \tag{27}$$

The Pareto distribution of the second kind is a mixture of exponential distribution when the mixture ratio distribution is a gamma distribution. That is, the failure time distribution is written in the following form:

$$F(t;a,c) = \int_0^\infty \{1-\exp(-\xi t)\}\frac{c^a\xi^{a-1}\exp(-c\xi)}{\Gamma(a)}d\xi = 1 - \frac{c^a}{(c+t)^a}. \tag{28}$$

For the EM algorithm of mixture-type SRMs, we also define the fault detection rate for each fault as a hidden variable.

Let $(T_1,\Xi_1),\ldots,(T_N,\Xi_N)$ be a set of failure time and its associated fault detection rate for all the failures. The complete LLF is given by

$$\begin{aligned} \mathrm{LLF}(\omega,a,c) =& N\log\omega - \omega + \sum_{i=1}^{N}\log\Xi_i - \sum_{i=1}^{N}\Xi_i T_i \\ &+ aN\log c + (a-1)\sum_{i=1}^{N}\log\Xi_i - c\sum_{i=1}^{N}\Xi_i - N\log\Gamma(a). \end{aligned} \tag{29}$$

Similar to gamma, we have the following EM-step formula from the MLEs of gamma distributions:

$$\omega \leftarrow \mathrm{E}[N|\mathcal{D};\omega',a',c'], \tag{30}$$

$$a \leftarrow \left\{a \,\middle|\, \log a - \psi(a) = \log\left(\frac{\mathrm{E}[\sum_{i=1}^{N}\Xi_i|\mathcal{D};\omega',a',c']}{\mathrm{E}[N|\mathcal{D};\omega',a',c']}\right) - \frac{\mathrm{E}[\sum_{i=1}^{N}\log\Xi_i|\mathcal{D};\omega',a',c']}{\mathrm{E}[N|\mathcal{D};\omega',a',c']}\right\} \tag{31}$$

$$b \leftarrow \frac{a\mathrm{E}[N|\mathcal{D};\omega',a',c']}{\mathrm{E}[\sum_{i=1}^{N}\Xi_i|\mathcal{D};\omega',a',c']}, \tag{32}$$

On the other hand, the formula for the expected value is given by

$$\mathrm{E}\left[\sum_{i=1}^{N}h(\Xi_i)\,\middle|\,\mathcal{D};\omega',\boldsymbol{\theta}'\right] = \sum_{i=1}^{k}\left\{\frac{x_i\int_{t_{i-1}}^{t_i}\tilde{h}(z;\boldsymbol{\theta}')dz}{\int_{t_{i-1}}^{t_i}\tilde{f}(z;\boldsymbol{\theta}')dz} + \frac{u_i\tilde{h}(t_i)}{\tilde{f}(t_i)}\right\} + \omega'\int_{t_k}^{\infty}\tilde{h}(z;\boldsymbol{\theta}')dz, \tag{33}$$

where $h(\cdot)$ is an arbitrary measurable function and

$$\tilde{h}(z;\boldsymbol{\theta}') = \int h(\xi)f(z;\xi)dG(\xi;\boldsymbol{\theta}'), \tag{34}$$

$$\tilde{f}(z;\boldsymbol{\theta}') = \int f(z;\xi)dG(\xi;\boldsymbol{\theta}'). \tag{35}$$

tnorm, lnorm: 'tnorm' and 'lnorm' are SRMs whose failure time distributions are truncated and log normal distributions, respectively. The failure time distributions for tnorm and lnorm are

$$\text{tnorm:} \qquad F(t) = \Phi\left(\frac{t-\mu}{\sigma}\right)/\{1-\Phi(-\mu/\sigma)\}, \tag{36}$$

$$\text{lnorm:} \qquad F(t) = \Phi\left(\frac{\log t-\mu}{\sigma}\right), \tag{37}$$

where $\Phi(\cdot)$ is the c.d.f. of the standard normal distribution. Since the EM algorithms for both models with failure time and count data were introduced in detail in the literature [34], this paper provides the EM-step formulas with the generalized data.

- EM-step formula for tnorm:

$$\tilde{\omega} \leftarrow N, \qquad \mu \leftarrow T^{(1)}/N, \qquad \sigma \leftarrow \sqrt{T^{(2)}/N - (T^{(1)}/N)^2} \tag{38}$$

where

$$N = \sum_{i=1}^{k}(x_i + u_i) + \tilde{\omega}\{\Phi(z_0) + \overline{\Phi}(z_k)\}, \tag{39}$$

$$T^{(1)} = \sum_{i=1}^{k}(x_i \tau_i^{(1)} + u_i t_i) + \tilde{\omega}\{\Phi^{(1)}(z_0) + \overline{\Phi}^{(1)}(z_k)\} \tag{40}$$

$$T^{(2)} = \sum_{i=1}^{k}(x_i \tau_i^{(2)} + u_i t_i^2) + \tilde{\omega}\{\Phi^{(2)}(z_0) + \overline{\Phi}^{(2)}(z_k)\}, \tag{41}$$

$$z_0 = -\mu/\sigma, \quad z_i = (t_i - \mu)/\sigma, \tag{42}$$

$$\tau_i^{(u)} = \frac{\overline{\Phi}^{(u)}(z_{i-1}) - \overline{\Phi}^{(u)}(z_i)}{\overline{\Phi}(z_{i-1}) - \overline{\Phi}(z_i)}, \tag{43}$$

- EM-step formula for lnorm:

$$\omega \leftarrow N, \qquad \mu \leftarrow T^{(1)}/N, \qquad \sigma \leftarrow \sqrt{T^{(2)}/N - (T^{(1)}/N)^2} \tag{44}$$

where

$$N = \sum_{i=1}^{k}(x_i + u_i) + \omega'\overline{\Phi}(z_k), \tag{45}$$

$$T^{(1)} = \sum_{i=1}^{k}(x_i \tau_i^{(1)} + u_i \log t_i) + \omega\overline{\Phi}^{(1)}(z_k), \tag{46}$$

$$T^{(2)} = \sum_{i=1}^{k}(x_i \tau_i^{(2)} + u_i (\log t_i)^2) + \omega\overline{\Phi}^{(2)}(z_k), \tag{47}$$

$$z_0 \to -\infty, \qquad z_i = (\log t_i - \mu)/\sigma, \tag{48}$$

$$\tau_i^{(u)} = \frac{\overline{\Phi}^{(u)}(z_{i-1}) - \overline{\Phi}^{(u)}(z_i)}{\overline{\Phi}(z_{i-1}) - \overline{\Phi}(z_i)}. \tag{49}$$

In the above formulas, $\overline{\Phi}(z)$ is the complementary c.d.f. of the standard normal distribution, and $\overline{\Phi}^{(1)}(z)$ and $\overline{\Phi}^{(2)}(z)$ are expressed with the p.d.f. of the standard normal distribution $\phi(z)$:

$$\overline{\Phi}^{(1)}(z) = \sigma\phi(z) + \mu\overline{\Phi}(z), \tag{50}$$

$$\overline{\Phi}^{(2)}(z) = (\sigma^2 z + 2\mu\sigma)\phi(z) + (\sigma^2 + \mu^2)\overline{\Phi}(z). \tag{51}$$

In addition, after the convergence, we take $\omega = \tilde{\omega}\overline{\Phi}(z_0)$ to obtain the ML estimate for ω in the case of tnorm.

tlogis, llogis tlogis and llogis are the SRMs with truncated and log logistic distributions, respectively. In particular, 'tlogis' is equivalent to the inflection S-shaped model [15]. The failure time distribution of tlogis is given by

$$F(t) = \Psi\left(\frac{t-\mu}{\psi}\right) / \{1 - \Psi(-\mu/\psi)\}, \tag{52}$$

where $\Psi(\cdot)$ is the c.d.f. of standard logistic distribution

$$\Psi(t) = \frac{1}{1+\exp(-t)}. \tag{53}$$

By taking into account the exponential of logistic distribution, we have the following failure time distribution of llogis:

$$F(t) = \Psi\left(\frac{\log t - \mu}{\psi}\right). \tag{54}$$

Since logistic distribution does not belongs to the exponential family of distributions, neither expectation nor maximization can be expressed as simple formulas. To construct the algorithm, we consider only one assumption; the number of all failures is not observed. Then, the EM-step formulas become

- The EM-step formula for tnorm

$$\tilde{\omega} \leftarrow \sum_{i=1}^{k}(x_i + u_i) + \tilde{\omega}' F(0; \boldsymbol{\theta}') + \tilde{\omega}' \overline{F}(t_k; \boldsymbol{\theta}') \tag{55}$$

$$\begin{aligned} \boldsymbol{\theta} \leftarrow \underset{\boldsymbol{\theta}}{\operatorname{argmax}} \Bigg\{ & \sum_{i=1}^{k}(x_i \log(F(t_i;\boldsymbol{\theta}) - F(t_{i-1};\boldsymbol{\theta})) - x_i \log x_i! + u_i \log f(t_i;\boldsymbol{\theta})) \\ & + (\tilde{\omega}' F(0;\boldsymbol{\theta}')) \log(F(0;\boldsymbol{\theta})) + (\tilde{\omega}' \overline{F}(t_k;\boldsymbol{\theta}')) \log(\overline{F}(t_k;\boldsymbol{\theta})) \Bigg\}. \end{aligned} \tag{56}$$

- The EM-step formula for lnorm

$$\omega \leftarrow \sum_{i=1}^{k}(x_i + u_i) + \omega' \overline{F}(t_k; \boldsymbol{\theta}') \tag{57}$$

$$\begin{aligned} \boldsymbol{\theta} \leftarrow \underset{\boldsymbol{\theta}}{\operatorname{argmax}} \Bigg\{ & \sum_{i=1}^{k}(x_i \log(F(t_i;\boldsymbol{\theta}) - F(t_{i-1};\boldsymbol{\theta})) - x_i \log x_i! + u_i \log f(t_i;\boldsymbol{\theta})) \\ & + (\omega' \overline{F}(t_k;\boldsymbol{\theta}')) \log(\overline{F}(t_k;\boldsymbol{\theta})) \Bigg\}. \end{aligned} \tag{58}$$

The second equations in both formulas indicate that $\boldsymbol{\theta}$ is updated by the MLEs when the number of all the failures is given by $\tilde{\omega}'$ and ω'. These algorithm are also stable if there exists a unique solution maximizing the right-hand side of the second term. Note that, after the convergence, the model parameter ω in tlogis can be obtained as $\omega = \tilde{\omega}\overline{F}(0; \boldsymbol{\theta})$.

txvmax, lxvmax, txvmin, lxvmin Suppose that the failure time caused by each failure follows an extreme value type I distribution. The extreme value type I distribution is called Gumbel distribution, and its definition is based on the limitation of the maximum value of random variables. Here, the c.d.f. of a standard Gumbel distribution is defined as

$$\Theta(t) = \exp\{-\exp(-t)\}. \tag{59}$$

Similar to tnorm, lnorm, tlogis, and llogis, we consider the truncation and logarithm of the extreme value distribution. In addition, since the extreme value distribution is not symmetric, we also consider the case of negative samples, i.e., the minimum value of random variables.

The failure time distributions of txvmax and lxvmax are, respectively,

$$F(t) = \Theta\left(\frac{t-\mu}{\theta}\right) / \{1 - \Theta(-\mu/\theta)\}, \tag{60}$$

$$F(t) = \Theta\left(\frac{\log t - \mu}{\theta}\right). \tag{61}$$

Similarly, the failure time distributions of txvmin and lxvmin are given by

$$F(t) = \overline{\Theta}\left(\frac{t+\mu}{\theta}\right) / \{1 - \overline{\Theta}(\mu/\theta)\}, \tag{62}$$

$$F(t) = \overline{\Theta}\left(\frac{\log t + \mu}{\theta}\right), \tag{63}$$

where $\overline{\Theta}(t) = 1 - \Theta(-t)$ corresponds to the c.d.f. of a standard extreme value type I distribution of the minimum. From Equation (63), we find that lxvmin is equivalent to the Weibull distribution.

Since the extreme value distribution is not an exponential family, we consider only one assumption; the number of all the failures is not observed. Then, the EM-step formulas are given by

- The EM-step formula for txvmax and txvmin

$$\tilde{\omega} \leftarrow \sum_{i=1}^{k}(x_i + u_i) + \tilde{\omega}' F(0; \boldsymbol{\theta}') + \tilde{\omega}'\overline{F}(t_k; \boldsymbol{\theta}') \tag{64}$$

$$\begin{aligned}\boldsymbol{\theta} \leftarrow \underset{\theta}{\operatorname{argmax}} \Bigg\{ &\sum_{i=1}^{k}(x_i \log(F(t_i; \boldsymbol{\theta}) - F(t_{i-1}; \boldsymbol{\theta})) - x_i \log x_i! + u_i \log f(t_i; \boldsymbol{\theta})) \\ &+ (\tilde{\omega}' F(0; \boldsymbol{\theta}')) \log(F(0; \boldsymbol{\theta})) + (\tilde{\omega}'\overline{F}(t_k; \boldsymbol{\theta}')) \log(\overline{F}(t_k; \boldsymbol{\theta})) \Bigg\}. \end{aligned} \tag{65}$$

- The EM-step formula for lxvmax and lxvmin

$$\omega \leftarrow \sum_{i=1}^{k}(x_i + u_i) + \omega'\overline{F}(t_k; \boldsymbol{\theta}') \tag{66}$$

$$\begin{aligned}\boldsymbol{\theta} \leftarrow \underset{\theta}{\operatorname{argmax}} \Bigg\{ &\sum_{i=1}^{k}(x_i \log(F(t_i; \boldsymbol{\theta}) - F(t_{i-1}; \boldsymbol{\theta})) - x_i \log x_i! + u_i \log f(t_i; \boldsymbol{\theta})) \\ &+ (\omega'\overline{F}(t_k; \boldsymbol{\theta}')) \log(\overline{F}(t_k; \boldsymbol{\theta})) \Bigg\}. \end{aligned} \tag{67}$$

4. Numerical Example

We investigated the numerical characteristics of the presented EM algorithms. Here, we compare the convergence property with the Nelder–Mead method and the quasi-Newton method (BFGS method). First, we check the trace of model parameters until they converge to MLE for the proposed method (EM algorithm), the Nelder–Mead method, and the BFGS method. In this experiment, we used the fault count data, which were collected from real software projects [53]. The statistics of fault count data are given as follow.

- Data label: SS1A
- Working days: 151
- The number of failures: 112
- LOC: Hundreds of thousands
- Software type: Operations

For the above failure count data, we estimated the parameters of 'exp'. The MLEs of exp are $\hat{\omega} = 354.75$ and $\hat{\beta} = 0.00251$, and the maximum LLF is $\mathrm{LLF}(\hat{\omega}, \hat{\beta}) = -180.79$.

Figures 1–3 show the trace of model parameters for EM algorithm, the Nelder–Mead method, and the BFGS method when the initial guesses are $\omega = 100$ and $\beta = 0.1$. We use the 'optim' function in R for the Nelder–Mead and BFGS methods. From these figures, we find that the EM algorithm stably updates the model parameters and converges to close parameters to the MLEs. However, the convergence speed is not fast, since the update becomes smaller as the parameters are close to the MLEs. The Nelder–Mead method provides the MLEs, but the trace of the algorithm is not stable. In particular, this algorithm sometimes takes invalid values that violate the parameter constraints, i.e., $\omega < 0$ or $\beta < 0$, while searching for the parameters. Figure 3 depicts the trace of parameters for the BFGS method. The convergence property is the worst among the three methods. Additionally, the algorithm fails to obtain the MLE.

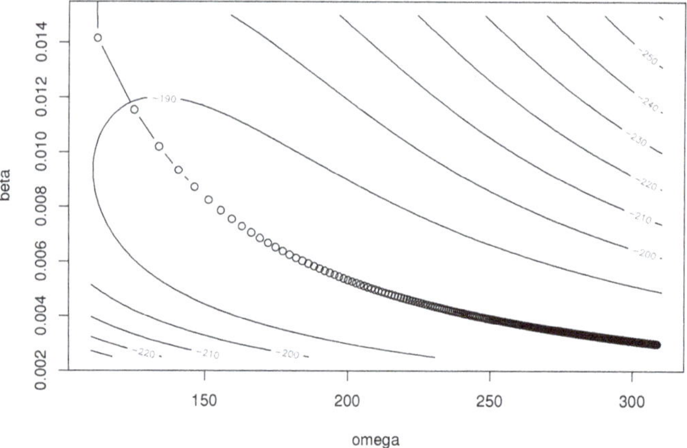

Figure 1. Trace of parameters in the EM algorithm.

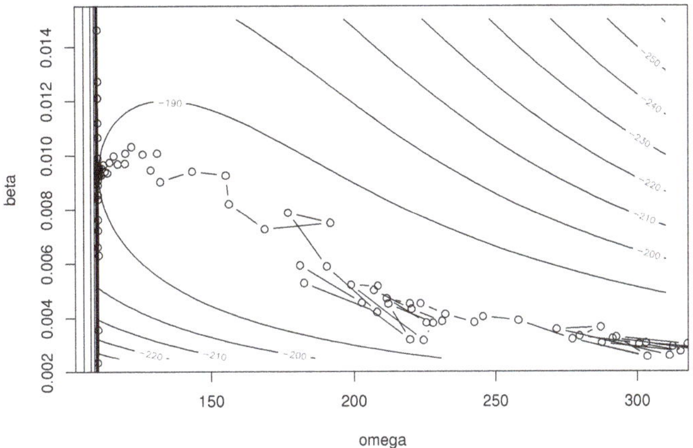

Figure 2. Trace of parameters in the Nelder–Mead method.

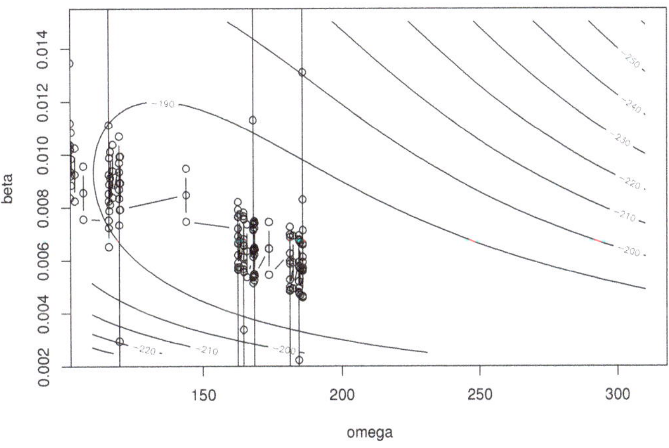

Figure 3. Trace of parameters in the BFGS method.

Next, we present the convergence properties for the proposed EM algorithm, the Nelder–Mead method, and the BFGS method quantitatively. Here, we use two additional fault count data that were collected from real software projects [53] as well as SS1A.

- Data label: SS1B
- Working days: 663
- The number of failures: 375
- LOC: Hundreds of thousands
- Software type: Operations

- Data label: SS1C
- Working days: 472
- The number of failures: 277
- LOC: Hundreds of thousands
- Software type: Operations

For three data sets—SS1A, SS1B, and SS1C—we applied the proposed EM algorithm, the Nelder–Mead method, and the BFGS method for 12 basic NHPP-based SRMs with 100 different initial parameters. In the experiment, the initial parameters were selected by random numbers. Tables 2–4 present the number of converged estimations, i.e., the number of times that the model parameters are successfully estimated for each NHPP-based SRMs and methods. If this value is 100, the method succeeded in obtaining the MLE for all of the initial parameters. On the other hand, if this value is 0, the estimation method fails to obtain the MLE for all of the initial parameters due to numerical computation errors such as overflow and underflow.

Table 2. The number of converged estimations (SS1A).

Model	EM	Nelder-Mead	BFGS
exp	100	19	19
gamma	100	34	34
pareto	100	57	57
tnorm	100	88	88
lnorm	100	98	98
tlogis	100	100	100
llogis	100	100	100
txvmax	100	100	100
lxvmax	99	99	99
txvmin	64	65	65
lxvmin	92	92	92

Table 3. The number of converged estimations (SS1B).

Model	EM	Nelder-Mead	BFGS
exp	100	7	7
gamma	100	7	6
pareto	100	18	18
tnorm	100	98	98
lnorm	87	87	87
tlogis	100	100	100
llogis	87	87	87
txvmax	99	99	99
lxvmax	0	0	0
txvmin	89	89	89
lxvmin	100	41	41

Table 4. The number of converged estimations (SS1C).

Model	EM	Nelder-Mead	BFGS
exp	100	9	8
gamma	100	13	12
pareto	100	47	47
tnorm	100	98	98
lnorm	96	97	97
tlogis	100	100	100
llogis	96	96	96
txvmax	99	99	99
lxvmax	0	0	0
txvmin	89	89	89
lxvmin	100	43	43

From these results, we find that the convergence rates of the proposed EM algorithms are 100% in the cases of exp, gamma, pareto, tnorm, and tlogis. Since the number of converged estimations of the Nelder–Mead is almost the same as that of BFGS for all cases, the convergence properties of their methods are the same if we use the 'optim' function in R. Additionally, since lxvmax did not fit SS1B and SS1C, all of the estimation methods failed to obtain the MLE. Furthermore, it was found that the numbers of converged estimates in the Nelder–Mead and BFGS methods are worse than that of the EM algorithm, specifically in the cases of exp, gamma, and pareto. Additionally, in the cases of tnorm and lnorm, the convergence property of EM is slightly superior to the other two methods. In exp, gamma, pareto, tnorm, and lnorm, the failure time distributions belong to the exponential family, and thus, their EM-step formulas do not include the numerical optimization step. That is, these EM-step formulas are 'pure' EM-step formulas. Therefore, the convergence properties outperform those of the Nelder–Mead and BFGS methods. On the other hand, in the cases of tlogis, llogis, txvmax, lxvmax, txvmin, and lvxmin, the failure time distributions are not in the exponential family, and we should use the numerical optimization step in their EM-step formulas. This is the reason why the convergence property of the presented EM algorithm is same as that in the Nelder–Mead and BFGS methods.

5. Conclusions

This paper derived EM-step formulas for 12 basic NHPP-based SRMs when the generalized failure count data are given. Since the generalized fault count data involve both time and count data formats, the presented EM algorithms can be applied to failure data experienced in practice. In addition, the convergence property of EM algorithm is better than or equivalent to other ordinary methods such as the Nelder–Mead and BFGS methods for practical software fault data. Thus, the presented algorithms are suitable for implementation in the automatic tool for software reliability evaluation. In fact, our research group has developed an AddIn of Microsoft Excel to estimate software reliability [54].

In the future, we will develop a reliability assessment tool by integrating a software repository such as GitHub, a bug tracking system, and a continuous integration system, and the tool will continuously monitor the reliability of software.

Author Contributions: Conceptualization, H.O. and T.D.; methodology, H.O.; software, H.O.; supervision, T.D.; validation, T.D.; writing—original draft, H.O. and T.D.; writing—review and editing, H.O. and T.D. All authors have read and agreed to the published version of the manuscript.

Funding: This research received no external funding.

Institutional Review Board Statement: Not applicable.

Informed Consent Statement: Not applicable.

Data Availability Statement: Not applicable.

Conflicts of Interest: The authors declare no conflict of interest.

Appendix A. Derivation of Equation (12)

For convenient, ω' and $\boldsymbol{\theta}'$ are written as ω and $\boldsymbol{\theta}$, respectively, and $\mathrm{E}[\cdot|\mathcal{D};\omega',\boldsymbol{\theta}']$ is simplified as $\mathrm{E}[\cdot|\mathcal{D}]$. Here we have

$$\mathrm{E}\left[\sum_{i=1}^{N} h(T_i)\middle|\mathcal{D}\right] = \sum_{i=1}^{n}\left\{\mathrm{E}\left[\sum_{j=s_{i-1}+1}^{s_{i-1}+x_i} h(T_j)\middle|\mathcal{D}\right] + b_i\mathrm{E}[h(T_{s_i})|\mathcal{D}]\right\} + \mathrm{E}\left[\sum_{i=s_n+1}^{N} h(T_i)\middle|\mathcal{D}\right], \tag{A1}$$

where $s_i = \sum_{j=1}^{i}(x_j + b_j)$.

According to the order statistics of failure times, the first term of the right-hand side of the above can be rewritten as follows.

$$\mathrm{E}\left[\sum_{j=s_{i-1}+1}^{s_{i-1}+x_i} h(T_j)\middle|\mathcal{D}\right] = \frac{\int_{t_{i-1}}^{t_i}\int_{z_1}^{t_i}\cdots\int_{z_{x_i}}^{t_i}\sum_{j=1}^{x_i} h(z_j)\prod_{j=1}^{x_i} f(z_j)dz_{x_i}\cdots dz_1}{\int_{t_{i-1}}^{t_i}\int_{z_1}^{t_i}\cdots\int_{z_{x_i}}^{t_i}\prod_{j=1}^{x_i} f(z_j)dz_{x_i}\cdots dz_1}. \tag{A2}$$

Since $T_{s_{i-1}+1}, \ldots, T_{s_{i-1}+x_i}$ are i.i.d. random variables, the multiple integrals of denominator in Equation (A2) is given by

$$\int_{t_{i-1}}^{t_i} \int_{z_1}^{t_i} \cdots \int_{z_{x_i}}^{t_i} \prod_{j=1}^{x_i} f(z_j) dz_{x_i} \cdots dz_1 = \frac{1}{x_i!} \left(\int_{t_{i-1}}^{t_i} f(z) dz \right)^{x_i}. \tag{A3}$$

Similarly, the numerator becomes

$$\int_{t_{i-1}}^{t_i} \int_{z_1}^{t_i} \cdots \int_{z_{x_i}}^{t_i} \sum_{j=1}^{x_i} h(z_j) \prod_{j=1}^{x_i} f(z_j) dz_{x_i} \cdots dz_1 = \frac{x_i}{x_i!} \int_{t_{i-1}}^{t_i} h(z) f(z) dz \left(\int_{t_{i-1}}^{t_i} f(z) dz \right)^{x_i - 1}. \tag{A4}$$

Henceforth we have

$$\mathrm{E}\left[\sum_{j=s_{i-1}+1}^{s_{i-1}+x_i} h(T_j) \middle| \mathcal{D} \right] = \frac{x_i \int_{t_{i-1}}^{t_i} h(z) f(z) dz}{\int_{t_{i-1}}^{t_i} f(z) dz}. \tag{A5}$$

The second term of the right-hand side of Equation (A1) is straightforwardly given by $h(t_i)$. The third term can be derived by a similar way to the first term. Taking account of the condition $N = \nu$, we have

$$\begin{aligned} \mathrm{E}\left[\sum_{i=s_n+1}^{N} h(T_i) \middle| \mathcal{D} \right] &= \frac{\sum_{\nu=s_k}^{\infty} e^{-\omega} \frac{\omega^{\nu}}{\nu!} \frac{\nu!(\nu-s_k)}{(\nu-s_k)!} \int_{t_k}^{\infty} h(z) f(z) dz \overline{F}(t_k)^{\nu-s_k-1}}{\sum_{\nu=s_k}^{\infty} e^{-\omega} \frac{\omega^{\nu}}{\nu!} \frac{\nu!}{(\nu-s_k)!} \overline{F}(t_k)^{\nu-s_k}} \\ &= \omega \int_{t_k}^{\infty} h(z) f(z) dz, \end{aligned} \tag{A6}$$

where $\overline{F}(t) = 1 - F(t)$.

References

1. Jelinski, Z.; Moranda, P.B. Software Reliability Research. In *Statistical Computer Performance Evaluation*; Freiberger, W., Ed.; Academic Press: New York, NY, USA, 1972; pp. 465–484.
2. Goel, A.L.; Okumoto, K. Time-Dependent Error-Detection Rate Model for Software Reliability and Other Performance Measures. *IEEE Trans. Reliab.* **1979**, *R-28*, 206–211. [CrossRef]
3. Lyu, M.R. (Ed.) *Handbook of Software Reliability Engineering*; McGraw-Hill: New York, NY, USA, 1996.
4. Musa, J.D.; Iannino, A.; Okumoto, K. *Software Reliability, Measurement, Prediction, Application*; McGraw-Hill: New York, NY, USA, 1987.
5. Pham, H. *Software Reliability*; Springer: Singapore, 2000.
6. Xie, M. *Software Reliability Modelling*; World Scientific: Singapore, 1991.
7. Rook, P. (Ed.) *Software Reliability Handbook*; Elsevier Science: London, UK, 1990.
8. Singpurwalla, N.D.; Wilson, S.P. *Statistical Methods in Software Engineering*; Springer: New York, NY, USA, 1997.
9. Li, Q.; Pham, H. A Generalized Software Reliability Growth Model With Consideration of the Uncertainty of Operating Environments. *IEEE Access* **2019**, *7*, 84253–84267. [CrossRef]
10. Li, Q.; Pham, H. NHPP software reliability model considering the uncertainty of operating environments with imperfect debugging and testing coverage. *Appl. Math. Model.* **2017**, *51*, 68–85. [CrossRef]
11. He, Y. NHPP software reliability growth model incorporating fault detection and debugging. In Proceedings of the 2013 IEEE 4th International Conference on Software Engineering and Service Science, Beijing, China, 23–25 May 2013; pp. 225–228.
12. Song, K.Y.; Chang, I.H.; Pham, H. NHPP Software Reliability Model with Inflection Factor of the Fault Detection Rate Considering the Uncertainty of Software Operating Environments and Predictive Analysis. *Symmetry* **2019**, *11*, 521. [CrossRef]
13. Sun, H.; Zhang, L.; Wu, J.; Wu, J.; Yang, H. A New Method of Model Combination Based on the NHPP Software Reliability Models. In *ICMSS 2018: Proceedings of the 2018 2nd International Conference on Management Engineering, Software Engineering and Service Sciences*; ACM: New York, NY, USA, 2018; pp. 153–158. [CrossRef]
14. Musa, J.D.; Okumoto, K. A Logarithmic Poisson Execution Time Model for Software Reliability Measurement. In *Proceedings of the 7th International Conference Software Engineering (ICSE-1084)*; IEEE CS Press: Los Alamitos, CA, USA; ACM: New York, NY, USA, 1984; pp. 230–238.
15. Ohba, M. Software Reliability Analysis. *IBM J. Res. Dev.* **1984**, *28*, 428–443. [CrossRef]
16. Ohba, M. Inflection S-Shaped Software Reliability Growth Model. In *Stochastic Models in Reliability Theory*; Osaki, S., Hatoyama, Y., Eds.; Springer: Berlin, Gremany, 1984; pp. 144–165.

17. Yamada, S.; Ohba, M.; Osaki, S. S-Shaped Reliability Growth Modeling for Software Error Detection. *IEEE Trans. Reliab.* **1983**, *R-32*, 475–478. [CrossRef]
18. Zhao, M.; Xie, M. On Maximum Likelihood Estimation for a General non-Homogeneous Poisson Process. *Scand. J. Stat.* **1996**, *23*, 597–607.
19. Pham, H.; Zhang, X. An NHPP Software Reliability Models and its Comparison. *Int. J. Reliab. Qual. Safe. Eng.* **1997**, *4*, 269–282. [CrossRef]
20. Shanthikumar, J.G. A General Software Reliability Model for Performance Prediction. *Microelectron. Reliab.* **1981**, *21*, 671–682. [CrossRef]
21. Langberg, N.; Singpurwalla, N.D. Unification of Some Software Reliability Models. *SIAM J. Sci. Comput.* **1985**, *6*, 781–790. [CrossRef]
22. Chen, Y.; Singpurwalla, N.D. Unification of Software Reliability Models by Self-Exciting Point Processes. *Adv. Appl. Probab.* **1997**, *29*, 337–352. [CrossRef]
23. Miller, D.R. Exponential Order Statistic Models of Software Reliability Growth. *IEEE Trans. Softw. Eng.* **1986**, *SE-12*, 12–24. [CrossRef]
24. Hossain, S.A.; Dahiya, R.C. Estimating the Parameters of a non-Homogeneous Poisson-Process Model for Software Reliability. *IEEE Trans. Reliab.* **1993**, *42*, 604–612. [CrossRef]
25. Knafl, G.; Morgan, J. Solving ML Equations for 2-parameter Poisson-Process Model for Ungrouped Software Failure Data. *IEEE Trans. Reliab.* **1996**, *45*, 42–53. [CrossRef]
26. Joe, H. Statistical Inference for General-Order-Statistics and Nonhomogeneous-Poisson-Process Software Reliability Models. *IEEE Trans. Softw. Eng.* **1989**, *15*, 1485–1490. [CrossRef]
27. Jeske, D.R.; Pham, H. On the maximum likelihood estimates for the Goel-Okumoto software reliability model. *Am. Stat.* **2001**, *55*, 219–222. [CrossRef]
28. Dempster, A.P.; Laird, N.M.; Rubin, D.B. Maximum likelihood from incomplete data via the EM algorithm. *J. R. Stat. Soc. B* **1977**, *B-39*, 1–38.
29. Wu, C.F.J. On the convergence properties of the EM algorithm. *Ann. Stat.* **1983**, *11*, 95–103. [CrossRef]
30. Okamura, H.; Dohi, T.; Osaki, S. EM algorithms for logistic software reliability models. In Proceedings of the 22nd IASTED International Conference on Software Engineering, Innsbruck, Austria, 17–19 February 2004; ACTA Press: Crete, Greece, 2004; pp. 263–268.
31. Okamura, H.; Murayama, A.; Dohi, T. EM Algorithm for Discrete Software Reliability Models: A Unified Parameter Estimation Method. In Proceedings of the 8th IEEE International Symposium High Assurance Systems Engineering, Tampa, FL, USA, 25–26 March 2004; pp. 219–228.
32. Okamura, H.; Watanabe, Y.; Dohi, T. Estimating a mixed software reliability models based on the EM algorithm. In Proceedings of the International Symposium on Empirical Software Engineering (ISESE2002), Nara, Japan, 3–4 October 2002; IEEE Computer Society Press: Los Alamitos, CA, USA, 2002; pp. 69–78.
33. Okamura, H.; Watanabe, Y.; Dohi, T. An iterative scheme for maximum likelihood estimation in software reliability modeling. In Proceedings of the 14th International Symposium on Software Reliability Engineering, Denver, CO, USA, 17–20 November 2003; pp. 246–256.
34. Okamura, H.; Dohi, T.; Osaki, S. Software reliability growth models with normal failure time distributions. *Reliab. Eng. Syst. Saf.* **2013**, *116*, 135–141. [CrossRef]
35. Ohishi, K.; Okamura, H.; Dohi, T. Gompertz software reliability model: Estimation algorithm and empirical validation. *J. Syst. Softw.* **2009**, *82*, 535–543. [CrossRef]
36. Okamura, H.; Dohi, T. Phase-type software reliability model: Parameter estimation algorithms with grouped data. *Ann. Oper. Res.* **2016**, *244*, 177–208. [CrossRef]
37. Okamura, H.; Dohi, T. Hyper-Erlang software reliability model. In Proceedings of the 2008 14th IEEE Pacific Rim International Symposium on Dependable Computing, Taipei, Taiwan, 15–17 December 2008.
38. Okamura, H.; Dohi, T. Building Phase-Type Software Reliability Model. In Proceedings of the 2006 17th International Symposium on Software Reliability Engineering, Raleigh, NC, USA, 7–10 November 2006; pp. 289–298.
39. Xiao, X.; Okamura, H.; Dohi, T. NHPP-based software reliability models using equilibrium distribution. *IEICE Trans. Fundam. Electron. Commun. Comput. Sci. (A)* **2012**, *E95-A*, 894–902. [CrossRef]
40. Kawasaki, M.; Okamura, H.; Dohi, T. A Comprehensive Evaluation of Software Reliability Modeling Based on Marshall-Olkin Type Fault-Detection Time Distribution. In Proceedings of the 2017 24th Asia-Pacific Software Engineering Conference (APSEC), Nanjing, China, 4–8 December 2017; pp. 486–494. [CrossRef]
41. Kimura, M.; Yamada, S. Software reliability management: Techniques and applications. In *Handbook of Reliability Engineering*; Pham, H., Ed.; Springer: London, UK, 2003; pp. 265–284.
42. Ledoux, J.; Rubino, G. A counting model for software reliability analysis. *Int. J. Model. Simul.* **1997**, *17*, 289–297. [CrossRef]
43. Okamura, H.; Dohi, T. Unification of software reliability models using Markovian arrival processes. In Proceedings of the 17th IEEE Pacific Rim International Symposium on Dependable Computing (PRDC-2011), Pasadena, CA, USA, 12–14 December 2011; IEEE CS Press: Los Alamitos, CA, USA, 2011; pp. 20–27.

44. Goel, A.L.; Okumoto, K. *An Imperfect Debugging Model for Reliability and Other Quantitative Measures of Software Systems*; Technical Report 78-1; Department of IE & OR, Syracuse University: Syracuse, NY, USA, 1978.
45. Littlewood, B. A Reliability Model for Systems with Markov Structure. *Appl. Stat.* **1975**, *24*, 172–177. [CrossRef]
46. Ledoux, J. Availability modeling of modular software. *IEEE Trans. Reliab.* **1999**, *48*, 159–168. [CrossRef]
47. Zeephongsekul, P.; Jayasinghe, C.L.; Fiondella, L.; Nagaraju, V. Maximum-Likelihood Estimation of Parameters of NHPP Software Reliability Models Using Expectation Conditional Maximization Algorithm. *IEEE Trans. Reliab.* **2016**, *65*, 1571–1583. [CrossRef]
48. Nagaraju, V.; Fiondella, L.; Zeephongsekul, P.; Jayasinghe, C.L.; Wandji, T. Performance Optimized Expectation Conditional Maximization Algorithms for Nonhomogeneous Poisson Process Software Reliability Models. *IEEE Trans. Reliab.* **2017**, *66*, 722–734. [CrossRef]
49. Okamura, H.; Dohi, T. Application of EM Algorithm to NHPP-Based Software Reliability Assessment with Ungrouped Failure Time Data. In *Stochastic Reliability and Maintenance Modeling: Essays in Honor of Professor Shunji Osaki on his 70th Birthday*; Dohi, T., Nakagawa, T., Eds.; Springer: London, UK, 2013; pp. 285–313. [CrossRef]
50. Littlewood, B. Rationale for a Modified Duane Model. *IEEE Trans. Reliab.* **1984**, *R-33*, 157–159. [CrossRef]
51. Goel, A.L. Software Reliability Models: Assumptions, Limitations and Applicability. *IEEE Trans. Softw. Eng.* **1985**, *SE-11*, 1411–1423. [CrossRef]
52. Gokhale, S.S.; Trivedi, K.S. Log-Logistic Software Reliability Growth Model. In Proceedings of the Third IEEE International High-Assurance Systems Engineering Symposium (HASE-1998), Washington, DC, USA, 13–14 November 1998; IEEE CS Press: Turku, Finland, 1998; pp. 34–41.
53. Musa, J. *Software Reliability Data*; Technical Report; Rome Air Development Center, Griffiss AFB: Rome, NY, USA, 1979.
54. Okamura, H.; Dohi, T. SRATS: Software reliability assessment tool on spreadsheet (Experience report). In Proceedings of the 24th International Symposium on Software Reliability Engineering (ISSRE 2013), Pasadena, CA, USA, 4–7 November 2013; IEEE CS Press: Los Alamitos, CA, USA, 2013; pp. 100–117.

Article

An Enhanced Evolutionary Software Defect Prediction Method Using Island Moth Flame Optimization

Ruba Abu Khurma [1], Hamad Alsawalqah [1], Ibrahim Aljarah [1,*], Mohamed Abd Elaziz [2,3] and Robertas Damaševičius [4,*]

1 King Abdullah II School for Information Technology, The University of Jordan, Amman 11942, Jordan; ruba_abukhurma@yahoo.com (R.A.K.); h.sawalqah@ju.edu.jo (H.A.)
2 Department of Mathematics, Faculty of Science, Zagazig University, Zagazig 44519, Egypt; abd_el_aziz_m@yahoo.com
3 School of Computer Science and Robotics, Tomsk Polytechnic University, 634050 Tomsk, Russia
4 Faculty of Applied Mathematics, Silesian University of Technology, 44-100 Gliwice, Poland
* Correspondence: i.aljarah@ju.edu.jo (I.A.); robertas.damasevicius@polsl.pl (R.D.)

Abstract: Software defect prediction (SDP) is crucial in the early stages of defect-free software development before testing operations take place. Effective SDP can help test managers locate defects and defect-prone software modules. This facilitates the allocation of limited software quality assurance resources optimally and economically. Feature selection (FS) is a complicated problem with a polynomial time complexity. For a dataset with N features, the complete search space has 2^N feature subsets, which means that the algorithm needs an exponential running time to traverse all these feature subsets. Swarm intelligence algorithms have shown impressive performance in mitigating the FS problem and reducing the running time. The moth flame optimization (MFO) algorithm is a well-known swarm intelligence algorithm that has been used widely and proven its capability in solving various optimization problems. An efficient binary variant of MFO (BMFO) is proposed in this paper by using the island BMFO (IsBMFO) model. IsBMFO divides the solutions in the population into a set of sub-populations named islands. Each island is treated independently using a variant of BMFO. To increase the diversification capability of the algorithm, a migration step is performed after a specific number of iterations to exchange the solutions between islands. Twenty-one public software datasets are used for evaluating the proposed method. The results of the experiments show that FS using IsBMFO improves the classification results. IsBMFO followed by support vector machine (SVM) classification is the best model for the SDP problem over other compared models, with an average G-mean of 78%.

Keywords: moth flame optimization; island-based model; feature selection; software defect prediction; software reliability

Citation: Khurma, R.A.; Alsawalqah, H.; Aljarah, I.; Elaziz, M.A.; Damaševičius, R. An Enhanced Evolutionary Software Defect Prediction Method Using Island Moth Flame Optimization. *Mathematics* **2021**, *9*, 1722. https://doi.org/10.3390/math9151722

Academic Editors: Tadashi Dohi and Shaoying Liu

Received: 28 June 2021
Accepted: 20 July 2021
Published: 22 July 2021

1. Introduction

The software industry has recently undergone further development in various aspects related to the software development life-cycle (SDLC). An important aspect to achieve during SDLC is reliability and error-free code. Software defect describes the error status that occurs at the program or system level which leads to erroneous results and unexpected actions and allows the system to behave in an unintended way [1]. There are several reasons behind software defects [2] such as incomplete or ambiguous requirements due to miscommunication and misinterpretation during requirements elicitation, errors in assumptions and preliminary specifications, lack of knowledge in the domain, developers with insufficient practical experience and technical skills, poor programming logic, and so forth. Software defects have many negative consequences for the quality of the software and the overall effectiveness of the system in terms of time, budget, risks, and resources [3]. For example, errors in the design stage may require a high cost of maintenance and restructuring. Poor quality software production will not satisfy customer requirements and will ultimately affect the company's reputation [4].

Defect prediction plays an important role in identifying error-prone modules and controlling the percentage of defects in the software, which improves the quality of the software. This will improve the testing process as it will focus on parts that are more likely to work incorrectly [5]. On the other hand, the distribution of errors in the code determines the refactoring candidates, which enhances the quality and the efficiency of the software product [6,7]. There are three categories of software defect prediction (SDP): prediction of the number of defects, prediction of the severity of defects, and prediction of whether the software module is defective or not. Among them, the last category is the most frequently used, where the SDP is formulated as a binary classification problem that deals with two classes called defect and non-defect [8].

In the literature, many machine learning algorithms have been proposed to predict software defects either through supervised or unsupervised learning [9–14]. Supervised learning is the most common machine learning method used to create SDP models, where the applied learning strategy is based on inferring a pattern from a set of instances (training data set). This pattern can then be applied to invisible instances (testing data set) to predict their class labels. Examples of supervised data mining methods used to reliably solve the software defects problem include decision trees (DT), artificial neural network (ANN), naïve Bayesian (NB), support vector machine (SVM), and random forest (RF) [15].

Feature selection (FS) is a data mining step to select the most informative features in the dataset. Its main target is to obtain a feature subset with a minimum length that, at the same time, achieves the maximum classification performance [16]. The FS process consists of search and evaluation sub-processes. The evaluation sub-process utilizes the dataset characteristics (e.g., filters) or classifier (e.g., wrappers) to evaluate a feature subset [17]. For applying the search in the FS process, many methods can be performed. Traditionally, brute force methods have been applied, but they are time-consuming. These are complete search methods because they generate the entire feature space and traverse all the feature subsets. Meta-heuristic methods such as swarm intelligence [18] algorithms generate random solutions and achieve promising results within less time [19]. Swarm intelligence methods have been used widely for enhancing the FS process, such as face recognition [20], machine scheduling [21], medical diagnosis [17,22], multi-objective power scheduling [23] and software defect prediction [24].

The moth flame optimization (MFO) algorithm is a swarm intelligence algorithm that is commonly used in many applications [25–27]. MFO generates a swarm of solutions to explore the search space. Furthermore, it adopts a spiral method to update the positions of moths and change their positions. The gradual degradation of the number of solutions improves the exploration/exploitation trade-offs. This supports the adaptive convergence behavior of the algorithm. However, MFO inherits the drawbacks of swarm intelligence algorithms, such as stagnation in local minima and premature convergence. To address these shortcomings, the improvement of the MFO algorithm has been proposed [28–31].

The island-based model has been integrated with many swarm intelligence algorithms. In this model, the members of the population are distributed among a set of sub-populations where they are managed separately using local rules. In a migration step, migrants interact with each other. Usually, this is done by exchanging the highly fit solutions between islands. This step increases the diversity among solutions and enhances the convergence trends. Three main factors affect the performance of the migration: the rate, the frequency, and the topology of migration. The rate of migration determines the number of exchanged solutions between islands. The frequency of migration indicates the number of invocations for the migration process. Lastly, the topology of migration defines the way the solutions are exchanged between islands. In the literature, there are many studies that integrate the island models with metaheuristic algorithms [32–35].

This paper proposes the island model to enhance the binary MFO (BMFO) algorithm. The new variant named IsBMFO is used to enhance the FS process and the prediction of software defects. The main objectives are enhancing the diversity of the solutions, alleviating the local minima problem, and enhancing convergence trends. The islands are generated from dividing the population into a group of islands. Each island consists of a group of solutions. Solutions are enhanced locally in each island, and then they are exchanged using a migration mechanism that adopts a random-ring topology. This topology exchanges the solution with the worst fitness in the destination island with the solution with the best fitness from the source island.

The remaining parts of this paper are arranged into sections as follows: Section 2 discusses related studies in the literature. Section 3 provides background about the applied classifiers, the MFO algorithm, and the island model. Section 4 describes the IsBMFO. Section 5 describes the experiments and the related discussions. Finally, Section 6 draws the conclusions of the paper and suggests some possible future works.

2. Related Works

Recently, the SDP problem has become a noteworthy research topic that has increasingly attracted the interest of researchers. Several methods from statistics, information theory, and machine learning fields have been used to predict defected models and reduce the cost of software production and maintenance [36,37].

In [38], the authors aimed to find the count of defects when the software process is not properly executed. For the classification of defects, the authors employed different DT algorithms such as C4.5 and ID3. Pattern mining methods were used to evaluate the defect patterns.

Can et al. [39] proposed a prediction model for software defects using particle swarm optimization (PSO) and SVM called the P-SVM. Specifically, the PSO was used for the optimization of parameters of the SVM. After identifying the optimal parameters of the SVM, it was used to predict the defects in the software. The experiments were performed over the JM1 dataset. P-SVM was compared with the SVM model, back propagation neural network (BPNN) model, and optimized SVM using the genetic algorithm (SVM-GA) model. The results proved the superiority of the P-SVM model.

Shuai et al. [40] proposed a cost-sensitive SVM (CSSVM) model which is based on dynamic SVM using the concept of cost-sensitivity. The model was optimized using the GA algorithm. The fitness function used the geometric accuracy metric. The results of the experiments showed that the GA-CSSVM achieved a higher area under the curve (AUC) value, indicating better prediction accuracy.

Agrawal and Tumar [41] proposed an FS approach based on a linear twin SVM (LTSVM) classifier to predict the defective software modules. They worked on determining the most important metrics set. The reduced metrics set, obtained after the FS process, was used to enhance the predictive power of their approach. The experiments on four PROMISE datasets showed the effectiveness of the LSTVM model.

In [42], the authors studied the software defect prediction using different methods such as DT, decision tables, RF, NN, NB, artificial immune recognition system, CLONALG, and Immunos. They used four software datasets from NASA. Principal component analysis (PCA) and correlation-based FS methods were applied for evaluation. The experiments proved that RF is the best predictor for large datasets while NP is the best predictor for small datasets. Moreover, the experiments showed that the Immunos-99 algorithm performed well when the FS method was applied, while the AIRSParallel algorithm performed better without applying FS methods.

Singh and Chung [15] applied common machine learning algorithms including artificial NN, PSO, DT, NB, and linear classifier. The authors used the KEEL tool and k-fold cross-validation method. The results on seven open-source NASA datasets proved the superiority of the linear classifier in terms of accuracy.

Recently, in [43], the authors used the oversampling technique SMOTE along with FS using PSO on object-oriented metrics. The obtained features were then utilized to train the datasets on SVM to predict defects. The experiments showed that SVM performed better when the dataset was balanced with SMOTE and PSO was used for selecting the feature set.

In [44], the authors studied the effect of 46 FS methods based on NB and DT classifiers over software defect datasets. The results proved that there is no model that can be considered the best FS method. This is because their performances depend on the applied classifiers, used evaluation metrics, and datasets.

Overall, in the literature, many studies used classification algorithms for classifying software defects datasets such as NB, KNN, C4.5, and SVM. Some of these studies proposed GA and PSO algorithms for optimizing the SVM. However, the number of works that addressed the problem of FS in the domain of software defect prediction is still few. This work focuses on identifying the features subset that is considered the optimal one for improving the efficiency of classifiers. Based on the no free lunch (NFL) theorem, no optimization algorithm is considered the best solution to solve every optimization problem. Hence, there is always room to develop, propose, and enhance optimization algorithms to tackle different optimization problems. MFO has remarkable proprieties among swarm intelligence algorithms. Therefore, in this study, we further enhance its performance to optimize FS and produce better results for software defect prediction.

3. Background

3.1. Classification Algorithms

3.1.1. K-Nearest Neighbor Classifier (k-NN)

This is a type of classification algorithm that belongs to a larger category of pattern recognition algorithms known as instance-based or lazy learning algorithms. Instead of conducting the generalization in an explicit training phase, they rely on computing the distance (similarities) between the unlabeled new query instance and its nearest k neighbors from the labeled training instances stored in memory. The basic idea for k-NN is that the nearby points in space are likely to have a similar class concept. In classification problems, the input to the k-NN is the k closest examples among the training examples, and the output is the labels of these examples. Assigning labels depends on the majority of votes obtained from the k closest neighbors for the required example. The comparison and the calculation of the closeness between points are done based on a predefined distance metric such as the Euclidean distance.

3.1.2. Support Vector Machines (SVM)

SVM is a supervised robust learning model that is based on a statistical learning framework. Given a set of training examples, the SVM maps these examples to one or the other category. This means that SVM is a binary classifier that applies an improbable linear method. The SVM tries to put the training examples in points in space in such a way to maximize the gap between the two categories. In addition, SVM can perform a non-linear classification using the kernel trick.

3.1.3. Naive Bayes Classifier (NB)

NB is a classification algorithm that applies the Bayes' theorem, and it is considered a probabilistic classifier. NB assumes strong independence between features. NB gives the probability of membership of an example to each class. NB is among the simplest of Bayesian network models that can achieve higher classification results.

3.2. Overview of Moth Flame Optimization Algorithm

The moth flame optimization algorithm (MFO) is a widely applied swarm intelligence algorithm [45] with remarkable results. The inspiration of MFO is from an insect called a moth. Moths move straight in nature by following a natural mechanism called transverse orientation. This mechanism enables moths to go far distances straight by keeping the same angle with a distant source of light such as the moon. However, the transverse orientation does not work correctly when the source light is near the moths. Consequently, moths are forced to enter a spiral path and move around the light. Figure 1 shows the movement of moths around a candle by following a spiral path.

Figure 1. The spiral path of moths around a candle.

The MFO identifies a set of solutions (population) where the solutions are called moths. The moths represent the possible solutions to the optimization problem. A specified fitness function is used to determine the fitness of each moth. Another component of the MFO is the flame. Both a moth and a flame are solutions; they differ in their update strategy. Moths are the identified solutions that are candidates to be the best solutions, but flames are the best achieved solutions. Each flame is replaced whenever a better solution is found so that the best solutions are never missed.

The spiral movement of moths around the flames is formulated in Equation (1), which describes the movement of moths in a spiral path around a candle where Mo_i is the ith moth, Fl_j is the jth flame, and Sp is the function of spiral path.

Equation (2) shows the logarithmic function used to formulate the spiral movement of moths, where Ds_i is the distance between the ith moth and the jth flame as shown in Equation (3), b is a constant value that determines the shape of the logarithmic spiral, and t is a random number in $[-1, 1]$. The parameter $t = -1$ represents the closest position of a moth to a flame, where $t = 1$ represents the farthest position between a moth and a flame. To increase exploitation, the t parameter is selected in the range $[r, 1]$, where r is decreased linearly across iterations from -1 to -2.

$$Moi = Sp(Mo_i, Fl_j) \tag{1}$$

$$Sp(Mo_i, Fl_j) = Ds_i \times e^{bt} \times cos(2\Pi) + Flj \tag{2}$$

$$Ds_i = |Mo_i - Fl_j| \tag{3}$$

Equation (4) shows the gradual decrease of the number of flames across the iterations, where Ct is the current number of iterations, Mfl is the maximum number of flames, and Mt is the maximum number of iterations.

$$FlameNumber = round(Mfl - Ct \times (Mfl - 1)/Mt) \tag{4}$$

3.3. Binary Moth Flame Optimization for Feature Selection

MFO was designed to solve continuous optimization problems. FS is a discrete problem in which the search space consists of two values, "0" or "1". For this reason, MFO needs some modification to be able to optimize in a binary feature space. In [46], the authors used the transfer functions to produce a binary optimizer from the original continuous version of the optimizer. A mapping procedure is used to convert the continuous update process into a binary process. Thus, the elements of the updated solutions are either "0" or "1".

In the proposed models, the sigmoid transfer function is used to produce a BMFO from the original MFO. The sigmoid function defines a probability for each element of the solution within a range [0, 1]. It was used in [47] to produce a binary variant of PSO. The velocity (step) is analogous to the first term of Equation (2) in the MFO algorithm. This term is redefined in Equation (5) as the probability for changing the position of moths. Each moth updates its position in the binary search space using Equation (7) based on the probability generated from Equation (6). Algorithm 1 shows the BMFO algorithm.

$$\Delta Mo = Ds_i \times e^{bt} \times cos(2\Pi) \tag{5}$$

$$Trf(\Delta Mo_t) = 1/(1 + e^{\Delta Mo_t}) \tag{6}$$

$$Mo_i^d(t+1) = \begin{cases} 0, & \text{if } rand < Trf(\Delta Mo_{t+1}) \\ 1, & \text{if } rand \geqslant Trf(\Delta Mo_{t+1}) \end{cases} \tag{7}$$

Algorithm 1 The pseudo-code of BMFO.

Input: Mt, n (# moths), d (# dimensions)
Output: near optimal moth
Initialization process for the moths
while $Ct \leq Mt$ **do**
  modify the number of flames using Equation (4)
  FMo = Fitness(Mo);
  if Ct == 1 **then**
    $Fl = sort(Mo)$;
    $FFl = sort(FMo)$;
  else
    $Fl = sort(Mo_{Ct-1}, Mo_{Ct})$;
    $FFl = sort(FMo_{Ct-1}, FMo_{Ct})$;
  end if
  for i = 1: n **do**
    for j = 1: d **do**
      Modify r and t;
      Compute Ds by Equation (3) based on the corresponding moth;
      Modify the step vector of a moth ΔMo using Equation (5).
      Compute the probabilities by Equation (6).
      Modify the position of a moth by Equation (7)
    end for
  end for
  $Ct = Ct + 1$;
end while

The fitness function is formulated in Equation (8), where Err is the error rate, $|Sf|$ is the number of selected features in the reduced data set, $|Cf|$ is the number of features in the original data set, and $\alpha \in [0,1]$, $\beta = (1-\alpha)$ are two parameters that indicate the significance of classification and the number of selected features according to [19].

$$Fitness = \alpha \times Err + \beta \times \frac{|Sf|}{|Cf|} \quad (8)$$

3.4. Fundamentals to Island-Based Model

The island model is an efficient method for structuring the population and increasing its heterogeneity [33,34]. This is applied by dividing the population into smaller subpopulations called (islands). The evolutionary algorithm is applied on each island either in a synchronous or asynchronous way. A migration process is applied after a period to allow solutions from different islands to exchange their positions. The exchange of solutions between islands improves exploration/exploitation trade-offs. This happens because the low-quality solutions with low-fitness values can approach the region where the global optima locate. Another advantage of the island model is that it enables the parallel implementation of the evolutionary algorithm on each island. This can minimize the computation time of complex optimization problems.

The island model has been applied with several evolutionary computation algorithms. The main purpose is to increase the population diversity and search the search space effectively. Examples of island-based models include the island differential evolution [48], island flower pollination algorithm [33], island ant colony [49], island bat algorithm [32], and island harmony search [34].

Several factors affect the island model such as the number of islands or the number of times the solutions are exchanged between islands. For integrating the island model with evolutionary algorithms, the partitioning and migration operators are used. Partitioning accounts for the number of islands (Is_n) and the size of the island (Is_s). In migration, the $Mr_m \times Is_s$ moths are to be swapped between islands after a predetermined number of iterations It_m, where Mr_m is the migration rate and Is_s is the island size.

The migration process can be performed in a synchronous or asynchronous way. In the synchronous way, the solutions are swapped between islands simultaneously. The asynchronous way enables solutions to change their islands to other ones after a specific time. Therefore, the migration times are different between islands. An important factor in migration is the topology. There are two migration typologies: either static or dynamic. The static typologies determine the paths between islands, so they are not changeable. The dynamic typologies determine the paths between islands during the execution time. The effectiveness of the island model is also affected by the migration process. This indicates which solutions will be selected to migrate between islands. A common migration policy is known as best–worst. It selects the best solution (with the highest fitness value) from the source island to be swapped with the worst solution (with the lowest fitness value) from the destination island [48]. Another known policy for applying migration is random-random. It selects a random solution from the source island to be swapped with a random solution from the destination island [50].

4. Island-Based MFO (IsMFO) Algorithm

This section proposes the island MFO algorithm. Figure 2 shows the overall methodology followed in this work. Initially, the population of moths is split into a set Is_n islands of moths. Each island is of size Is_s moths. The MFO runs independently and asynchronously on each island. The number of times the algorithm runs depends on the migration frequency Fr_m. The moths are exchanged based on random-ring migration topology, and the number of moths to be exchanged depends on the migration rate Mr_m. The migration policy used is the best–worst. This technique is applied more than one time until reaching the maximum iteration.

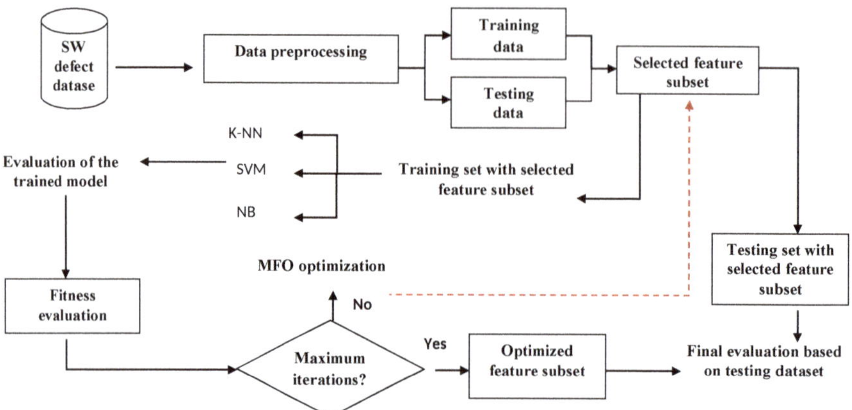

Figure 2. Architecture of the proposed methodology.

The IsBMFO flowchart is shown in Figure 3, and the pseudo-code is shown in Algorithm 2.

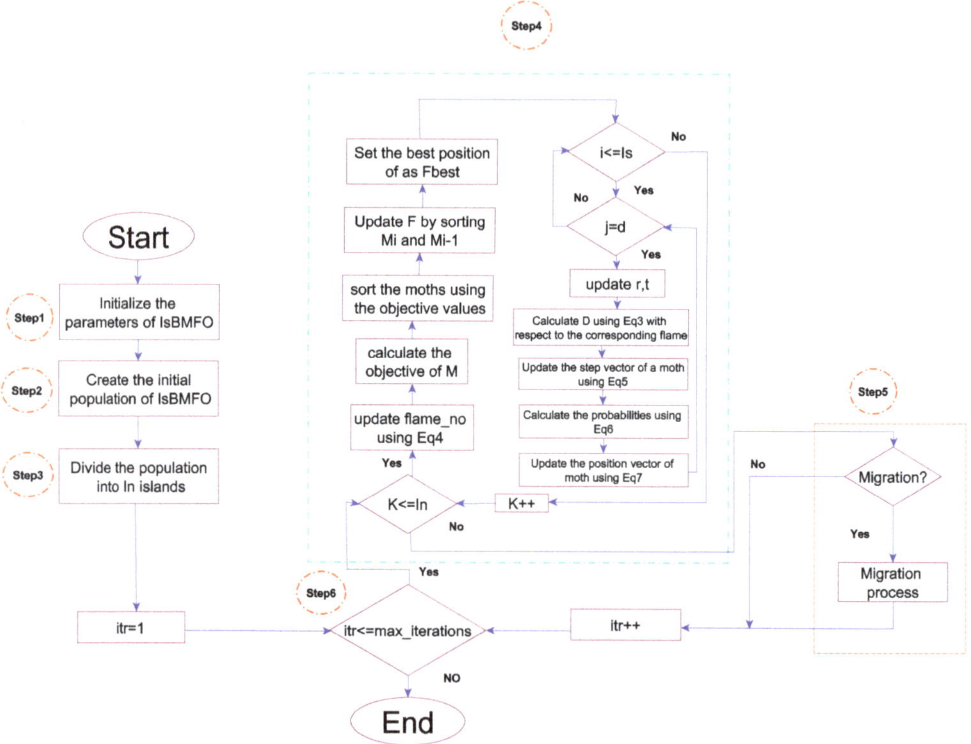

Figure 3. The flowchart of the proposed IsBMFO algorithm.

Algorithm 2 The IsBMFO pseudo-code.

```
———Identification of the IsBMFO parameters———
Set the IsBMFO parameters Mt, n, d, Is_n, Is_s, Mr_m, Fr_m
———Initialize the IsBMFO positions———
Initialize the positions of moths
0: ——Split IsBMFO into a group of islands———
  Flag(y) = False, ∀y = (1,2....n)
  for K = 1 : Is_n do
    for i = 1 : Is_s do
      select y , where y ∈ (1, 2,..., n)
      while Flage( y) is true do
        select y , where y ∈ 1, 2, . . . , S_n
      end while
      Add x_y to island Is_k
    end for
  end for
  while Ct ≤ Mt do
    ——Improvement step———
    for i = 1: Is_n do
      Update flame no using Equation (4)
      FMo = Fitness(Mo);
      if Ct == 1 then
        Fl = sort(Mo);
        FFl = sort(FMo);
      else
        Fl = sort(Mo_{Ct−1}, Mo_{Ct});
        FFl = sort(FMo_{Ct−1}, FMo_{Ct});
      end if
      for i = 1: Is do
        for j = 1: d do
          Modify r and t;
          Compute Ds by Equation (3) based on the corresponding moth;
          Modify the step vector of a moth ΔMo by Equation (5).
          Compute the probabilities by Equation (6).
          Modify the position of a moth by Equation (7)
        end for
      end for
    end for
    ——- —— Migration of moths———
    if t mod Fr_m = 0 then
      for y = 1, .., Is_n do
        k = 1
        while k ≤ Mr_m × Is do
          xWorst(k, y + 1) = xBest(k, y)
        end while
      end for
    end if
    Ct = Ct + 1
  end while
```

The IsBMFO steps are explained next:

Step 1: This is the initialization step for the BMFO parameters. These include the # dimensions (d), # moths (n), and the # iterations ($\#Mt$). The fitness function $f(Mo)$ and the representation of a moth Mo = ($mo_1, mo_2, \ldots, mo_d$) are also defined. The island model parameters should be identified as follows:

- Island number (Is_n): this determines the number of sub-populations that is less than or equal to n.
- The size of island (Is_s): the population size for each island can be computed using the formula $Is_s = n/Is_n$ since all islands are homogeneous.
- The frequency of migration (Fr_m): this indicates the required iterations number to call the migration function.
- The rate of migration (Mr_m): this indicates the number of moths swapped between islands based on Is_s, where $Mr_m \times Is_s \leq Is_s$.

Step 2: Identifies the solutions in the population of IsBMFO. In this step, IsBMFO follows the same process as in the MFO. The random moths are $Mo = (mo_1, mo_2, ..., mo_n)$, and the fitness function (i.e., $f(mo)$) for each moth (mo_j, where $j \in (1, 2, ..., n)$) is computed.

Step 3: Split the IsBMFO population into a set of islands Is_n of size Is_s for each one as shown in Figure 4. The island vector is Is = (Is_1, Is_2, . . . , Is_n), where each variable $Is_j \in (1, 2, ..., Is_n)$. As an example, assume Is_n = 4 and Is_s = 3 are the division of IsBMFO population of size n=12. Assume that Is = (3, 4, 2, 1, 4, 2, 4, 1, 3, 2, 1, 3), then island Is_1 = (M_4, M_8, M_{11}), island Is_2 = (M_3, M_6, M_{10}), island Is_3 = (M_1, M_9, M_{12}), and island Is_4 = (M_2, M_5, M_7). Remember that each moth is assigned randomly to an island.

Step 4: The step of improvement includes updating the flames number, calculating the objective values of moths, and sorting of moths based on their fitness values. In this stage, the moth is updated based on the computed distance between a moth and the flame corresponding to it.

Step 5: Migration process of IsBMFO. The main target of the migration process is to exchange the moths between islands. After a predefined iteration number specified by (Fr_m), the migration process is applied as shown in Algorithm 2. A specific number of moths are exchanged on each island based on the migration rate Mr_m, where $Mr_m \times Is_s \leq Is_s$. The migration uses the best–worst policy and a random ring topology. The best–worst policy selects the best $Mr_m \times Is_s$ moths from an island to replace the worst $Mr_m \times Is_s$ moths on a neighboring island. In random-ring topology, the islands are rearranged in a random way to compose a ring $(Is_j, Is_{j+1}, ..., Is_k, Is_j)$ in which the island neighbor to Is_j is island Is_{j+1}, and the island neighbor to Is_{j+1} is island Is_{j+2}, etc.

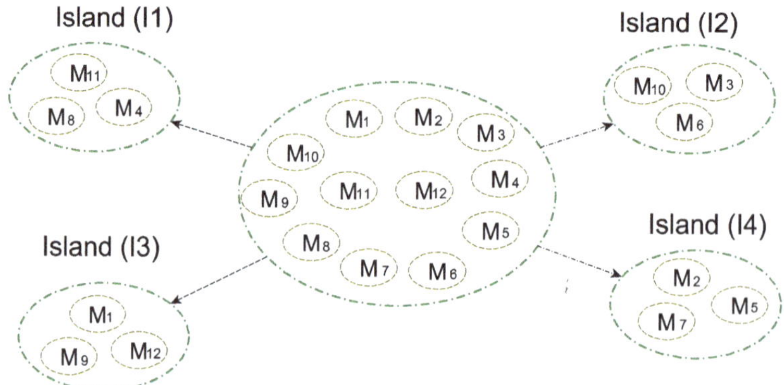

Figure 4. An illustration of island-based model.

5. Experimental Results

5.1. Model Evaluation Metrics

The basic evaluation metric that is used to evaluate the proposed software defect prediction algorithm is the confusion matrix. Table 1 shows the confusion matrix.

Table 1. Confusion Matrix.

		Actual Labels	
		Defect	Non-Defect
Predicted labels	Defect	*TruePos*	*FalsePos*
	Non-defect	*FalseNeg*	*TrueNeg*

From the confusion metric, other evaluation metrics can be deduced, such as:

1. *Recall*: The ratio of correctly classified defected instances, as in Equation (9):

$$Recall = \frac{TruePos}{TruePos + FalseNeg} \tag{9}$$

2. *Precision*: The ratio of the correctly classified defected instances among the retrieved instances. It can be calculated by Equation (10):

$$Precision = \frac{TruePos}{TruePos + FalsePos} \tag{10}$$

3. *G-mean*: The recall of each class, as in Equation (11):

$$G\text{-}mean = \sqrt{\frac{TruePos}{TruePos + FalseNeg} \times \frac{TrueNeg}{FalsePos + TrueNeg}} \tag{11}$$

5.2. Datasets Specifications

The methodology is verified by a series of 21 public benchmark software datasets. Table 2 describes the datasets. Eleven of these datasets are downloaded from the NASA corpus (cleaned versions) https://figshare.com/articles/dataset/MDP_data_sets_D_and_D_-_zipped_up/6071675 (accessed on 28 May 2021), while the remaining datasets are from the PROMISE software engineering corpus http://promise.site.uottawa.ca/SERepository/ (accessed on 28 May 2021). NASA collected datasets from real software projects with different specifications such as the programming language, the code size, and software measures. The datasets consist of a set of features that have values and a goal field that describes the instance as defect or non-defect. These features describe the program from different sides including the lines of code measure (program length, count of lines of comments, count of lines of comments), McCabe metrics, base Halstead measures, derived Halstead measures, unique operators, unique operands, total operators, total operands, cyclomatic complexity, essential complexity, design complexity, and a branch-count. The PROMISE datasets were collected from open-source software projects.

Table 2. Description of datasets.

No.	Name	Features	Instances	Defects	Non-Defects	Defect Ratio	Non-Defect Ratio
D_1	cm1	38	327	42	285	12.8	87.2
D_2	jm1	22	7782	1672	6110	21.5	78.5
D_3	kc1	22	1183	314	869	26.5	73.5
D_4	kc3	40	194	36	158	18.6	81.4
D_5	mc1	39	1988	46	1942	2.3	97.7
D_6	mw1	38	253	27	226	10.7	89.3
D_7	pc1	38	705	61	644	8.7	91.3
D_8	pc2	37	745	16	729	2.1	97.9
D_9	pc3	38	1077	134	943	12.4	87.6
D_10	pc4	38	1287	177	1110	13.8	86.2
D_11	pc5	39	1711	471	1240	27.5	72.5
D_12	ant-1.7	21	745	166	579	22.3	77.7
D_13	camel-1.6	21	965	188	777	19.5	80.5
D_14	ivy-2.0	21	352	40	312	11.4	88.6
D_15	jedit-4.3	21	492	11	481	2.2	97.8
D_16	log4j-1.2	21	205	189	16	92.2	7.8
D_17	lucene-2.4	21	340	203	137	59.7	40.3
D_18	poi-3.0	21	442	281	161	63.6	36.4
D_19	tomcat-6	20	858	77	781	9	91
D_20	xalan-2.6	21	885	411	474	46.4	53.6
D_21	xerces-1.4	21	588	437	151	74.3	25.7

5.3. Results and Discussion

The methodology for applying training and testing in the experiments is based on a hold-out strategy in which each data set is split in a random way into 80% for training and 20% for testing. The experiments were repeated 30 times to obtain significant results. All experiments were conducted using a personal computer with AMD Athlon Dual-Core QL-60 CPU at 1.90 GHz and 2 GB of memory. The EvoloPy-FS [51] was used to run the experiments. It is a framework in Python for applying binary swarm intelligence algorithms to solve FS problems. It is open-source and available at (www.evo-ml.com accessed on 28 May 2021). The population size and the maximum iterations were set to 10 and 100, respectively [52].

Figure 5a illustrates the average recall obtained from applying the classifiers NB, KNN, and SVM without FS, with BMFO-FS, and with IsBMFO-FS. As can be seen, there was a dynamic increase in the values of recall. The lower values from the three classifiers were achieved when the classifiers were applied to the datasets without implementing FS. There was an increase in the recall values of the three classifiers when BMFO-FS was implemented. The best recall results were achieved when the IsBMFO-FS was implemented. This can be explained by the FS process having an effective influence on the classifiers' performance. Furthermore, the island-based affected the performance of the classifiers and enhanced the optimizer job in the feature search space. In three experiments, the SVM classifier achieved the best performance, followed by the NB classifier. The lowest recall results were obtained by the KNN classifier. This can be explained by the SVM having the capability to distinguish between classes more than the KNN and NB. Furthermore, the integration of the FS process and the island enhancement helped to increase its efficiency. Figures 5b and 6 show the results of the precision and gmean. As can be seen, the precision and gmean were increased dramatically when FS and FS with the island enhancement were applied to the BMFO.

Figures 7a,b and 8 show the recall, precision, and recall results obtained from applying IsBMFO-FS to all the datasets. It can be noticed that the SVM classifier achieved the best results on most of the datasets. On the other hand, lower results were achieved by the NB and KNN classifiers. It can be noticed that the performance results of the NB and KNN were similar.

Figure 9 shows the convergence behavior of the three classifiers KNN, NB, and SVM with the proposed IsBMFO. It can be seen that the convergence behavior of the classifier SVM was better than the NB and KNN on 71% of the datasets. This can be seen in the tails of the convergence curves that reached low values of fitness. This means that IsBMFO with the SVM classifier can reach the global best in the time the other classifiers fall in the local minima. In addition, the classifier NB achieved better convergence scales compared with KNN on fifteen datasets. The convergence scales of the three classifiers were similar on six datasets: mw1, pc2, pc3, ant-1.7, xalan-2.6, and xerces-1.4.

Table 3 shows p-values of the Wilcoxon test based on fitness. This statistical test takes into consideration all runs to determine if the IsBMFO-SVM is meaningfully different from other methods. Table 3 shows the superiority of IsBMFO-SVM over IsBMFO-NB and IsBMFO-KNN.

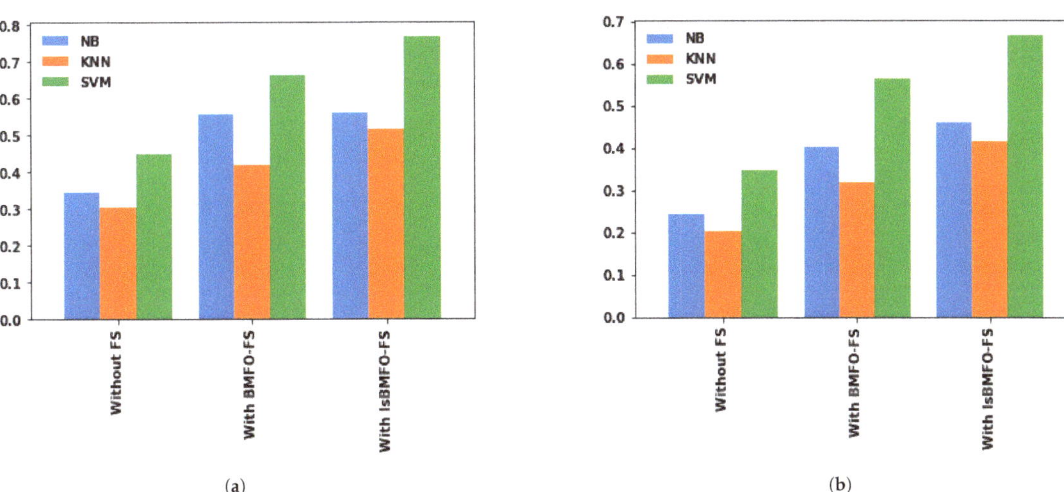

Figure 5. Results of applying classifiers without FS, with BMFO-FS, and with IsBMFO-FS on all datasets. Average recall (**a**) and average precision (**b**).

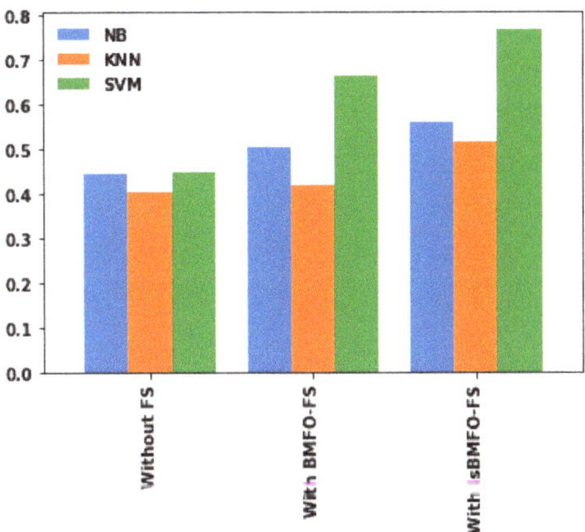

Figure 6. Average gmean results of applying classifiers without FS, with BMFO-FS, and with IsBMFO-FS on all datasets.

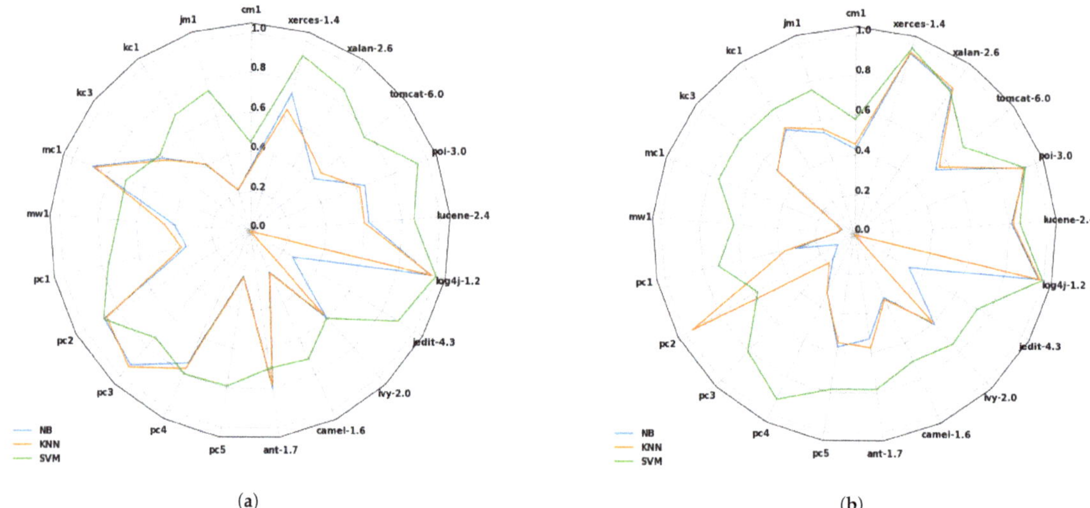

(a) (b)

Figure 7. Results of applying classifiers with IsBMFO-FS on all datasets. Average recall (**a**) and Average precision (**b**).

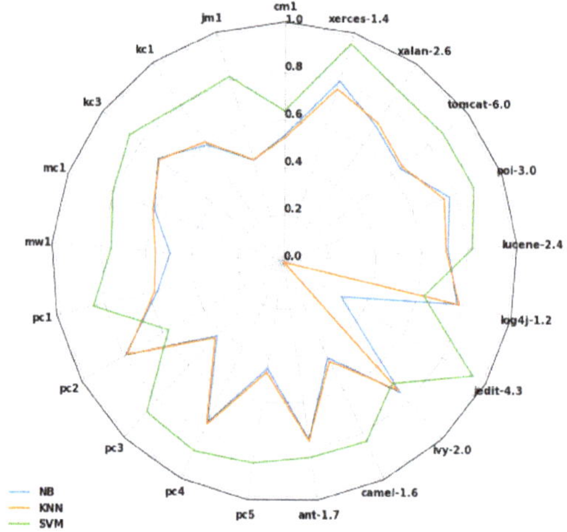

Figure 8. Gmean results of applying classifiers with IsBMFO-FS on all datasets.

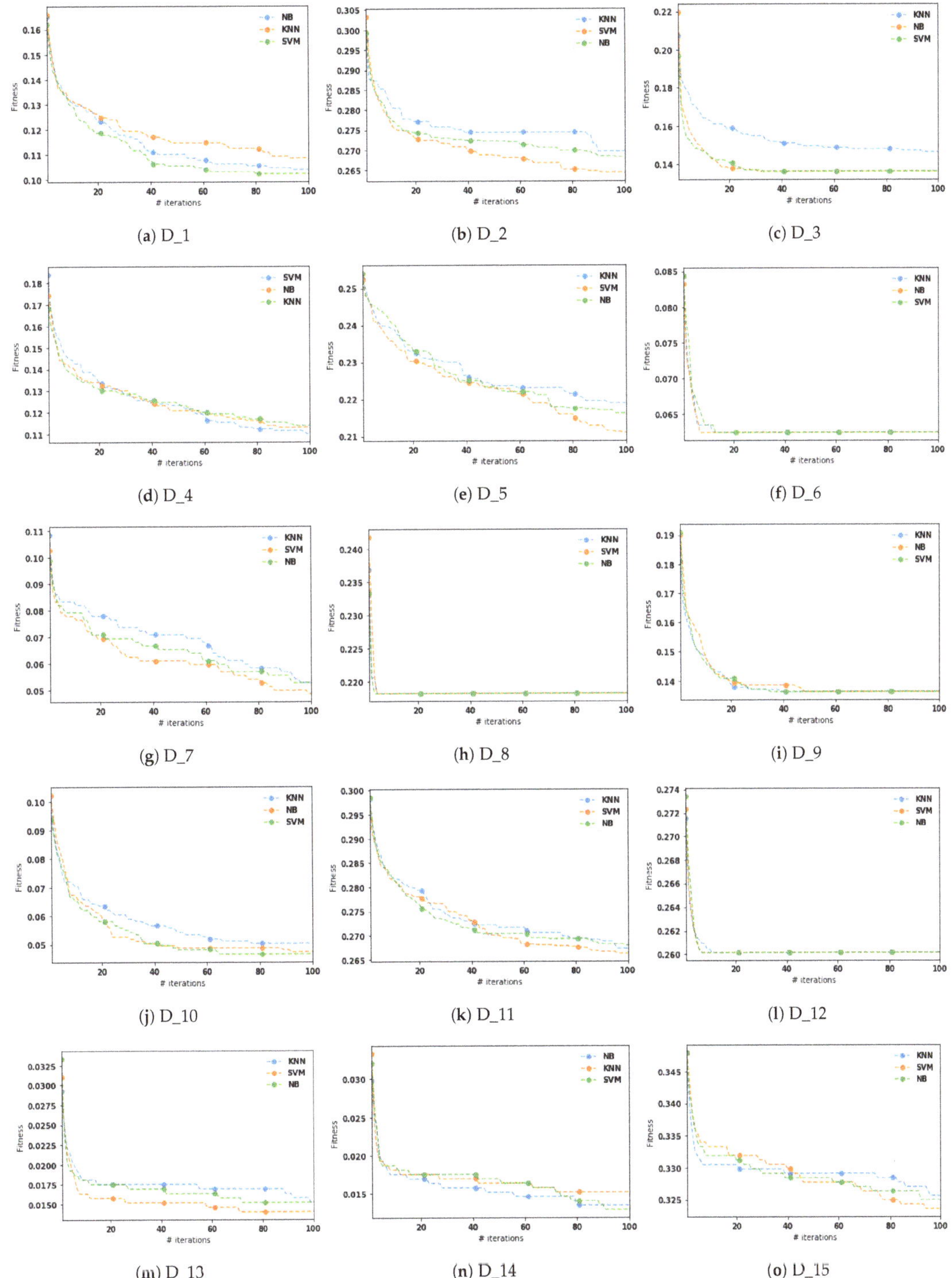

(**a**) D_1 (**b**) D_2 (**c**) D_3

(**d**) D_4 (**e**) D_5 (**f**) D_6

(**g**) D_7 (**h**) D_8 (**i**) D_9

(**j**) D_10 (**k**) D_11 (**l**) D_12

(**m**) D_13 (**n**) D_14 (**o**) D_15

Figure 9. *Cont.*

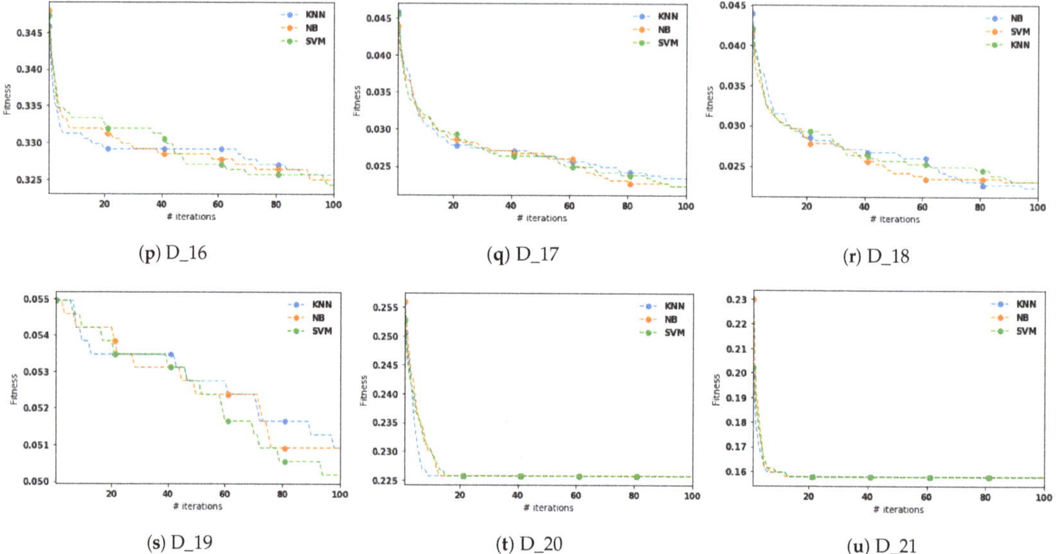

(**p**) D_16 (**q**) D_17 (**r**) D_18

(**s**) D_19 (**t**) D_20 (**u**) D_21

Figure 9. Convergence curves for IsBMFO with the three classifiers KNN, NB, and SVM.

Table 3. p-values of the Wilcoxon test for the IsBMFO-SVM and other classifiers using fitness ($p > 0.05$ are underlined).

Datasets	IsBMFO-KNN	IsBMFO-NB
D_1	2.44×10^{-5}	1.56×10^{-5}
D_2	4.31×10^{-5}	2.24×10^{-4}
D_3	$\underline{1.21 \times 10^{-1}}$	$\underline{1.37 \times 10^{-1}}$
D_4	1.28×10^{-4}	1.61×10^{-10}
D_5	1.31×10^{-4}	1.81×10^{-9}
D_6	1.23×10^{-10}	1.62×10^{-10}
D_7	2.38×10^{-13}	1.30×10^{-12}
D_8	2.46×10^{-10}	2.23×10^{-10}
D_9	4.52×10^{-6}	5.14×10^{-7}
D_10	4.95×10^{-11}	2.25×10^{-11}
D_11	3.14×10^{-11}	3.56×10^{-10}
D_12	6.63×10^{-4}	9.22×10^{-14}
D_13	9.41×10^{-13}	5.56×10^{-12}
D_14	$\underline{1.78 \times 10^{-1}}$	$\underline{1.12 \times 10^{-1}}$
D_15	$\underline{2.32 \times 10^{-1}}$	2.35×10^{-5}
D_16	2.23×10^{-11}	3.25×10^{-12}
D_17	5.18×10^{-12}	2.12×10^{-13}
D_18	4.13×10^{-7}	3.82×10^{-7}
D_19	3.66×10^{-9}	8.4×10^{-12}
D_20	2.61×10^{-6}	3.13×10^{-7}
D_21	2.65×10^{-11}	2.14×10^{-11}

5.4. Analytical Description of the Relevant Features

This section presents an analytical description of the most informative features. These features are obtained by the IsBMFO-SVM approach. Referring to Table 4, it shows the # all features in each dataset (AF), the number of selected features (SF), the feature reduction ratio (FRR), and the selected relevant features (RF) in the dataset. For the FFR, it is calculated by Equation (12).

$$\text{FFR} = \frac{\text{AF} - \text{SF}}{\text{AF}} \quad (12)$$

As can be seen, the FFR ranged between 48% on poi-3.0 and jedit-4.3 datasets to 74% on pc1 dataset. The average FRR on all the datasets is 62%. This ratio indicates that the proposed IsBMFO-SVM can reduce the dimensionality of the datasets by more than half. This supports the proposed IsBMFO-SVM, which also outperformed other approaches in terms of recall, precision, gmean, and convergence scales.

Table 4. Relevant features in software datasets.

Datasets	AF	SF	FFR%	RF
cm1	38	15	61%	F2, F3, F7, F9, F11, F14, F17, F19, F23, F25, F26, F32, F33, F36, F38
jm1	22	9	59%	F1, F2, F4, F6, F7, F11, F13, F16, F19
kc1	22	8	64%	F3, F7, F9, F10, F14, F15, F18, F21
kc3	40	16	60%	F2, F4, F7, F8, F9, F11, F17, F19, F23, F26, F28, F31, F33, F35, F39, F40
mc1	39	12	69%	F3, F7, F8, F12, F13, F15, F20, F24, F28, F29, F32, F38
mw1	38	13	66%	F1, F5, F9, F11, F14, F16, F19, F20, F22, F25, F27, F30, F31
pc1	38	10	74%	F1, F6, F13, F15, F17, F21, F24, F27, F29, F35
pc2	37	14	62%	F1, F3, F4, F8, F13, F14, F17, F25, F27, F29, F30, F34, F36, F37
pc3	38	11	71%	F2, F3, F6, F9, F17, F20, F22, F26, F31, F34, F37
pc4	38	11	71%	F1, F5, F8, F9, F14, F22, F23, F26, F31, F35, F38
pc5	39	12	69%	F2, F4, F8, F10, F12, F15, F19, F25, F29, F30, F32, F37
ant-1.7	21	7	67%	F1, F5, F7, F10, F16, F19, F21
camel-1.6	21	9	57%	F2, F5, F6, F9, F11, F12, F14, F17, F20
ivy-2.0	21	10	52%	F3, F7, F10, F11, F13, F15, F16, F19, F20, F21
jedit-4.3	21	11	48%	F1, F4, F5, F7, F9, F13, F14, F15, F18, F20, F21
log4j-1.2	21	8	62%	F6, F9, F10, F13, F15, F17, F20, F21
lucene-2.4	21	9	57%	F2, F4, F7, F10, F11, F15, F18, F19, F21
poi-3.0	21	11	48%	F2, F5, F7, F8, F9, F10, F11, F14, F17, F19, F21
tomcat-6	20	7	65%	F4, F5, F8, F11, F13, F19, F20
xalan-2.6	21	8	62%	F1, F3, F6, F10, F11, F12, F19, F20
xerces-1.4	21	10	52%	F1, F4, F6, F7, F8, F11, F14, F16, F20, F21

6. Conclusions and Future Trends

This paper proposes the island model to enhance the BMFO for solving the FS problem in the domain of software defect prediction. The new variant is called (IsBMFO). The island model divides the population of moths into a set of islands and applies a migration process to share features between islands. This concept can improve the diversity of solutions and control the convergence of the algorithm. In IsMFO, different copies of MFO are applied separately on each island in an asynchronous way. Three measurements are used to evaluate the proposed approach, recall, precision, and G-mean, in addition to the convergence scales and statistical rank test. The experiments compared the average recall, precision, and gmean obtained from applying the classifiers NB, KNN, and SVM without FS, with BMFO-FS, and with IsBMFO-FS. There was a dynamic increase in the values of the evaluation measures. The lower values from the three classifiers were achieved when the classifiers were applied to the datasets without implementing FS. The best results were achieved when the IsBMFO-FS was implemented. In three experiments, the SVM classifier achieved the best performance, followed by the NB classifier. The lowest results were obtained by the KNN classifier. Furthermore, the convergence behavior of the classifier SVM was better than the NB and KNN on 71% of the datasets.

The best achieved results were obtained by the IsBMFO-SVM model. These results demonstrate that the proposed model can serve as an effective predictive model for the software defect problem.

For future works, we suggest applying the proposed model on other classification problems such as for medical diagnosis. Furthermore, the island-based enhancement can be investigated with other metaheuristic algorithms.

Author Contributions: Data curation, R.A.K. and I.A.; Formal analysis, R.A.K. and I.A.; Funding acquisition, R.D.; Investigation, R.A.K., H.A. and I.A.; Methodology, R.A.K. and H.A.; Resources, R.A.K. and I.A.; Software, R.A.K., H.A. and I.A.; Supervision, I.A.; Validation, R.A.K., M.A.E., I.A. and R.D.; Visualization, R.A.K. and H.A.; Writing—original draft, R.A.K., H.A. and I.A.; Writing—review & editing, R.A.K., I.A., M.A.E., H.A. and R.D. All authors have read and agreed to the published version of the manuscript.

Funding: This research received no external funding.

Data Availability Statement: Data are available from the corresponding author upon reasonable request.

Conflicts of Interest: The authors declare no conflict of interest.

References

1. Levendel, Y. Reliability analysis of large software systems: Defect data modeling. *IEEE Trans. Softw. Eng.* **1990**, *16*, 141–152. [CrossRef]
2. Ehrlich, W.K.; Iannino, A.; Prasanna, B.; Stampfel, J.P.; Wu, J.R. How faults cause software failures: Implications for software reliability engineering. In Proceedings of the 1991 International Symposium on Software Reliability Engineering, Austin, TX, USA, 17–18 May 1991; IEEE Computer Society: Washington, DC, USA, 1991; pp. 233–234.
3. Laprie, J.C. Dependability of computer systems: Concepts, limits, improvements. In Proceedings of the IEEE Sixth International Symposium on Software Reliability Engineering (ISSRE'95), Toulouse, France, 24–27 October 1995; pp. 2–11.
4. Mandeville, W.A. Software costs of quality. *IEEE J. Sel. Areas Commun.* **1990**, *8*, 315–318. [CrossRef]
5. Singpurwalla, N.D. Determining an optimal time interval for testing and debugging software. *IEEE Trans. Softw. Eng.* **1991**, *17*, 313–319. [CrossRef]
6. Mens, T.; Tourwé, T. A survey of software refactoring. *IEEE Trans. Softw. Eng.* **2004**, *30*, 126–139. [CrossRef]
7. Alsawalqah, H.; Hijazi, N.; Eshtay, M.; Faris, H.; Radaideh, A.A.; Aljarah, I.; Alshamaileh, Y. Software defect prediction using heterogeneous ensemble classification based on segmented patterns. *Appl. Sci.* **2020**, *10*, 1745. [CrossRef]
8. Wahono, R.S. A systematic literature review of software defect prediction. *J. Softw. Eng.* **2015**, *1*, 1–16.
9. Li, Z.; Jing, X.; Zhu, X. Progress on approaches to software defect prediction. *IET Softw.* **2018**, *12*, 161–175. [CrossRef]
10. Son, L.H.; Pritam, N.; Khari, M.; Kumar, R.; Phuong, P.T.M.; Thong, P.H. Empirical study of software defect prediction: A systematic mapping. *Symmetry* **2019**, *11*, 212. [CrossRef]
11. Shen, Z.; Chen, S. A Survey of Automatic Software Vulnerability Detection, Program Repair, and Defect Prediction Techniques. *Secur. Commun. Netw.* **2020**, *2020*, 8858010. [CrossRef]
12. Li, N.; Shepperd, M.; Guo, Y. A systematic review of unsupervised learning techniques for software defect prediction. *Inf. Softw. Technol.* **2020**, *122*, 106287. [CrossRef]
13. Aljarah, I.; Mafarja, M.; Heidari, A.A.; Faris, H.; Mirjalili, S. Multi-verse optimizer: Theory, literature review, and application in data clustering. In *Nature-Inspired Optimizers*; Springer: Cham, Switzerland, 2020; pp. 123–141.
14. Mafarja, M.; Heidari, A.A.; Faris, H.; Mirjalili, S.; Aljarah, I. Dragonfly algorithm: Theory, literature review, and application in feature selection. In *Nature-Inspired Optimizers*; Springer: Cham, Switzerland, 2020; pp. 47–67.
15. Singh, P.D.; Chug, A. Software defect prediction analysis using machine learning algorithms. In Proceedings of the 2017 IEEE 7th International Conference on Cloud Computing, Data Science & Engineering-Confluence, Noida, India, 12–13 January 2017; pp. 775–781.
16. Khurma, R.A.; Aljarah, I.; Sharieh, A. Rank based moth flame optimisation for feature selection in the medical application. In Proceedings of the 2020 IEEE Congress on Evolutionary Computation (CEC), Glasgow, UK, 19–24 July 2020; pp. 1–8.
17. Khurma, R.A.; Aljarah, I.; Sharieh, A. An Efficient Moth Flame Optimization Algorithm using Chaotic Maps for Feature Selection in the Medical Applications. In Proceedings of the 9th International Conference on Pattern Recognition Applications and Methods (ICPRAM), Valletta, Malta, 22–24 February 2020; pp. 175–182.
18. Faris, H.; Aljarah, I.; Alqatawna, J. Optimizing feedforward neural networks using krill herd algorithm for e-mail spam detection. In Proceedings of the 2015 IEEE Jordan Conference on Applied Electrical Engineering and Computing Technologies (AEECT), Amman, Jordan, 3–5 November 2015; pp. 1–5.
19. Khurma, R.A.; Aljarah, I.; Sharieh, A. A Simultaneous Moth Flame Optimizer Feature Selection Approach Based on Levy Flight and Selection Operators for Medical Diagnosis. *Arab. J. Sci. Eng.* **2021**, 1–26. [CrossRef]
20. Agarwal, V.; Bhanot, S. Firefly inspired feature selection for face recognition. In Proceedings of the 2015 IEEE Eighth International Conference on Contemporary Computing (IC3), Noida, India, 20–22 August 2015; pp. 257–262.

21. Jouhari, H.; Lei, D.; Al-qaness, M.A.A.; Abd Elaziz, M.; Damaševičius, R.; Korytkowski, M.; Ewees, A.A. Modified Harris Hawks optimizer for solving machine scheduling problems. *Symmetry* **2020**, *12*, 1460. [CrossRef]
22. Sahlol, A.T.; Elaziz, M.A.; Jamal, A.T.; Damaševičius, R.; Hassan, O.F. A novel method for detection of tuberculosis in chest radiographs using artificial ecosystem-based optimisation of deep neural network features. *Symmetry* **2020**, *12*, 1146. [CrossRef]
23. Makhadmeh, S.N.; Al-Betar, M.A.; Alyasseri, Z.A.A.; Abasi, A.K.; Khader, A.T.; Damaševičius, R.; Mohammed, M.A.; Abdulkareem, K.H. Smart home battery for the multi-objective power scheduling problem in a smart home using grey wolf optimizer. *Electronics* **2021**, *10*, 447. [CrossRef]
24. Anbu, M.; Mala, G.A. Feature selection using firefly algorithm in software defect prediction. *Clust. Comput.* **2019**, *22*, 10925–10934. [CrossRef]
25. Khurma, R.; Castillo, P.; Sharieh, A.; Aljarah, I. Feature Selection using Binary Moth Flame Optimization with Time Varying Flames Strategies. In *Volume 1: ECTA, INSTICC, Proceedings of the 12th International Joint Conference on Computational Intelligence, Budapest, Hungary, 2–4 November 2020*; SciTePress: Setúbal, Portugal, 2020; pp. 17–27. [CrossRef]
26. Hussien, A.G.; Amin, M.; Abd El Aziz, M. A comprehensive review of moth-flame optimisation: Variants, hybrids, and applications. *J. Exp. Theor. Artif. Intell.* **2020**, *32*, 705–725. [CrossRef]
27. Shehab, M.; Abualigah, L.; Al Hamad, H.; Alabool, H.; Alshinwan, M.; Khasawneh, A.M. Moth–flame optimization algorithm: Variants and applications. *Neural Comput. Appl.* **2020**, *32*, 9859–9884. [CrossRef]
28. Kaur, K.; Singh, U.; Salgotra, R. An enhanced moth flame optimization. *Neural Comput. Appl.* **2020**, *32*, 2315–2349. [CrossRef]
29. Khurmaa, R.A.; Aljarah, I.; Sharieh, A. An intelligent feature selection approach based on moth flame optimization for medical diagnosis. *Neural Comput. Appl.* **2021**, *33*, 7165–7204. [CrossRef]
30. Xu, Y.; Chen, H.; Luo, J.; Zhang, Q.; Jiao, S.; Zhang, X. Enhanced Moth-flame optimizer with mutation strategy for global optimization. *Inf. Sci.* **2019**, *492*, 181–203. [CrossRef]
31. Khan, M.A.; Sharif, M.; Akram, T.; Damaševičius, R.; Maskeliūnas, R. Skin lesion segmentation and multiclass classification using deep learning features and improved moth flame optimization. *Diagnostics* **2021**, *11*, 811. [CrossRef]
32. Al-Betar, M.A.; Awadallah, M.A. Island bat algorithm for optimization. *Expert Syst. Appl.* **2018**, *107*, 126–145. [CrossRef]
33. Al-Betar, M.A.; Awadallah, M.A.; Doush, I.A.; Hammouri, A.I.; Mafarja, M.; Alyasseri, Z.A.A. Island flower pollination algorithm for global optimization. *J. Supercomput.* **2019**, *75*, 5280–5323. [CrossRef]
34. Al-Betar, M.A.; Awadallah, M.A.; Khader, A.T.; Abdalkareem, Z.A. Island-based harmony search for optimization problems. *Expert Syst. Appl.* **2015**, *42*, 2026–2035. [CrossRef]
35. Awadallah, M.A.; Al-Betar, M.A.; Bolaji, A.L.; Doush, I.A.; Hammouri, A.I.; Mafarja, M. Island artificial bee colony for global optimization. *Soft Comput.* **2020**, *24*, 13461–13487. [CrossRef]
36. Gupta, A.; Suri, B.; Kumar, V.; Misra, S.; Blažauskas, T.; Damaševičius, R. Software code smell prediction model using Shannon, Rényi and Tsallis entropies. *Entropy* **2018**, *20*, 372. [CrossRef] [PubMed]
37. Kumari, M.; Misra, A.; Misra, S.; Sanz, L.F.; Damasevicius, R.; Singh, V.B. Quantitative quality evaluation of software products by considering summary and comments entropy of a reported bug. *Entropy* **2019**, *21*, 91. [CrossRef] [PubMed]
38. Naidu, M.S.; Geethanjali, N. Classification of defects in software using decision tree algorithm. *Int. J. Eng. Sci. Technol.* **2013**, *5*, 1332–1340.
39. Can, H.; Xing, J.; Zhu, R.; Li, J.; Yang, Q.; Xie, L. A new model for software defect prediction using particle swarm optimization and support vector machine. In Proceedings of the 2013 IEEE 25th Chinese Control and Decision Conference (CCDC), Guiyang, China, 25–27 May 2013; pp. 4106–4110.
40. Shuai, B.; Li, H.; Li, M.; Zhang, Q.; Tang, C. Software defect prediction using dynamic support vector machine. In Proceedings of the 2013 IEEE Ninth International Conference on Computational Intelligence and Security, Emeishan, China, 14–15 December 2013; pp. 260–263.
41. Agarwal, S.; Tomar, D. A feature selection based model for software defect prediction. *Int. J. Adv. Sci. Technol.* **2014**, *65*, 39–58. [CrossRef]
42. Abaei, G.; Selamat, A. A survey on software fault detection based on different prediction approaches. *Viet. J. Comput. Sci.* **2014**, *1*, 79–95. [CrossRef]
43. Malhotra, R.; Nishant, N.; Gurha, S.; Rathi, V. Application of Particle Swarm Optimization for Software Defect Prediction Using Object Oriented Metrics. In Proceedings of the 2021 IEEE 11th International Conference on Cloud Computing, Data Science & Engineering (Confluence), Noida, India, 28–29 January 2021; pp. 88–93.
44. Balogun, A.O.; Basri, S.; Mahamad, S.; Abdulkadir, S.J.; Capretz, L.F.; Imam, A.A.; Almomani, M.A.; Adeyemo, V.E.; Kumar, G. Empirical Analysis of Rank Aggregation-Based Multi-Filter Feature Selection Methods in Software Defect Prediction. *Electronics* **2021**, *10*, 179. [CrossRef]
45. Mirjalili, S. Moth-flame optimization algorithm: A novel nature-inspired heuristic paradigm. *Knowl. Based Syst.* **2015**, *89*, 228–249. [CrossRef]
46. Mirjalili, S.; Lewis, A. S-shaped versus V-shaped transfer functions for binary particle swarm optimization. *Swarm Evol. Comput.* **2013**, *9*, 1–14. [CrossRef]
47. Kennedy, J.; Eberhart, R.C. A discrete binary version of the particle swarm algorithm. In Proceedings of the 1997 IEEE International conference on Systems, Man, and Cybernetics. Computational Cybernetics and Simulation, Orlando, FL, USA, 12–15 October 1997; Volume 5, pp. 4104–4108.

48. Kushida, J.i.; Hara, A.; Takahama, T.; Kido, A. Island-based differential evolution with varying subpopulation size. In Proceedings of the 2013 IEEE 6th International Workshop on Computational Intelligence and Applications (IWCIA), Hiroshima, Japan, 13 July 2013; pp. 119–124.
49. Michel, R.; Middendorf, M. An island model based ant system with lookahead for the shortest supersequence problem. In Proceedings of the International Conference on Parallel Problem Solving from Nature, Amsterdam, The Netherlands, 27–30 September 1998; Springer: Berlin/Heidelberg, Germany, 1998; pp. 692–701.
50. Araujo, L.; Merelo, J.J. Diversity through multiculturality: Assessing migrant choice policies in an island model. *IEEE Trans. Evol. Comput.* **2010**, *15*, 456–469. [CrossRef]
51. Khurma, R.A.; Aljarah, I.; Sharieh, A.; Mirjalili, S. Evolopy-fs: An open-source nature-inspired optimization framework in python for feature selection. In *Evolutionary Machine Learning Techniques*; Springer: Singapore, 2020; pp. 131–173.
52. Khurma, R.A.; Sabri, K.E.; Castillo, P.A.; Aljarah, I. Salp Swarm Optimization Search Based Feature Selection for Enhanced Phishing Websites Detection. In Proceedings of the Applications of Evolutionary Computation: 24th International Conference, EvoApplications 2021, Held as Part of EvoStar 2021, Virtual Event, 7–9 April 2021; Springer Nature: Basingstoke, UK, 2021; Volume 12694, pp. 146–161.

Article

Performance of Enhanced Multiple-Searching Genetic Algorithm for Test Case Generation in Software Testing

Wanida Khamprapai [1,2], Cheng-Fa Tsai [2,*], Paohsi Wang [3] and Chi-En Tsai [4]

1 Department of Tropical Agriculture and International Cooperation, National Pingtung University of Science and Technology, Pingtung 91201, Taiwan; wanida.kpp@gmail.com
2 Department of Management Information Systems, National Pingtung University of Science and Technology, Pingtung 91201, Taiwan
3 Department of Food and Beverage Management, Cheng Shiu University, Kaohsiung 83347, Taiwan; k0627@gcloud.csu.edu.tw
4 Department of Multimedia Business Unit II, Realtek Semiconductor Corporation, Hsinchu 30076, Taiwan; t82327@gmail.com
* Correspondence: cftsai@mail.npust.edu.tw; Tel.: +886-08-770-3201 (ext. 7906)

Abstract: Test case generation is an important process in software testing. However, manual generation of test cases is a time-consuming process. Automation can considerably reduce the time required to create adequate test cases for software testing. Genetic algorithms (GAs) are considered to be effective in this regard. The multiple-searching genetic algorithm (MSGA) uses a modified version of the GA to solve the multicast routing problem in network systems. MSGA can be improved to make it suitable for generating test cases. In this paper, a new algorithm called the enhanced multiple-searching genetic algorithm (EMSGA), which involves a few additional processes for selecting the best chromosomes in the GA process, is proposed. The performance of EMSGA was evaluated through comparison with seven different search-based techniques, including random search. All algorithms were implemented in EvoSuite, which is a tool for automatic generation of test cases. The experimental results showed that EMSGA increased the efficiency of testing when compared with conventional algorithms and could detect more faults. Because of its superior performance compared with that of existing algorithms, EMSGA can enable seamless automation of software testing, thereby facilitating the development of different software packages.

Keywords: search-based test case generation; genetic algorithm; branch coverage; object-oriented

Citation: Khamprapai, W.; Tsai, C.-F.; Wang, P.; Tsai, C.-E. Performance of Enhanced Multiple-Searching Genetic Algorithm for Test Case Generation in Software Testing. *Mathematics* **2021**, *9*, 1779. https://doi.org/10.3390/math9151779

Academic Editor: Vassilis C. Gerogiannis

Received: 22 June 2021
Accepted: 23 July 2021
Published: 27 July 2021

Publisher's Note: MDPI stays neutral with regard to jurisdictional clai-ms in published maps and institutio-nal affiliations.

1. Introduction

Software testing is an important process in the software development life cycle. It is performed to investigate the quality of software and to evaluate the risks in software implementation. Software testing involves both valid and invalid inputs and includes the processes of executing the developed software and checking for the expected responses. Several techniques can be used to automatically produce inputs that conform to the behavior of the software being tested, and these techniques provide high coverage in a given branch, line, condition, or path. Various techniques have been proposed to reduce the cost, resources, and time involved in the testing process.

The genetic algorithm (GA) is a popular and efficient search-based technique for test case generation. GAs have been widely used to create suitable test cases [1–4]. Suitable test case generation helps to reduce costs in software testing given the huge cost of creating test cases, which accounts for more than 50% of the total cost of developing a program [5]. Researchers have investigated methods to enhance the solution efficiencies of GAs. Multiple-searching genetic algorithm (MSGA) [6] is a successfully solved optimal solution with high probability for routing in network system. MSGA is attractive to utilize in other fields. From previous work [7], MSGA can generate test cases for small to medium scale software but cannot increase the percentage of coverage for complex software. This means test cases generated with MSGA cannot increase the number of executed statements or source code in complex software. Therefore, while MSGA may be suitable for generating test cases for small to medium scale software, it may not be flexible enough for test case generation for complex software. Some algorithms may be suitable for generating test cases for small to medium scale software but may not succeed in complex cases. For this reason, we present a new algorithm for improving MSGA to make it suitable for generating test cases. We expect that the test case generation using our algorithm will also detect more errors or faults in the software and therefore reduce the cost of software testing by creating the minimum number of test cases while getting the maximum coverage. Further, our algorithm can create test cases for complex software. In this study, we used MSGA to generate test cases for software testing because MSGA can reach the global optimum faster than a traditional GA [7]. In addition, we refactored the algorithm to solve the problem of executing the source code for more access to the statements.

In this study, a new algorithm called the enhanced multiple-searching genetic algorithm (EMSGA), which is an improved MSGA incorporating some additional processes, was developed. The genetic operators constitute the basic mechanism of the GA, namely selection, crossover, and mutation. Additional processes in EMSGA include the evaluation of chromosomes and selection of the best chromosomes to add to the next generation. In the original MSGA, all the chromosomes that are executed with the genetic operators are added to the next generation. EMSGA was expanded in EvoSuite, and its effectiveness was compared with that of MSGA and seven other techniques available in EvoSuite. The SF110 corpus and nine open-source Java projects developed by Google and the Apache Software Foundation were employed as case studies for generating test cases using the aforementioned algorithms.

The remainder of this paper is organized as follows. Section 2 discusses previous research works related to this study. Section 3 describes search-based techniques for generating test cases, including representation and fitness functions. The proposed algorithm is also introduced in this section. Section 4 presents the problem instances and tools used to evaluate EMSGA. Section 5 presents the experimental results. Section 6 reports threats to the validity of the algorithm. A discussion of the results is presented in Section 7. Finally, Section 8 concludes the paper.

2. Related Work

In software engineering, GA has been successful in many areas, such as software design, effort estimation, and maintenance. For software design [8], GA can help migration from structure programming to object-oriented programming, and the results are better than greedy algorithm and Monte Carlo. In software effort estimation, GA is stable, has higher accuracy than a random approach, and consists of an exhaustive framework [9]. Furthermore, GA is utilized to manage maintenance packages taking into account the cost-effectiveness of the package and to reduce human bias [10].

Various search-based techniques are available for test case generation. GA is one of the most widely used techniques. Many GAs have been remodeled for increased search efficiency. For example, a population aging process was added in a traditional GA without modifying any original parameters of the GA to reduce the number of test cases and increase the test coverage [4]. The features of GA and ant colony optimization (ACO) were combined to increase the efficiency and health of test cases [11]. GA and negative selection algorithms were merged to reduce the generation of duplicate test cases [12]. The results of the studies indicate that these improved algorithms are capable of efficiently generating test cases, even though the algorithms were originally improved for other applications. MSGA is an improved GA for network systems. Even though it was improved for and utilized in another field, we believe that an enhanced version of MSGA can increase the efficiency of test case generation.

EMSGA reuses and refactors existing algorithms. The reusable nature of this algorithm [13] helps to increase the reliability of results, provides faster algorithm development, and reduces costs. Algorithm refactoring is caused by insufficient existing algorithms to perform certain tasks. Consequently, algorithms are improved to suit the task. Algorithm refactoring is challenging in terms of selecting some parts of an algorithm to improve the performance or adding some processes to make it suitable for solving a given problem. Several studies have examined refactoring. For example, Liu et al. (2020) [14] studied automated refactoring for real-time systems to help reduce the effort required by programmers to isolate portions for the execution of real-time systems under limitations. Several researchers have used the SF110 corpus and EvoSuite to compare newly developed algorithms and existing algorithms. For example, the EvoTLBO algorithm was extended into EvoSuite to compare the results with traditional GA and monotonic GA using 50 random classes from SF110 [15]; EvoSuite and SF110 were utilized to compare the performance of memetic algorithm with traditional GA [16]; and nontrivial classes were selected from SF110 to compare the efficiency of the DynaMOSA algorithm with the many objective sorting algorithm (MOSA), the whole suite approach with archive (WSA), and the traditional whole suite approach (WS) [17]. The SF110 corpus is considered as a benchmark for test generation [18]. The SF110 corpus contains 110 Java projects from SourceForge, 100 random projects, and the 10 most popular projects in SourceForge. EvoSuite is an automatic test generation tool for Java classes based on GA. In the present study, the SF110 corpus and EvoSuite were considered sufficient to measure the effectiveness of the proposed algorithm for test case generation. EMSGA was tested using SF110, and its effectiveness was compared with that of seven algorithms available in EvoSuite.

3. Search-Based Test Case Generation

The search-based technique is widely used for test case generation [19–22]. The following subsections describe some of the most well-known search-based techniques before introducing the proposed EMSGA.

3.1. Representation

A population of candidate solutions is represented as a test suite [17,22], which is a collection of test cases $T = \{t_1, t_2, \ldots, t_n\}$. Each test case is composed of various statements $t = \langle s_1, s_2, \ldots, s_l \rangle$, where l is the total number of statements. A statement [23] can be a variable declaration or a method call and can be of several different types, namely a primitive, a constructor, a method, an array, or an assignment.

Figure 1 presents the generated test cases from Java code by considering the required variables and methods to generate statements for testing the class under test. When considering Java code, the integer array variable is a required variable to maintain the numbers for sorting. Therefore, the integer array variable is declared in the test case. The number of statements depends on the instruction to be used for each test. The length of either the test case or the chromosome depends on the number of statements. The population evolves iteratively to yield better solutions. The processes are repeated until a stopping criterion is satisfied.

3.2. Fitness Function

In software testing, a fitness function is used to evaluate the ability of the generated test suites to execute the source code of the program. Typically, fitness functions are assessed based on the branch coverage metric. Complete branch coverage refers to all control structures being executed and all lines of code being tested. This metric is defined as follows [24,25]:

$$f(T) = |M| - |M_T| + \sum_{b \subset B} d(b, T), \tag{1}$$

where $|M|$ denotes the total number of methods, $|M_T|$ is the number of methods executed in test suite T, and $d(b, T)$ represents the branch distance for each branch b on test suite

T that b is an element of in a set of branches B. The branch distance $d(b, T)$ is defined as follows:

$$d(b,T) = \begin{cases} 0 & \text{if the branch has been covered,} \\ d_{min}(b,T) & \text{if the predicate has been executed at least twice,} \\ 1 & \text{otherwise.} \end{cases} \quad (2)$$

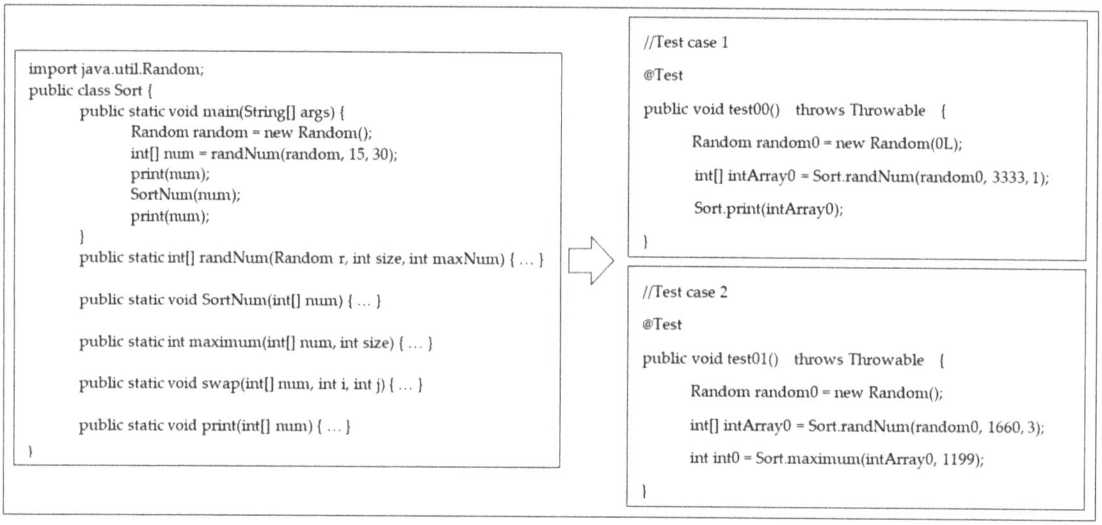

```
import java.util.Random;
public class Sort {
        public static void main(String[] args) {
                Random random = new Random();
                int[] num = randNum(random, 15, 30);
                print(num);
                SortNum(num);
                print(num);
        }
        public static int[] randNum(Random r, int size, int maxNum) { … }

        public static void SortNum(int[] num) { … }

        public static int maximum(int[] num, int size) { … }

        public static void swap(int[] num, int i, int j) { … }

        public static void print(int[] num) { … }
}
```

```
//Test case 1
@Test
public void test00()  throws Throwable  {
        Random random0 = new Random(0L);
        int[] intArray0 = Sort.randNum(random0, 3333, 1);
        Sort.print(intArray0);
}
```

```
//Test case 2
@Test
public void test01()  throws Throwable  {
        Random random0 = new Random();
        int[] intArray0 = Sort.randNum(random0, 1660, 3);
        int int0 = Sort.maximum(intArray0, 1199);
}
```

Figure 1. Generated test cases from source code.

3.3. Genetic Algorithms

GAs [4,26] solve problems through the use of three basic operators: selection, crossover, and mutation. In GA, a chromosome is defined as a set of parameters that represent a proposed solution to the problem that the GA is being used to solve. The selection operator selects certain chromosomes as parent chromosomes. Chromosomes are selected on the basis of their fitness values. Chromosomes with higher fitness values have a higher chance of being selected. The crossover and mutation operators are applied to the parent chromosomes to produce offspring for the next generation. The crossover operator exchanges certain genes of two chromosomes. The mutation operator changes the value of some genes in a few chromosomes.

Several researchers have proposed techniques to improve the traditional GA for enhancing its solution efficiency and enabling its application in complex problems. These efforts have relied on adjustments of factors or integration of GAs with other strategies. For example, the monotonic GA [26] involves additional processes after the mutation process in the traditional GA. These additional processes measure the fitness values to determine the best offspring or the best parent for the next population; in contrast, the traditional GA increases the number of mutated offspring in the next population and then calculates the fitness values of all chromosomes. Another improved version of GA is the steady-state GA [27,28], in which the fitness values of the mutated offspring are determined and then the offspring is compared with the parent. If the offspring is better than the best parent, the offspring replaces the parent in the current population. The advantages of monotonic GA and steady-state GA are similar, namely removing duplicate chromosomes and ensuring the best chromosome is not discarded. A breeder GA [29] differs from the traditional GA in that it uses the principle of breeding, which involves selecting the fittest chromosomes and reproducing using those chromosomes. The breeder GA is more precise as it utilizes the science of breeding [30]. A cellular GA [31] is an improved GA that selects the best offspring after the crossover operator has been applied. The best offspring is mutated, and the fitness value is determined. The selection of cellular GA is restricted to the overlapping neighborhood producing slow solutions [32,33]. Table 1 summarizes the characteristic of each GA.

Table 1. Comparison of GA-based characteristics.

Algorithm	Characteristic of Algorithm
Traditional GA	Applies only three basic operators: selection, crossover, and mutation
Monotonic GA	Still applies three basic operators but adds some processes to select the best chromosome for the next generation.
Steady-state GA	Adds some processes to select the best chromosome. Similar to the monotonic GA but replaces the best chromosome in the current population.
Breeder GA	Applies the principle of breeding to select chromosomes before performing the basic operators.
Cellular GA	Performs mutation operator on only one crossed chromosome. Chromosomes are selected for mutation by choosing at random.

3.4. Chemical Reaction Optimization (CRO)

Chemical reaction optimization (CRO) [34] is a search-based technique that combines the advantages of GA and simulated annealing. CRO solves problems using a set of molecules. Each molecule possesses a molecular structure, potential energy, and kinetic energy. The molecular structure represents a possible solution that does not have any specific format. The potential energy is the fitness value of the corresponding molecule. The kinetic energy quantifies the tolerance of the worst solution. The iterative processes of CRO are similar to those of GA. A basic CRO involves four types of reactions: on-wall ineffective collision and decomposition are reactions where a single molecule hits a wall of the surface, and intermolecular ineffective collision and synthesis are reactions where multiple molecules collide with each other.

On-wall ineffective collision represents a local search. There is minimal change in the structure or properties of the molecule during this process. Decomposition is a type of collision that produces two or more new molecules. This process represents a global search. Intermolecular ineffective collision is the collision of multiple molecules, which produces minimal changes in the structure or properties of the molecules, similar to on-wall ineffective collision. Two or more collided molecules undergo small changes in structure or properties. Synthesis is a reaction that represents a global search. In this reaction, multiple colliding molecules fuse into a single molecule.

3.5. Random Search

Random search is the simplest search-based technique. It involves iterative searches until an optimal solution is obtained. In each iteration, the solution is incremented with a random vector. The fitness value of the modified solution is determined. If the modified solution is better than the previous solution, the former replaces the latter. Otherwise, the previous solution is retained. Random search is often utilized for comparison with other techniques [35]. This technique can effectively solve large-scale problems [36].

4. Proposed Algorithm: Enhanced Multiple-Searching Genetic Algorithm (EMSGA)

In the multiple-searching genetic algorithm (MSGA) introduced by Tsai et al. [6], two types of chromosomes are created to prevent the search from falling into a local optimum. The MSGA utilizes the candidate mechanism to create more chromosomes with the same features, resulting in better chromosomes. The MSGA has been successfully used to find the optimal multicast route in network systems. We believe that the MSGA can also be integrated with other strategies to increase search ability. Therefore, we propose EMSGA, a regeneration MSGA with the addition of a feature-selection strategy. After the mutation operator is employed and the fitness value is determined, only chromosomes from the best offspring or the best parent will be selected to be included in the next-generation population. If the mutated offspring are better than the parents, then they replace the parents in the next generation. Otherwise, the parents are retained. Choosing the best chromosome increases the chances of reaching the optimal solution. Generally, two mutated offspring are added to the next-generation population, and the parents are discarded. The processes involved in EMSGA are similar to those in MSGA, with the exception of the aforementioned best chromosome selection mechanism after the mutation process (Figure 2). Algorithm 1 shows the pseudocode of EMSGA.

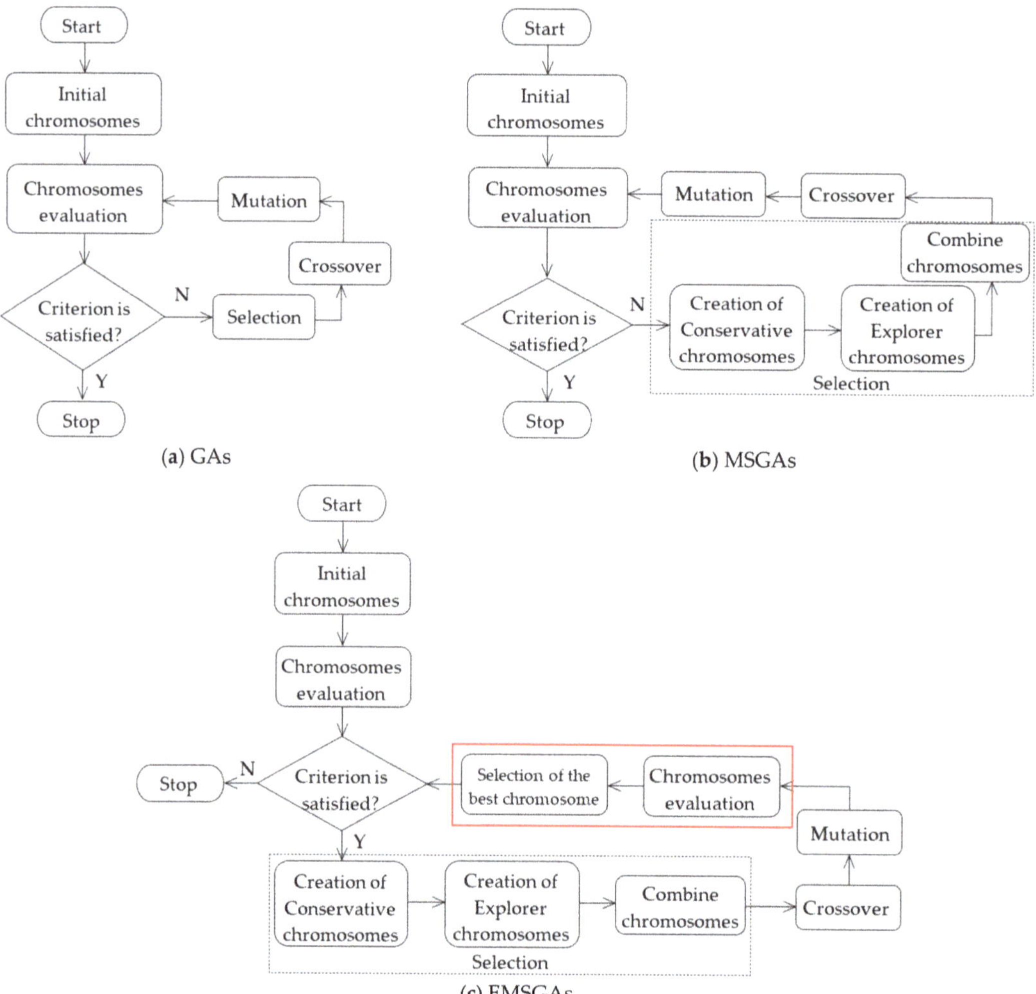

Figure 2. Flowcharts of GA (**a**) [7], MSGA (**b**) [7], and EMSGA (**c**). The red box indicates the additional processes in EMSGA. The black dashed box displays the additional processes in MSGA.

Algorithm 1 Pseudocode for EMSGA

```
Procedure EMSGA()
Create initial chromosomes
Evaluate fitness value of initial chromosomes and order by descending
while not terminal condition do
  Select half chromosomes with the highest fitness value //Conservative chromosomes
  Call procedure CreateExplorerChromosomes(Conservative chromosomes)
  Evaluate fitness value of explorer chromosomes
  Combine Conservative and Explorer chromosomes
  Call procedure Crossover(all chromosomes)
  //Mutation of EMSGA is the same as traditional GA
  Mutate Conservative chromosomes with M1
  Mutate Explorer chromosomes with M2
  Evaluate fitness value of the mutated chromosomes
  if the offspring is better than the best parent then
      Add offspring in the next population
  else
      Add parent in the next population
  end if
end while
return chromosomes
end procedure

Procedure CreateExplorerChromosomes(Conservative chromosomes)
for each conservative chromosome i
  for each gene of conservative chromosome j of i
      Keep jth gene of ith conservative chromosome to jth candidate gene set
  end for
end for
Creates explorer chromosomes with a number equal to the number of conservative
chromosomes
for each explorer chromosome i
  for each candidate gene set j
      Select one gene from jth candidate gene set
      Preserve the selected gene in jth gene of ith explorer chromosome
  end for
end for
return explorer chromosomes
end procedure

Procedure Crossover(all chromosomes)
Set a random number r
if r is less than crossover probability then
  for half of all chromosomes from i = 1 to (population size / 2)
      Select ith chromosome and (population size − i + 1)th chromosome
      Split the selected chromosomes with crossover method
      Cross both chromosomes
  end for
end if
return chromosomes
end procedure
```

The EMSGA process starts with the creation of initial chromosomes. Then, the fitness value of the population is determined, and half of the chromosomes with the highest fitness values are retained. The rest of the chromosomes are discarded. The preserved chromosomes are called the conservative chromosomes. Next, the candidate mechanism is utilized to build the explorer chromosomes by selecting the genes of the conservative chromosomes.

The candidate mechanism is created to gather genes of all conservative chromosomes that are in the same position into the same candidate gene set. Each candidate gene set selects only one gene to create as a gene of explorer chromosome. Figure 3 illustrates the method for creating an explorer chromosome. Thereafter, crossover and mutation are performed on the conservative and explorer chromosomes separately. Both types of chromosomes are assigned the same crossover probability. The mutation probabilities are defined differently. At the end of each iteration, the chromosomes are evaluated in terms of the fitness value, and the best chromosomes are selected and added to the next-generation population.

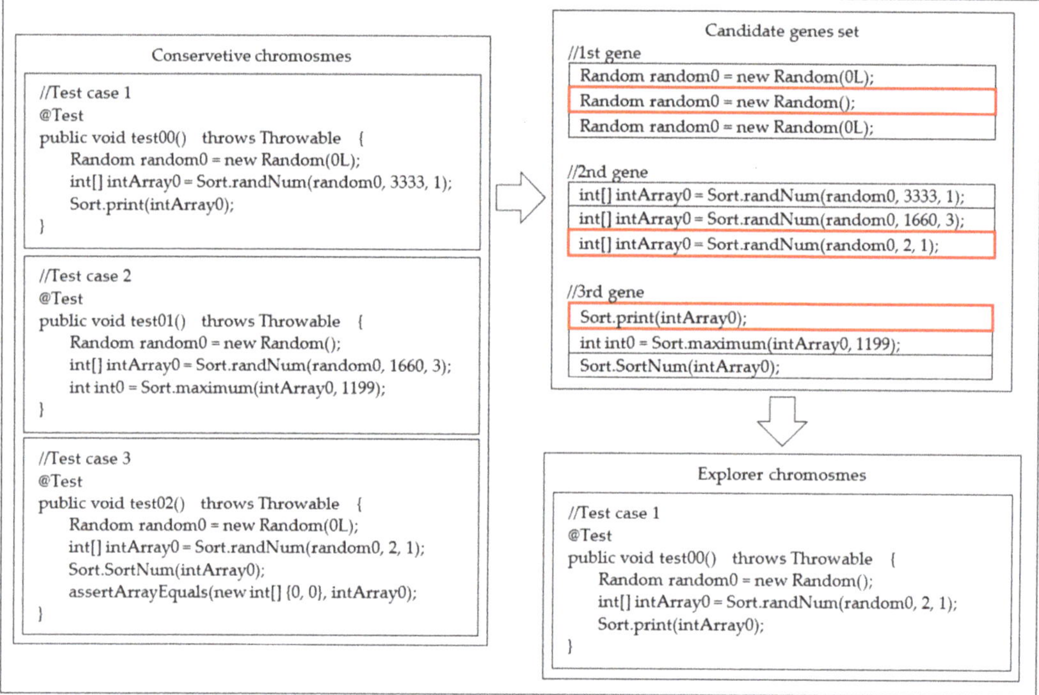

Figure 3. Mechanism of creating explorer chromosome. Red boxes demonstrate which one gene from each candidate gene set was chosen.

5. Experimental Evaluation

The aim of this study was to evaluate the capability of EMSGA to generate test cases and to compare the feasibility and effectiveness of EMSGA with those of other algorithms.

5.1. Problem Instances

The selection of problem instances is important for any empirical study on automatic test case generation. This study utilized the SF110 corpus (the details of SF110 are available online: https://www.evosuite.org/experimental-data/sf110/ (accessed on 4 March 2020)) [18] and nine open-source Java projects developed by Google and the Apache Software Foundation to evaluate EMSGA. The SF110 corpus is widely used as a benchmark [17,24,37]. It contains 110 projects that were written with the Java language. Not all classes in the SF110 corpus were employed in this experiment. Only 203 classes were chosen based on the selection in a previous study [38]. Furthermore, nine problem instances from Google and the Apache Software Foundation were chosen uniformly and at random based on their sizes and functionalities (Table 2), consisting of a total of 1382 classes. EvoSuite was applied to a total of 203 + 1382 = 1585 classes.

Table 2. Details of open-source Google and Apache projects. Note: the second column lists the number of non-commenting source lines of code reported by JavaNCSS (http://www.kclee.de/clemens/java/javancss/ (accessed on 10 December 2020)). The fourth column lists the number of branches reported by EvoSuite.

Problem Instances	No. of Lines	No. of Classes	No. of Branches
Java Certificate Transparency	955	30	178
Commons CLI	1480	22	961
Commons Codec	5545	68	3050
Commons Email	1505	20	209
Commons Jelly	4688	95	636
Commons Math3	65,389	918	28,450
Commons Numbers	317	5	225
Joda-Time	19,441	166	9924
Truth	4117	58	223
Total	103,437	1382	43,856

5.2. Test Generation Tool

The testing tool employed EvoSuite (EvoSuite can be downloaded from http://www.evosuite.org (accessed on 20 February 2020)) [24] to generate test cases for Java code. EvoSuite is widely used in software testing [3,39,40]. It utilizes search-based methods, including genetic algorithms, to generate test cases using Java bytecode. Furthermore, EvoSuite supports various coverage criteria to determine the quality of a solution.

In the experiment, the proposed algorithm was implemented as an extension to the EvoSuite. To extend the new algorithm in Evosuite, a developer must create a new class in the client module and extend the abstract class *GeneticAlgorithm*. The EMSGA class implemented the basic methods for GA that EvoSuite prepares. In addition, the EMSGA class added some processes for creating two types of chromosomes and selected the best chromosome. Test cases of each algorithm were automatically generated, and problem instances were executed through EvoSuite. The performance of EMSGA was compared with that of the MSGA, traditional GA, monotonic GA, steady-state GA, breeder GA, cellular GA, CRO, and random search. These search-based methods are provided in EvoSuite. The coverage achieved by the algorithms was assessed in terms of the branch coverage metric. Search budget configuration uses EvoSuite's default of 60 s [41]. Search budget is the time for generating test cases of the algorithm each time. The experiment was independently repeated 10 times.

The parameter settings influence the performance of search-based methods. The EvoSuite guides the default values (e.g., selection function, crossover function, crossover probability, mutation function, mutation probability, population size, and chromosome length) for test case generation. The default values of EvoSuite are the approximate values that are suitable for generating test cases that are based on GA. Table 3 shows the default values in EvoSuite. The same parameter setting may not be enough to fully extract the efficiency of the algorithm [42]. As Arcuri and Fraser (2013) [43] pointed out, the default values of EvoSuite are sufficient to evaluate the performance of algorithms for test case generation, whereas the suitable parameter setting is time-consuming and may or may not produce good results for algorithms. In addition, Črepinšek et al. (2014) [44] perceptively stated that all algorithms should be examined under the same conditions.

Therefore, the default values for all nine algorithms were used in the experiment. EMSGA assigns different mutation probabilities to the conservative and explorer chromosomes. If the explorer chromosomes are defined as having a higher mutation probability than the conservative chromosomes, the optimal solution can be obtained [6]. Several researchers have set the probability as $1/l$ for the mutation operator, where l is the chromosome length [43,45,46]. Accordingly, mutation probabilities of $1/l$ and 0.75 (default) were used for the conservative and the explorer chromosomes, respectively, in this study.

Table 3. Default values of parameters in Evosuite.

Parameters	Default Values
Population size	50
Chromosome length	40
Selection function	Rank
Crossover function	Single point relative
Crossover probability	0.75
Mutation function	Uniform
Mutation probability	0.75
Search budget	60 s

The experiment involved 1585 × 9 × 10 = 142,650 runs of EvoSuite with the aforementioned settings. The search in each run was limited to 60 s. The experiment required at least 142,650/(60 × 24) = 99.0625 days of computational time. It was conducted on a Windows 10 Professional (Seattle, WA, USA) ×64 system having an Intel® Core i7 CPU with 3.40 GHz and 16 GB of RAM.

5.3. Experimental Analysis

The coverage achieved was evaluated based on the branch criterion, number of test cases (#T), and mutation score. All the experimental results were analyzed via nonparametric Mann–Whitney U tests with a significance level (p-value) of 0.05, the Vargha–Delaney $\hat{A}_{12}$ effect size, and a 95% confidence interval for the branch coverage achieved. Boxplots and marginal distribution plots were created using RStudio Version 1.1.383.

6. Experimental Results

The experimental results for EMSGA and the competing algorithms are presented and analyzed in this section. The experimental results are tabulated in Table 4, which shows the standard deviation (σ), a 95% confidence interval (CI) of the branch coverage, the p-value for the Mann–Whitney U tests, and the Vargha-Delaney $\hat{A}_{12}$ effect size.

Table 4. Results of test case generation using each algorithm.

Algorithm	Branch Coverage			Mut. Score			#T	p-Value	$\hat{A}_{12}$ (EMSGA:Others)
	Avg.	σ	CI	Avg.	σ	CI			
EMSGA	0.5900	0.0032	(0.5877, 0.5923)	0.4174	0.0038	(0.4146, 0.4201)	180.49351	-	-
MSGA	0.5846	0.0033	(0.5823, 0.5870)	0.4166	0.0043	(0.4135, 0.4196)	181.5325	0.00578	0.87
GA	0.5829	0.0040	(0.5801, 0.5858)	0.4159	0.0046	(0.4127, 0.4192)	177.8818	0.00168	0.92
Monotonic GA	0.5855	0.0063	(0.5810, 0.5901)	0.4162	0.0050	(0.4127, 0.4198)	182.3091	0.03752	0.74
Steady-State GA	0.5699	0.0036	(0.5673, 0.5725)	0.4168	0.0023	(0.4152, 0.4185)	178.7455	0.00018	1
Breeder GA	0.5821	0.0059	(0.5779, 0.5864)	0.4167	0.0040	(0.4138, 0.4195)	180.0545	0.00466	0.88
Cellular GA	0.5588	0.0034	(0.5563, 0.5612)	0.4056	0.0044	(0.4024, 0.4087)	174.2039	0.00018	1
CRO	0.5717	0.0040	(0.5688, 0.5746)	0.4120	0.0062	(0.4076, 0.4164)	177.9416	0.00018	1
Random search	0.5683	0.0036	(0.5657, 0.5709)	0.4127	0.0025	(0.4109, 0.4144)	179.5857	0.00018	1

EMSGA achieved the highest branch coverage (0.5900). This means test cases of EMSGA can execute 59% of the source code of the class test. The branch coverage of EMSGA obtained that similar to the monotonic GA. However, EMSGA generated fewer test cases than the monotonic GA due to the limited search budget. Each algorithm had 60 s to search for the optimal test cases for each class. Although EMSGA generated fewer cases, the branch coverage of EMSGA was higher. This means that EMSGA is more efficient than monotonic GA. In terms of the mutation score, EMSGA achieved the best performance. The mutation score represents the number of faults that can be detected in the test cases, which is a measure of the quality of the test cases generated by each algorithm [47]. The $\hat{A}_{12}$ measure is a comparison of effect size between the EMSGA and the others; if $\hat{A}_{12} > 0.5$, it means EMSGA can beat that algorithm more than 50% of the time. For example, $\hat{A}_{12} = 0.74$ means EMSGA can beat the monotonic GA 74% of the time. The values of this metric for all the algorithms were found to be greater than 0.5. This means that the EMSGA can generate higher-quality test cases than the other algorithms.

Considering the values of all the metrics, EMSGA clearly outperformed MSGA in most categories. Furthermore, specifically in terms of the $\hat{A}_{12}$ measure, EMSGA performed significantly better than MSGA (average $\hat{A}_{12}$ effect size was 0.93). In the Mann–Whitney U tests, EMSGA exhibited a p-value of less than 0.05. From a comparison between EMSGA and MSGA, it can be concluded that EMSGA possesses a more effective best chromosome selection process due to the addition of genetic operators and is hence more efficient than the traditional MSGA. The higher mutation score implies that EMSGA is better at detecting faults than the other algorithms.

The distributions of the average branch coverage and average mutation scores obtained from the 1585 classes during the execution of the test cases generated by each algorithm are shown in Figure 4. The length of the box indicates the distribution of values between the 25% and 75% quantiles. The horizontal line in the box represents the median value. The dot in the box represents the mean value. The vertical lines indicate the smallest and largest values outside the middle 50%. The dots outside the box denote the outlier values. Despite the similar distributions of coverage and mutation score for all the algorithms, outliers of mutation score were observed across all the algorithms (see Figure 4b) except EMSGA and random search. This suggests that EMSGA and random search can detect up to 100% of the faults, while the other algorithms can detect approximately 80–90% of the faults (the outliers represent the undetected faults). Considering the distribution of coverage (see Figure 4a), EMSGA exhibited a higher average coverage than random search. Furthermore, EMSGA presented a narrower distribution, that is, less scattered data.

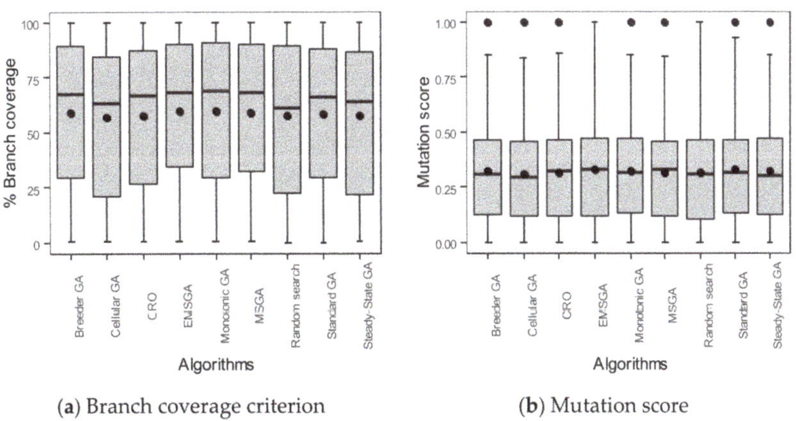

(**a**) Branch coverage criterion (**b**) Mutation score

Figure 4. Coverage and mutation scores achieved by each algorithm.

Figure 5 presents the distributions of the branch coverage, number of test cases, and mutation score achieved by each algorithm. Each marginal distribution displays the average of each metric (dashed line) and the marginal density. The marginal density is the solid line on the right side of each marginal distribution plot that indicates the distribution of results. The average branch coverage of all the algorithms was 57.71% (Figure 5a). Five algorithms achieved values exceeding the average, namely EMSGA, MSGA, standard GA, monotonic GA, and breeder GA. In terms of the number of test cases (Figure 5b) as well, four algorithms achieved values better than the average (179.19 test cases), namely EMSGA, MSGA, monotonic GA, breeder GA, and random search. All algorithms exhibited mutation scores above the average (0.41). Thus, EMSGA achieved values exceeding the average for all three evaluation metrics. The ratio of classes reached branch coverage within each 10% branch coverage interval, as shown in Figure 6. For example, 35% of all classes that were tested in the test cases generated by EMSGA achieved a branch coverage between 81% and

100%. From the experimental results, it is evident that EMSGA is feasible and effective for generating test cases.

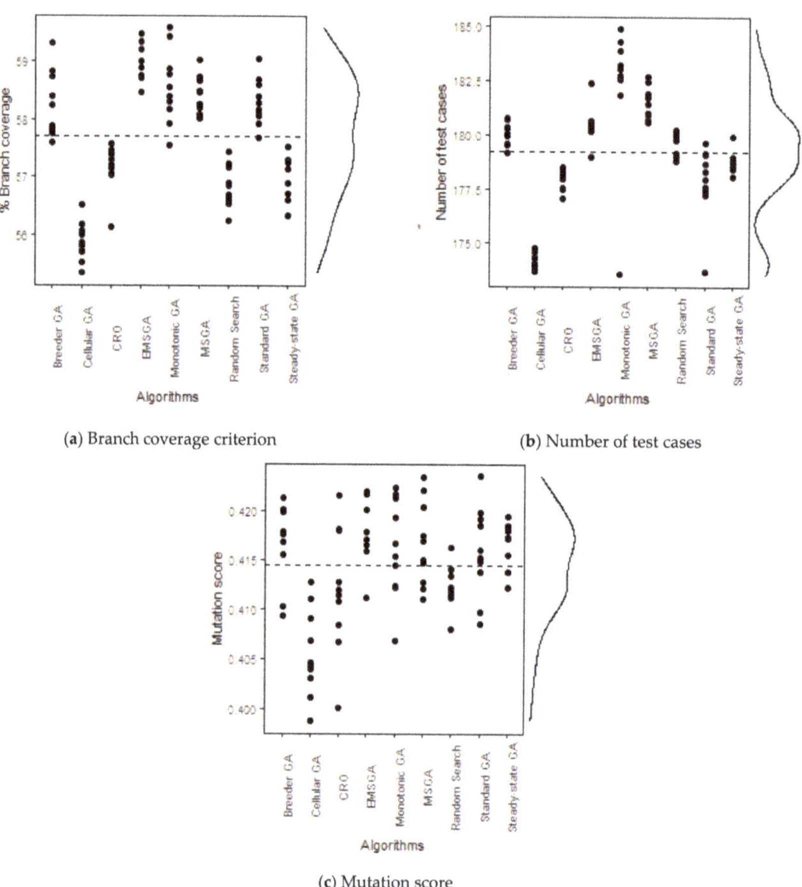

(**a**) Branch coverage criterion

(**b**) Number of test cases

(**c**) Mutation score

Figure 5. Average values of metrics for each algorithm.

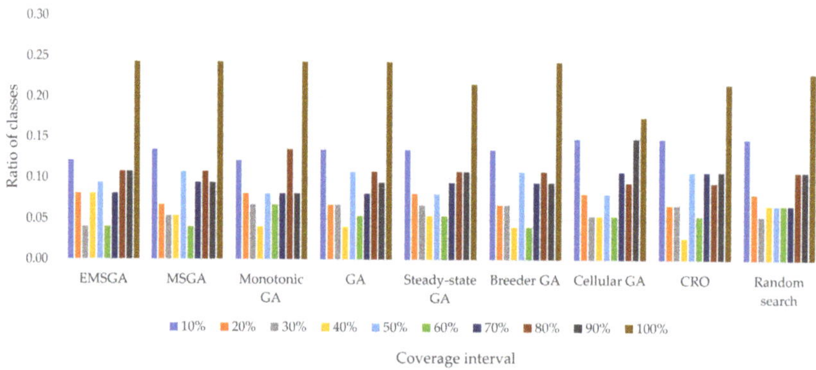

Figure 6. Proportion of classes for different branch coverage intervals.

Figure 7 displays the association between the number of test cases and the achieved branch coverage when problem instances were executed using test cases of each algorithm. Several problem instances indicated the EMSGA achieved greater or equal branch coverage while the number of test cases was less than the others. The problem instance Truth is a small-scale program, and the test cases of all algorithms executed a similar number of source code.

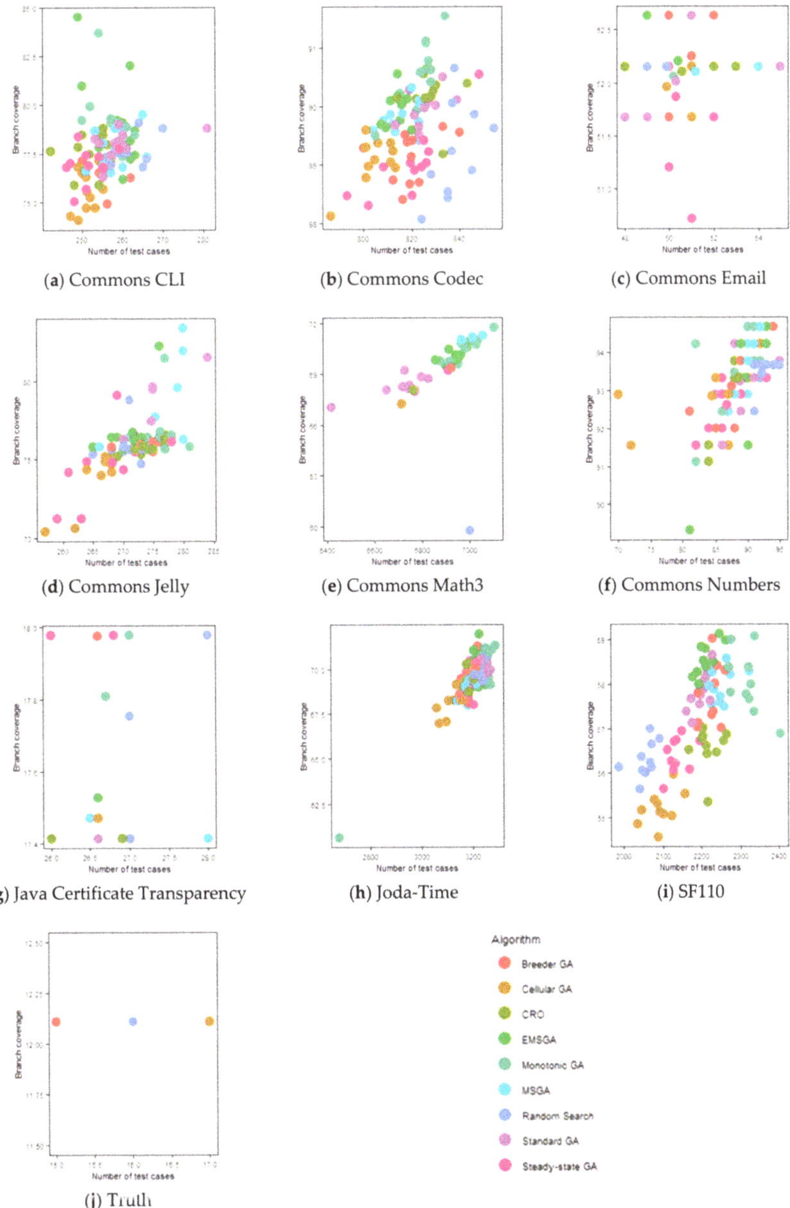

(**a**) Commons CLI (**b**) Commons Codec (**c**) Commons Email (**d**) Commons Jelly (**e**) Commons Math3 (**f**) Commons Numbers (**g**) Java Certificate Transparency (**h**) Joda-Time (**i**) SF110 (**j**) Truth

Figure 7. Problem instances that were evaluated with each algorithm.

7. Threats to Validity

Based on the results obtained, threats to internal validity are related to factors affecting the behavior of the software under test [48]. One such factor observed in the experiment was the number of test cases generated by all algorithms. Single testing might be inadequate for summarizing the performance of the algorithms in terms of generating test cases. In this experiment, each algorithm was run 10 times with the same tools. Furthermore, all parameters were defined with the same default values.

Threats to external validity are related to the generalization of the results beyond the scope of experimental analysis [22]. The SF110 corpus and nine open-source Java projects developed by Google and the Apache Software Foundation were utilized as case studies, which required a large number of experiments to be conducted. In this study, a total of 1585 classes were used, which included 203 classes from the SF110 corpus chosen based on previous studies [37] and all classes of the nine open-source Java projects. The reported results are limited to the search-based techniques employed in the experiments.

8. Discussion

EMSGA modifies the MSGA processes by comparing the parent and offspring and choosing the better chromosomes for the next generation. The selection of the better chromosome as input to the next generation allows for approaching the optimal solution. Our experimental results are in accordance with the results of previous experiments, which indicates that the branch coverage increases when a better chromosome is selected. For example, the monotonic GA achieved better results than the traditional GA [15,22]. Our results show that EMSGA can achieve a higher branch coverage, generate more test cases, and obtain a higher mutation score than MSGA.

One of the contributions of this research is our examination of the efficiency of EMSGA by extending it to EvoSuite, which is an automatic tool for generating test cases. The results of this application provide the number of test cases, the percentage of coverage, and mutation score. The results also indicate that EMSGA achieves a similar coverage with fewer test cases compared with monotonic GA. This is probably because the population of EMSGA contains two types of chromosomes, namely conservative and explorer chromosomes. The explorer chromosomes are created from high-fitness chromosomes. The main objective of software testing is to minimize the number of test cases and increase the coverage. The number of test cases affects the software development cost [5,49]. Although EMSGA produces fewer test cases than monotonic GA does in 60 s, the former achieves a higher coverage for the same number of test cases. A comparison of the efficiency between the existing algorithms in EvoSuite and EMSGA suggests that, in test case generation, the branch coverage may not be enough to clearly demonstrate the difference between results. The finding is consistent with Campos et al. (2018) [21], who indicated that the efficiencies of algorithms in EvoSuite may provide little difference in results for generating test cases. This could be due to a limitation on setting parameters, such as population size, basic function, or timing. In particular, as Fraser and Arcuri (2015) [50] pointed out, achieving a certain percentage of branch coverage and mutation score for a limited time may lead to higher mutation scores, but the coverage may be lower. The above experimental results also show that we can obtain higher mutation scores while having coverage very close to other algorithms. These findings lead us to believe that EMSGA has the potential to generate more test cases within a limited time and increase its coverage. Arcuri and Fraser [47] reported that the performance of a search-based technique depends on the parameter settings. A possible alternative is to find the best value of the parameters suitable for generating test cases [22], although the default values of EvoSuite are sufficient for evaluating algorithms in terms of test case generation. Therefore, appropriate values for EMSGA should be determined to generate the maximum number of test cases. Furthermore, EMSGA should be examined for other test coverage criteria.

9. Conclusions

This paper proposes an enhanced MSGA (EMSGA) to generate test cases for software testing. In EMSGA, the selection process involves creating two types of chromosomes to obtain better chromosomes before performing crossover and mutation operations. The performance of EMSGA on the basis of branch coverage, number of test cases, and mutation score was compared with that of other algorithms available in EvoSuite. The results show that EMSGA is more efficient than MSGA as well as the other algorithms. In addition, EMSGA can detect more faults in programs than the other algorithms. Therefore, because of its superior performance, EMSGA is expected to enable seamless automation of software testing, thereby facilitating the development of different software packages in the future.

Author Contributions: Conceptualization, W.K. and C.-F.T.; methodology, W.K. and C.-F.T.; software, W.K.; validation, P.W. and C.-E.T.; formal analysis, W.K.; investigation, W.K. and C.-F.T.; resources, W.K.; data curation, P.W. and C.-E.T.; writing—original draft preparation, W.K.; writing—review and editing, W.K. and C.-F.T.; visualization, W.K.; supervision, C.-F.T.; project administration, C.-F.T.; funding acquisition, C.-F.T. All authors have read and agreed to the published version of the manuscript.

Funding: This research was funded by the Ministry of Science and Technology, Taiwan, grant numbers MOST-108-2637-E-020-003, MOST-108-2321-B-020-003, and MOST-109-2637-E-020-003.

Institutional Review Board Statement: Not applicable.

Informed Consent Statement: Not applicable.

Data Availability Statement: The proposed algorithm in this study including source code and results are available on request from the corresponding author.

Acknowledgments: The authors would like to express their sincere gratitude to the anonymous reviewers for their useful comments and suggestions for improving the quality of this paper. We also thank the staff of the Department of Tropical Agriculture and International Cooperation, Taiwan; Department of Management Information Systems, Taiwan; National Pingtung University of Science and Technology, Taiwan; and the Ministry of Science and Technology, Taiwan. It is their kind help and support that have made to complete this research.

Conflicts of Interest: The authors declare no conflict of interest.

References

1. Khan, R.; Amjad, M.; Srivastava, A.K. Optimization of Automatic Generated Test Cases for Path Testing Using Genetic Algorithm. In Proceedings of the 2nd International Conference on Computational Intelligence & Communication Technology, Ghaziabad, India, 12–13 February 2016; pp. 32–36. [CrossRef]
2. Jatana, N.; Suri, B. Particle Swarm and Genetic Algorithm applied to mutation testing for test data generation: A comparative evaluation. *J. King Saud Univ. Comput. Inf. Sci.* **2020**, *32*, 514–521. [CrossRef]
3. Aleti, A.; Grunske, L. Test data generation with a Kalman filter-based adaptive genetic algorithm. *J. Syst. Softw.* **2015**, *103*, 343–352. [CrossRef]
4. Yang, S.; Man, T.; Xu, J.; Zeng, F.; Li, K. RGA: A lightweight and effective regeneration genetic algorithm for coverage-oriented software test data generation. *Inf. Softw. Technol.* **2016**, *76*, 19–30. [CrossRef]
5. Kumar, D.; Mishra, M.M. The Impacts of Test Automation on Software's Cost, Quality and Time to Market. *Procedia Comput. Sci.* **2016**, *79*, 8–15. [CrossRef]
6. Tsai, C.F.; Tsai, C.W.; Wu, H.C. A novel algorithm for multimedia multicast routing in a large scale network. *J. Syst. Softw.* **2004**, *72*, 431–441. [CrossRef]
7. Khamprapai, W.; Tsai, C.F.; Wang, P. Analyzing the Performance of the Multiple-Searching Genetic Algorithm to Generate Test Cases. *Appl. Sci.* **2020**, *10*, 7264. [CrossRef]
8. Selim, M.; Siddik, M.S.; Gias, A.U.; Abdullah-Al-Wadud, M.; Khaled, S.M. A Ge-netic Algorithm for Software Design Migration fromStructured to Object Oriented Paradigm. In Proceedings of the 8th International Conference on Computer Engineering and Application (CEA 2014), Tenerife, Spain, 10–12 January 2014; pp. 187–192.
9. Murillo-Morera, J.; Quesada-López, C.; Castro-Herrera, C.; Jenkins, M. A genetic algorithm based framework for software effort prediction. *J. Softw. Eng. Res. Dev.* **2017**, *5*, 4. [CrossRef]
10. Bennett, T.E.; Brown, M.S.; Pelosi, M. A Genetic Algorithm for the Generation of Software Maintenance Release Plans without Human Bias. *J. Softw. Eng. Practice* **2015**, *1*, 6–21.

11. Khari, M.; Kumar, P.; Shrivastava, G. Enhanced approach for test suite optimisation using genetic algorithm. *Int. J. Comput. Aided Eng. Technol.* **2019**, *11*, 653–668. [CrossRef]
12. Mohi-Aldeen, S.M.; Mohamad, R.; Deris, S. Optimal path test data generation based on hybrid negative selection algorithm and genetic algorithm. *PLoS ONE* **2020**, *15*, e0242812. [CrossRef]
13. Rathee, A.; Chhabra, J.K. A multi-objective search based approach to identify reusable software components. *J. Comput. Lang.* **2019**, *52*, 26–43. [CrossRef]
14. Liu, Y.; An, K.; Tilevich, E. RT-Trust: Automated refactoring for different trusted execution environments under real-time constraints. *J. Comput. Lang.* **2020**, *56*, 100939. [CrossRef]
15. Shahabi, M.M.D.; Badiei, S.P.; Beheshtian, S.E.; Akbari, R.; Moosavi, M.R. EVOTLBO: A TLBO based Method for Automatic Test Data Generation in EvoSuite. *Int. J. Adv. Comput. Sci. Appl.* **2017**, *8*, 214–226. [CrossRef]
16. Fraser, G.; Arcuri, A.; McMinn, P. A Memetic Algorithm for whole test suite generation. *J. Syst. Softw.* **2015**, *103*, 311–327. [CrossRef]
17. Panichella, A.; Kifetew, F.M.; Tonella, P. A large scale empirical comparison of state-of-the-art search-based test case generators. *Inf. Softw. Technol.* **2018**, *104*, 236–256. [CrossRef]
18. Fraser, G.; Arcuri, A. A Large-Scale Evaluation of Automated Unit Test Generation Using EvoSuite. *ACM Trans. Softw. Eng. Methodo* **2014**, *24*, 1–42. [CrossRef]
19. Wang, R.; Sato, Y.; Liu, S. Mutated Specification-Based Test Data Generation with a Genetic Algorithm. *Mathematics* **2021**, *9*, 331. [CrossRef]
20. Rani, S.; Suri, B.; Goyal, R. On the Effectiveness of Using Elitist Genetic Algorithm in Mutation Testing. *Symmetry* **2019**, *11*, 1145. [CrossRef]
21. Tian, T.; Gong, D.; Kuo, F.C.; Liu, H. Genetic algorithm based test data generation for MPI parallel programs with blocking communication. *J. Syst. Softw.* **2019**, *155*, 130–144. [CrossRef]
22. Campos, J.; Ge, Y.; Albunian, N.; Fraser, G.; Eler, M.; Arcuri, A. An empirical evaluation of evolutionary algorithms for unit test suite generation. *Inf. Softw. Technol.* **2018**, *104*, 207–235. [CrossRef]
23. Fraser, G.; Zeller, A. Mutation-Driven Generation of Unit Tests and Oracles. *IEEE Trans. Softw. Eng.* **2012**, *38*, 278–292. [CrossRef]
24. Fraser, G.; Arcuri, A. Whole test suite generation. *IEEE Softw. Eng.* **2012**, *39*, 276–291. [CrossRef]
25. Aleti, A.; Moser, I.; Grunske, L. Analysing the fitness landscape of search-based software testing problems. *Autom. Softw. Eng.* **2017**, *24*, 603–621. [CrossRef]
26. Whitley, D. Next Generation Genetic Algorithms: A User's Guide and Tutorial. In *Handbook of Metaheuristics*, 3rd ed.; Gendreau, M., Potvin, J.Y., Eds.; Springer: Cham, Switzerland, 2019; Volume 272, pp. 245–274. [CrossRef]
27. Sundar, S. A Steady-State Genetic Algorithm for the Dominating Tree Problem. In Proceedings of the 10th International Conference on Simulated Evolution and Learning, Dunedin, New Zealand, 15–18 December 2014; pp. 48–57. [CrossRef]
28. Agapie, A.; Wright, A.H. Theoretical analysis of steady state genetic algorithms. *Appl. Math.* **2014**, *59*, 509–525. [CrossRef]
29. Muhlenbein, H.; Schlierkamp-Voosen, D. Predictive Models for the Breeder Genetic Algorithm I. Continuous Parameter Optimization. *Evol. Comput.* **1996**, *1*, 25–49. [CrossRef]
30. Mühlenbein, H.; Schlierkamp-Voosen, D. The Science of Breeding and Its Application to the Breeder Genetic Algorithm. *Evol. Comput.* **1994**, *1*, 335–360. [CrossRef]
31. Dorronsoro, B.; Alba, E. A Simple Cellular Genetic Algorithm for Continuous Optimization. In Proceedings of the IEEE International Conference on Evolutionary Computation, Vancouver, BC, Canada, 16–21 July 2006; pp. 2838–2844. [CrossRef]
32. Pedemonte, M.; Panizo-LLedot, A.; Bello-Orgaz, G.; Camacho, D. Exploring Multi-objective Cellular Genetic Algorithms in Community Detection Problems. In *Intelligent Data Engineering and Automated Learning*; Analide, C., Novais, P., Camacho, D., Yin, H., Eds.; Springer: Cham, Switzerland, 2020; Volume 12490, pp. 223–235. [CrossRef]
33. Whitley, D. A genetic algorithm tutorial. *Stat. Comput.* **1994**, *4*, 65–85. [CrossRef]
34. Lam, A.Y.S.; Li, V.O.K. Chemical Reaction Optimization: A tutorial. *Memetic. Comp.* **2012**, *4*, 3–17. [CrossRef]
35. Marrison, C.I.; Stengel, R.F. The use of random search and genetic algorithms to optimize stochastic robustness functions. In Proceedings of the 1994 American Control Conference, Baltimore, MD, USA, 29 June–1 July 1994; pp. 1484–1489. [CrossRef]
36. Zabinsky, Z.B. Random search algorithms. In *Wiley Encyclopedia of Operations Research and Management Science*; Cochran, J.J., Cox, L.A., Keskinocak, P., Kharoufeh, J.P., Smith, J.C., Eds.; John Wiley & Sons: New York, NY, USA, 2010; pp. 1–13. [CrossRef]
37. Rojas, J.M.; Fraser, G.; Arcuri, A. Seeding strategies in search-based unit test generation. *Softw. Test. Verif. Reliab.* **2016**, *26*, 366–401. [CrossRef]
38. Panichella, A.; Kifetew, F.M.; Tonella, P. Automated Test Case Generation as a Many-Objective Optimisation Problem with Dynamic Selection of the Targets. *IEEE Softw. Eng.* **2016**, *44*, 122–158. [CrossRef]
39. Grano, G.; Palomba, F.; Nucci, D.D.; Lucia, A.D.; Gall, H.C. Scented since the beginning: On the diffuseness of test smells in automatically generated test code. *J. Syst. Softw.* **2019**, *156*, 312–327. [CrossRef]
40. Ma, P.; Cheng, H.; Zhang, J.; Xuan, J. Can This Fault Be Detected: A Study on Fault Detection via Automated Test Generation. *J. Syst. Softw.* **2020**, *170*, 110769. [CrossRef]
41. Fraser, G. A Tutorial on Using and Extending the EvoSuite Search-Based Test Generator. In Proceedings of the 10th International Symposium, Montpellier, France, 8–9 September 2018; pp. 106–130. [CrossRef]

42. Hansen, N.; Auger, A.; Finck, S.; Ros, R. Real-Parameter Black-Box Optimization Benchmarking 2010: Experimental Setup. 2010. Available online: https://hal.inria.fr/inria-00462481 (accessed on 31 May 2021).
43. Arcuri, A.; Fraser, G. Parameter tuning or default values? An empirical investigation in search-based software engineering. *Empir. Softw. Eng.* **2013**, *18*, 594–623. [CrossRef]
44. Črepinšek, M.; Liu, S.; Mernik, M. Replication and comparison of computational experiments in applied evolutionary computing: Common pitfalls and guidelines to avoid them. *Appl. Soft. Comput.* **2014**, *19*, 161–170. [CrossRef]
45. Aston, E.; Channon, A.; Belavkin, R.V.; Gifford, D.R.; Krašovec, R.; Knight, C.G. Critical Mutation Rate has an Exponential Dependence on Population Size for Eukaryotic-length Genomes with Crossover. *Sci. Rep.* **2017**, *7*, 1–12. [CrossRef] [PubMed]
46. Deb, K.; Deb, D. Analysing mutation schemes for real-parameter genetic algorithms. *Int. J. Artif. Intell. Soft Comput.* **2014**, *4*, 1–28. [CrossRef]
47. Jia, Y.; Merayo, M.; Harman, M. Introduction to the special issue on Mutation Testing. *Softw. Test. Verif. Reliab.* **2015**, *25*, 461–463. [CrossRef]
48. Luo, Q.; Moran, K.; Poshyvanyk, D.; Penta, M.D. Assessing Test Case Prioritization on Real Faults and Mutants. In Proceedings of the IEEE International Conference on Software Maintenance and Evolution, Madrid, Spain, 23–29 September 2018; pp. 240–251. [CrossRef]
49. Ammann, P.; Offutt, J. *Introduction to Software Testing*, 2nd ed.; Cambridge University Press: New York, NY, USA, 2016; pp. 18–19.
50. Fraser, G.; Arcuri, A. Achieving scalable mutation-based generation of whole test suites. *Empir. Softw. Eng.* **2015**, *20*, 783–812. [CrossRef]

Article

Deep Cross-Project Software Reliability Growth Model Using Project Similarity-Based Clustering

Kyawt Kyawt San [1], Hironori Washizaki [1,*], Yoshiaki Fukazawa [1], Kiyoshi Honda [2], Masahiro Taga [3] and Akira Matsuzaki [3]

[1] Department of Computer Science and Engineering, Waseda University, Shinjuku-ku, Tokyo 169-8555, Japan; kks@fuji.waseda.jp (K.K.S.); fukazawa@waseda.jp (Y.F.)
[2] Department of Information Systems, Osaka Institute of Technology, Hirakata City, Osaka 573-0196, Japan; kiyoshi.honda@oit.ac.jp
[3] e-Seikatsu Co., Ltd., Minato-ku, Tokyo 106-0047, Japan; masahiro.taga@e-seikatsu.co.jp (M.T.); akira.matsuzaki@e-seikatsu.co.jp (A.M.)
* Correspondence: washizaki@waseda.jp

Abstract: Software reliability is an essential characteristic for ensuring the qualities of software products. Predicting the potential number of bugs from the beginning of a development project allows practitioners to make the appropriate decisions regarding testing activities. In the initial development phases, applying traditional software reliability growth models (SRGMs) with limited past data does not always provide reliable prediction result for decision making. To overcome this, herein, we propose a new software reliability modeling method called a deep cross-project software reliability growth model (DC-SRGM). DC-SRGM is a cross-project prediction method that uses features of previous projects' data through project similarity. Specifically, the proposed method applies cluster-based project selection for the training data source and modeling by a deep learning method. Experiments involving 15 real datasets from a company and 11 open source software datasets show that DC-SRGM can more precisely describe the reliability of ongoing development projects than existing traditional SRGMs and the LSTM model.

Keywords: software reliability; deep learning; long short-term memory; project similarity and clustering; cross-project prediction

Citation: San, K.K.; Washizaki, H.; Fukazawa, Y.; Honda, K.; Taga, M.; Matsuzaki, A. Deep Cross-Project Software Reliability Growth Model Using Project Similarity-Based Clustering. *Mathematics* **2021**, *9*, 2945. https://doi.org/10.3390/math9222945

Academic Editors: Tadashi Dohi and Shaoying Liu

Received: 16 October 2021
Accepted: 10 November 2021
Published: 18 November 2021

1. Introduction

Reliability is one of the most significant attributes in enhancing the quality of the product in the software development process [1–3]. Assessing software reliability is vital to delivering a failure-free software system. Despite the enormous amount of testing, a number of software defects always occur in the product [4]. Software Reliability Growth Models (SRGMs) express the number of potential errors or defects that might happen in the future by analyzing past data, such as the cumulative number of errors, test cases, error rate, and detection time [5]. Therefore, the application of SRGMs helps to optimize resource planning and achieve highly reliable systems.

SRGMs are not always a reliable indicator in evaluating the situation of an ongoing software project and may even lead to an incorrect plan for testing resources [6]. New ongoing projects often do not have enough past data, which are needed in SRGM model fitting. In most studies, SRGM model fitting relies on past data to predict the future for the same project. Cross-project prediction is feasible in such cases requiring past data by applying other projects. However, if a source project is dissimilar to the target project, it affects prediction performance and leads to unstable future prediction results. One challenge in the cross-project prediction is that the distribution of the source and target project usually differ significantly [7,8].

To adopt a more reliable cross-project method of software reliability growth modeling while eliminating the unrelated data from all source projects for each target project, this study introduces a new SRGM method which can be utilized at the beginning stage of ongoing projects by processing only the project data with the most common features of the target project. For a target project with an insufficient amount of data, this method acquires the required information and features from similar projects to use in building the model. More specifically, a clustering method, k-means, is applied according to the features of projects such as the correlation of datasets and the number of bugs to create a new training data source. According to the identified clusters, the included datasets are combined. Prediction modeling is performed by a deep long short-term memory (LSTM) model using the merged dataset.

The goals of the study are to:

- Identify the correlation among projects by bug occurrence patterns and the same attributes of the projects.
- Determine groups of similar projects from a defect prediction viewpoint.
- Adopt a new approach for SRGM for the initial or ongoing stage of software development projects.

Although the idea of taking previous similar projects as a basis for the prediction of errors is common to cross-project prediction methods, our method has a novelty in using deep learning in combination with cluster-based project selection.

Here, we apply our proposed method, named Deep Cross-Project Software Reliability Growth Model (DC-SRGM), to 15 cloud service development projects of a company, e-Seikatsu, and 11 open source software (OSS) projects. Then we compare the performance of DC-SRGM with traditional models and the deep learning LSTM models. In our case study, DC-SRGM achieves the best scores in most cases. Hence, it can be regarded as an effective SRGM capable of improving deep learning LSTM models. Additionally, it significantly outperforms conventional SRGMs. Therefore, the DC-SRGM method allows software developers and managers to understand project situations in an ongoing stage with limited historical data.

The contributions of this work are as follows:

- A new SRGM method that uses a combination of deep learning and a cluster-based project selection method.
- Experimental comparison to two different models using 15 empirical projects and 11 open source projects to verify the prediction accuracy of the proposed model compared with two other models.
- Analysis of effective metrics, clustering factors, and suitable time to create reliability growth models.

The rest of the paper is organized as follows. Section 2 reviews the background and the related works. Section 3 presents the proposed DC-SRGM. Section 4 explains the experimental setup, data, and design. Section 5 reports the results and evaluations. Section 6 describes the threats to validity. Finally, Section 7 provides conclusions and future work.

This paper is extended from our previous study [9]. We conducted additional experiments to investigate the impact of clustering factors, another similarity score using dynamic time warping, applied at different time points of ongoing projects and predictions across organizations.

2. Background and Related Work

Studies have been conducted on SRGMs and their adoption for current project prediction as well as cross-project prediction. In this section, we firstly show related works on SRGMs in general. Secondly, we explain the current project prediction as the context of this study. Finally, we present related works on cross-project prediction and their limitations to motivate our method.

2.1. Software Reliability Growth Model

The widely used Software Reliability Models (SRMs) [10] are Software Reliability Growth Models (SRGMs) that are used for modeling the failure or defect arrival pattern [11] based on failure data regardless of the source code characteristics. Many SRGMs have been studied to measure the failure process. These models require external parameters to be estimated by the least-squares or maximum likelihood estimation to build the relevant parameters [1]. N. Ullah et al. [11] studied different SRGMs using defect data in industrial and open source software and performed a comparative analysis between them. To evaluate the qualities of development projects monitored by SRGM applications, K. Honda et al. [6] analyzed the tendencies for unstable situations in the results of different SRGM models. K. Okumoto et al. [4] applied SRGM in developing a reliability assessment automated tool.

SRGM processes are usually performed with data from testing. Detecting and resolving failures or defects would enable software systems to be more stable and reliable. To understand the underlying condition of the system, such processes are often described using a mathematical expression, usually based on parameters such as the number of failures or failure density [12]. Studies report many ways to create models based on the model's assumption of failure occurrence patterns.

Similar to previous studies [6,13], we focused on the Logistic model, which is the most suitable concerning fitness for the collected experimental datasets. We employed the model using the number of detected bugs and detected time. The Logistic model can be expressed as

$$N(t) = \frac{N_{max}}{1 + exp{-A(t-B)}} \tag{1}$$

where $N(t)$ is the number of bugs detected by time t. The parameters, N_{max}, A and B were estimated using Nonlinear Least Square Regression (NLR) function [6].

2.2. Current Project Prediction

SRGMs can be applied to current ongoing projects to allow project managers or other stakeholders to assess the release readiness and consider optimal testing resource allocations. Current project prediction applies existing project data as a training source and then makes predictions for future days. Therefore, prediction models in this study are created using only 50 percent data points of the target project's existing data. Then these models are used to predict the subsequent days for the rest of the data points. Each data point refers to the cumulative number of bugs that have been reported by the corresponding time. We considered an RNN-based LSTM as well as the Logistic model as prediction models for current project prediction.

A Recurrent Neural Network (RNN) connects neurons with one or more feedback loops, which is capable of modeling sequential data in sequence recognition and prediction [14,15] because it includes high-dimensional hidden states with nonlinear dynamics. These hidden states perform as the memory of the network, and its current state is conditioned on its previous one [16]. A simple RNN structure has an input layer, recurrent hidden and output layers, which accept the input sequences through time. Consequently, RNNs are capable of storing, remembering, and processing data from past periods, which enables the RNN to elucidate sequential dependencies [14]. However, it comes with the challenges that the memory produced from the recurrent connections may be limited to learning long-range sequential data.

An RNN-based LSTM network is designed to solve that problem. The LSTMs are capable of bridging very long-time lags with an advanced RNN architecture, with self-connected units [14,17,18]. The inputs and outputs of hidden units are controlled by gates, which maintain the extracted features from previous time steps [14,18]. LSTM contains an input gate, forget gate, cell state, output gate, and output response. The input gate and forget gate manage the information flow into and out of the cell, respectively. The output gate decides what information is passed to the output cell. The memory cell has a self-connected recurrent edge of weight, ensuring that the gradient can pass across many time steps without exploding [19]. The advantage of an LSTM model is it can keep information over long periods by removing or adding information to the state.

For current project prediction, traditional SRGMs such as the Logistic model cannot realize underlying project conditions if they are applied at the initial stage with limited historical data. As a result of the preliminary experiment using one of the industrial projects of the company, we confirmed that the Logistic model did not work well, as shown in Figure 1a.

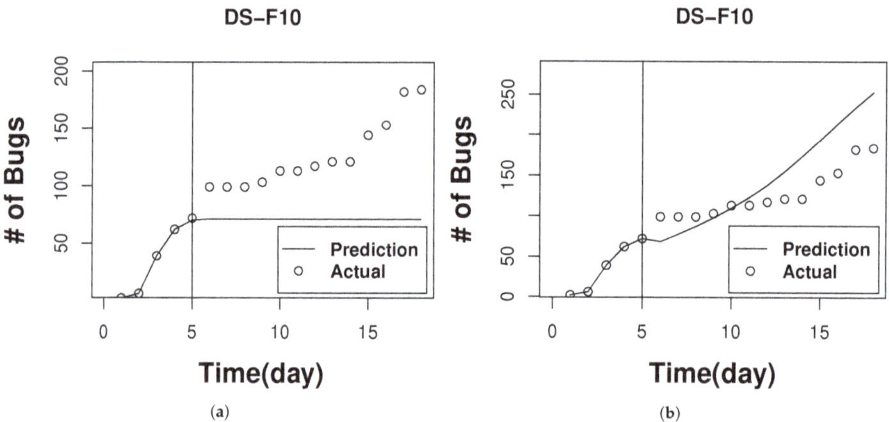

Figure 1. Applying the Logistic model and LSTM model on day 5 for ongoing project F10. (**a**) Logistic Model. (**b**) LSTM Model.

Therefore, we applied an advanced technique LSTM model with the same amount of data during model construction. At each step, the input layer receives a vector of the number of bugs and passes the data to hidden layers, with four LSTM neurons in each. An output layer generates a single output that gives the predictions for the next time step. Although improvements occur (Figure 1b), the LSTM model does not always provide accurate results at the beginning in cases with very little data that has different reliability growth patterns.

2.3. Cross-Project Prediction

Ongoing projects have limited data for use as historical defect data. One alternative is to employ a cross-project prediction, which utilizes external projects to construct a prediction model for the current project [3,20]. In the literature, cross-project prediction is a very well-studied subject by utilizing project data of different organizations. K. Honda et al. [5] proposed a cross-project SRGM model to compare software products within the same company. However, they did not implement cross-project applications of SRGMs for ongoing projects. Remarkably, there are a few studies in SRGM modeling using cross-project data.

The mismatch between the randomly selected source projects and the target project affects the cross-project prediction performance and creates unstable results. Earlier studies in [21,22] implied that usage of cross-company data without any modification degrades the accuracy of prediction models. Irrelevant source project data may decrease the efficiency of the cross-project prediction model. To overcome this issue, C. Liu et al. [23] considered the Chidamber and Kemerer (CK) metric suite [24] and size metrics to implement a cross-project model, which detects change-proneness class files. Source projects were selected by the best-matched distribution.

To choose appropriate training data, X. Zhang et al. [7] investigated the efficiencies of nine different relevancy filtering methods. A cross-project defect prediction model was constructed with a random forest classifier on the PROMISE repository. M. Jureczko et al. [25]

also studied a similar project clustering approach using k-means and hierarchical clustering by a stepwise linear regression in the PROMISE data repository. They confirmed that k-means could successfully identify similar project clusters from a defect prediction viewpoint. The above studies with cross-project prediction focused on the clustering or filtering approaches and employed a specific classifier to label defective modules or classes. None of these methods dealt with the observed time series failure data.

J. Wang et al. [1] proposed an encoder–decoder-based deep learning model RNN and performed analysis between non-parameter models and parameter models. They applied the cumulative executive time and the accumulated number of defects. However, a cross-project prediction model was not implemented.

In addition, most of the past studies have not investigated sufficiently in SRGMs modeling that utilizes cross-project prediction. This study conducted projects reliability assessment by SRGM modeling with a sophisticated method rather than traditional approaches using cross-project data, which were carefully selected with a project similarity method.

In earlier studies, cross-project predictions models have been utilized to resolve the requirement of huge historical data. However, one challenge in the cross-project prediction is that the distribution of the source and target project usually differ significantly [7,8]. If the training data contain all the source project data, a poor prediction quality can result. Ideally, one defect prediction model should work well for all projects that belong to a group [25].

3. Deep Cross-Project Software Reliability Model

To eliminate the unrelated data from all source projects for each target project, we propose the Deep Cross-Project Software Reliability Growth Model (DC-SRGM), which processes only the project data with the most common features of the target project. DC-SRGM utilizes a cross-project prediction method that uses other projects; data as a training data source with the advantage of LSTM modeling for time series data.

Figure 2 overviews the proposed model DC-SRGM. It includes three processes, similarity scoring, clustering-based project selection, and prediction modeling. Figure 3 details the process of selecting the most appropriate projects that share common characteristics with the target project. The core feature of DC-SRGM is that it filters irrelevant projects from training data sources and only selects projects with the most common characteristics as the target project.

3.1. Similarity Scoring

Each project has its own features, such as the project size and the number of bugs [3]. Identifying similarities among the datasets is the basis used to eliminate differences between the data across projects. Otherwise, inappropriate source data may be chosen. To exclude irrelevant projects from training data sources, the clustering factors include project similarity scores. In DC-SRGM, cross-correlation is applied to identify the correlation of projects against the target project. Furthermore, Dynamic Time Warping (DTW) is considered as a comparative similarity measurement.

Cross-correlation: A measure of the similarity among the projects by aligning two time series. The coefficients identify the connections between different time series of datasets [26]. In given time series datasets for cumulative numbers of bugs, each dataset is considered as one time series. The cross-correlation function of each pair taken from two datasets is calculated.

Dynamic Time Warping (DTW): A well-known technique to measure the optimal alignment or similarity between time series sequences of different lengths concerning the shape of information and patterns [27]. It calculates the minimal distance to observe dissimilarities among the datasets according to the scale and distribution of the project. Here, it is used to compare the performances of DC-SRGM with different similarity measurements.

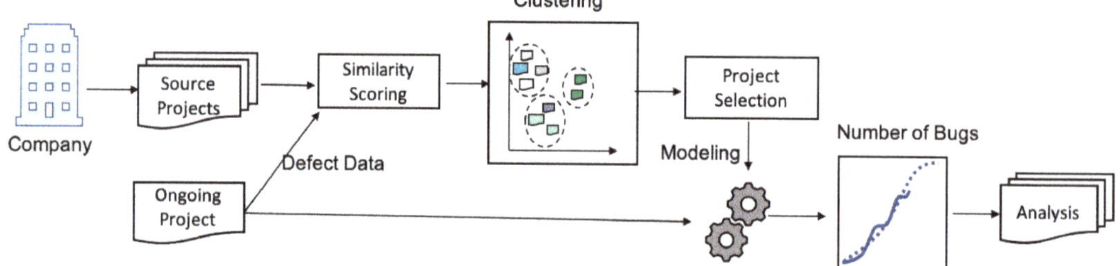

Figure 2. Overview of the DC-SRGM model.

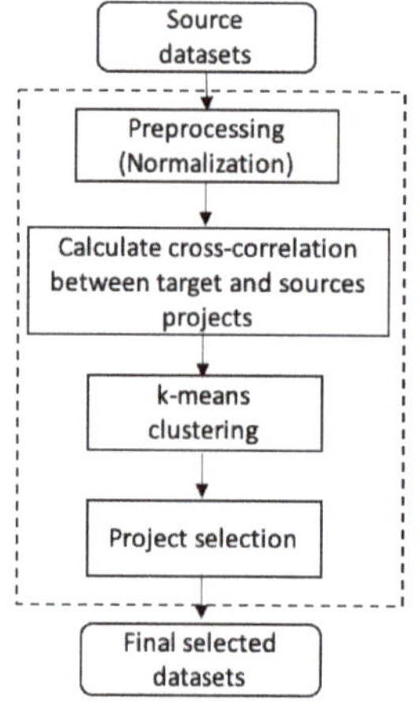

Figure 3. Project selection process.

3.2. Project Clustering

Project clustering groups similar projects together using the k-means algorithm with the following clustering factors:

- Cross-correlation similarity scores between the number of bug growth patterns;
- Normalized values of the maximum number of bugs;
- Normalized values of the maximum number of days.

Clustering results usually indicate three groups. Each group includes projects with characteristics similar to the target project according to the cross-correlation scores and the distribution of the projects, such as the number of bugs and the number of days.

3.3. Selection

To investigate whether a cluster for SRGM modeling exists, a prediction model is created by the datasets from each same cluster. According to our initial analysis, the cluster from the number of bugs prediction viewpoint exists only in the group with the target project itself. Each group shares the most common attributes of the projects, such as failure occurrence patterns, and only those within the same group are appropriate to model for each project. In addition, only a cluster that belongs to the target project is selected. All the containing projects in that cluster are combined, but the target project itself is excluded when merging the data. Eventually, the merged group of projects eliminating the irrelevant training data is used for model training.

3.4. Training and Prediction

To employ the LSTM model, the input to the network at each time step is a vector of the number of bugs, and the single output is the number of bugs for the next step. Figure 4 shows the process of LSTM training at each time step. Because the ranges of the input values can vary, the values of bugs are scaled into the range of zero to one. By considering the prediction process as a time series, the input layer receives the values of the number of bugs for nine days, and the single output node produces the prediction for the next day. By shifting by one in each step, the model is trained to the maximum days of the training dataset. The model is trained with 300 epochs because the results are similar to those using 500 epochs. The stochastic gradient descent method is employed using the mean squared error loss function.

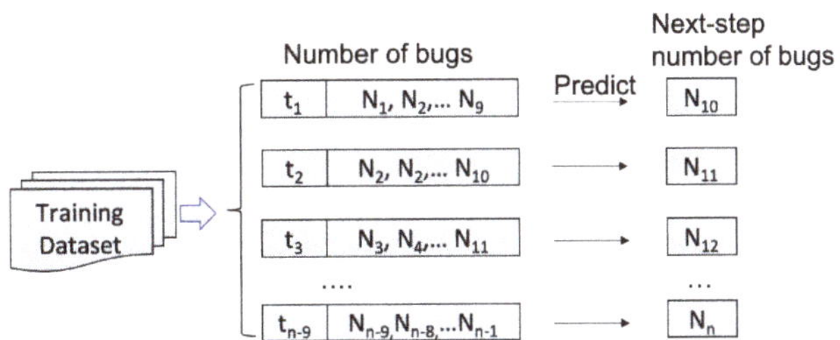

Figure 4. Model training process.

For a target project prediction, the trained model uses fifty percent of the data points of its project to predict the following fifty percent of the data points because we considered a project to be ongoing.

4. Experiment Methodology

Experiments were conducted to answer the following research questions RQ1–RQ5. Figure 3 overviews the evaluation design for each research question. RQ1 compares two different types of current project prediction: LSTM and Logistic models using only the first half of the current project data to predict the second half of the same project, and DC-SRGM using past projects' data for training and the first half of the current project data as input for prediction of the second half of the same project. Furthermore, RQ2–RQ5 address only DC-SRGM using past projects' data for training and the first half of the current project data as input for prediction of the second half with different settings. We explained this distinction as follows in Section 4.

- **RQ1: Is DC-SRGM more effective in ongoing projects than other models?**
  This question evaluates the effectiveness of the DC-SRGM model compared to the Logistic model and LSTM model (Figure 5, RQ1). That is, does the proposed method correctly describe ongoing projects' reliability despite insufficient data to apply in a prediction model? Specifically, we used a case study to compare the performance of different models for 15 industrial projects with a duration longer than 14 days and 11 OSS projects. Because the target is an ongoing project, the first half of its data is used to obtain the similarity scores as well as for input data. Then the models are used to predict the second half of the target data. The results should reveal whether cluster-based similar project selection improves the LSTM model performance relative to that of a traditional Logistic model.
- **RQ2: What factors influence the performance of DC-SRGM?**
  This question examines the performance of DC-SRGM upon applying a different clustering factor to the similarity scores of the projects. Domain experts indicated that the projects are clustered according to the project domain type, and the same types of projects are applied as the training source projects for modeling. We compared the prediction results with similarity scores in terms of AE values to reveal how different

clustering factors influence the prediction results. This RQ helps to assess whether DC-SRGM can be utilized when the same type of other projects is not available.

- **RQ3: Do different similarity measurements affect the prediction quality of DC-SRGM?**
  This question investigates the performances of DC-SRGM based on cross-correlation and Dynamic Time Warping (DTW) to determine the impact of the similarity measurement techniques on the model (Figure 3, RQ3). We analyzed the effect of the similarity measurement on DC-SRGM by comparing the performance of two methods in terms of AE values by model. In general, AE > 0.10 indicated a satisfactory model.
- **RQ4: Can DC-SRGM precisely describe an ongoing project's status?**
  This question explores the relation of the amount of utilized project data and the model's prediction capability for new initial stage projects. It aims to determine if there is a suitable time for managers to begin to evaluate projects with acceptable accuracy by DC-SRGM. Therefore, we applied the DC-SRGM model at different time points in ongoing projects to assess the prediction performance and the impact of the target project's past data usage.
- **RQ5: Can DC-SRGM trained with OSS datasets indicate the industrial projects' situation?**
  Even if previous source projects' data are unavailable, this question evaluates whether DC-SRGM created with OSS datasets can predict the conditions for an industrial project. We used open source datasets to create DC-SRGM with the same setting and procedure performed on industrial datasets. Then the results are compared to those predicted using industrial datasets.

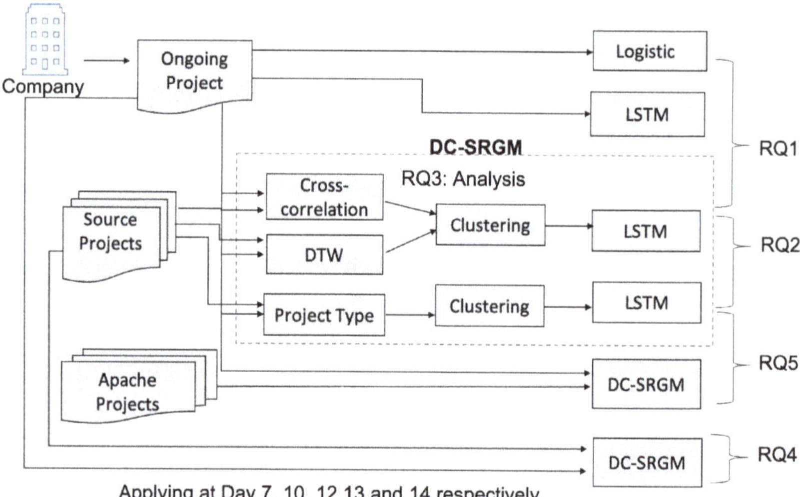

Figure 5. Overview of the experiment design (Research Questions).

4.1. Initial Analysis

To identify similar groups, the initial analysis used cosine similarity and DTW. However, the similarity measurements and the prediction performance were not correlated. Therefore, the k-means clustering method was applied. Then the optimum number of clusters, k, was determined by the Elbow method. Initially, the clustering produced biased results on the number of days. After adding cross-correlation coefficients in clustering factors, projects with similar characteristics were classified well.

4.2. Performance Measure

We evaluated the prediction capability in terms of accuracy by considering the ratio between the difference in the error values and the prediction over a time period, namely average error (AE) [1]. AE is defined as:

$$AE = \frac{1}{n}\sum_{i=1}^{n}\left|\frac{U_{ij} - D_j}{D_j}\right| \tag{2}$$

where U_{ij} denotes the cumulative number of predicted bugs by time t_j, D_j represents the cumulative number of detected bugs by time t_j, and n is the project size [1]. A value closer to zero indicates a better prediction accuracy.

We employed the Friedman test with the Nemenyi test as a post hoc test to evaluate the statistically significant difference in performances between DC-SRGM and the baseline methods because it is better suited for non-normal distributions.

4.3. Data Collection

The datasets were from 15 industrial projects' data with a duration longer than 14 days from real cloud services development projects. Each dataset consisted of the time series number of bugs per testing day. The domains of the projects were property information management, customer relationship management, contract management, money receipt/payment management, and content management systems [6]. To derive more generalized results, we aimed to include as many software projects as possible. Thus, 11 datasets from Apache open source projects were also collected from apache.org using a bug tracking system, JIRA, to study reliability growth modeling. All the issues reported in two minor versions, which were declared as bugs or defects excluding any other categories, were collected for each project. Tables 1 and 2 describe details of each dataset.

Table 1. Industrial project details.

Project	Days	# of Bugs
F01	19	91
F02	22	137
F03	12	47
F04	17	259
F05	19	188
F06	26	263
F07	15	146
F08	17	97
F09	16	99
F10	18	184
F11	14	74
F12	25	351
F13	22	187
F14	34	331
F15	18	752

Table 2. OSS projects details.

Project	Days	# of Bugs	Studied Version	
Camel	36	32	2.15.1	2.15.2
Ignite	48	149	2.5	2.6
Jclouds	175	25	2.1.0	2.1.1
Karaf	56	64	4.1	4.2
Lucene	91	6	6.6.0	6.6.1
Maven	160	22	3.5.1	3.5.2
Shiro	30	6	1.3.0	1.3.1
Spark	99	185	2.3.1	2.3.2
Syncope	80	36	2.0.2	2.0.3
Tez	120	27	0.6.0	0.6.1
Zookeeper	86	14	3.4.12	3.4.13

5. Experiment Results and Discussions

5.1. Project Clustering Result of Industrial Datasets

In terms of the application of DC-SRGM targeting the industrial datasets, Table 3 summarizes the clustering factors, which are the cross-correlation similarity score, the maximum number of bugs, and maximum number of days. Table 4 summarizes the project clustering results in the industrial datasets. The number of projects in each group differs slightly based on the similarity scores between the candidate target and source datasets for each target dataset. Table 4 details each cluster, including the range of the number of bugs, number of days, and the overall number of bugs of the included projects. "Grad" indicates a gradual increase in the detected number of bugs. "Expo" refers to an exponential rise in bug growth. "Expo and Grad" denotes both an exponential and gradual increase in the number of bugs.

Table 3. Summary of the clustering factors.

Similarity	Max Bugs	Max Days
0∼1	47∼752	14∼36

Table 4. Summary of the clustering results. Projects are generally clustered into three groups according to similarity scores and the project scales. Grad, Expo and Grad, and Expo indicate the growth of the number of bugs is gradually increasing, exponentially increasing and gradually increasing, and exponentially increasing.

Cluster	Clustered Projects	Max Bugs	Max Days	Growth	Type
C1	F01, F02, F04, F05, F07, F08, F09, F10, F11	91∼188	14∼22	Grad	Similarity
C2	F12, F15	540∼752	18∼24	Expo	# Bugs
C3	F03, F06, F13, F14	47∼331	22∼36	Expo and Grad	# Days

Table 5 shows the clustering results by project, where "Cluster" represents the cluster containing the target project. Projects applied for model building are presented in Table 4 according to the expressed cluster name. "Actual Growth" describes the bug growth of each project. "Prediction Result" shows the growth of the number of bugs by the prediction model created by clustered projects.

Table 5. Summary of the clustering results by project. Grad, Expo and Grad, Expo, and Const indicate that the number of bugs is gradually increasing, exponentially increasing and gradually increasing, exponentially increasing, and constantly increasing.

Project	Cluster	Actual Growth	Prediction Result
F01	C1	Grad	Grad
F02	C1	Grad	Grad
F03	C3	Grad	Grad
F04	C1	Grad	Grad
F05	C1	Grad	Grad
F06	C3	Expo	Expo
F07	C1	Grad and Expo	Grad
F08	C2	Grad	Grad
F09	C3	Expo and Grad	Grad
F10	C4	Grad	Grad
F11	C5	Grad	Grad
F12	C2	Expo	Expo and Grad
F13	C3	Expo and Grad	Expo and Grad
F14	C3	Expo and Grad	Const
F15	C2	Expo	Expo

In this study, since the maximum number of bugs, the maximum number of days, and cross-correlation scores for the connections between projects are used as clustering factors, the obtained clusters are basically three main groups depending on these factors, their similar attributes, and data patterns. The first cluster denotes a group with moderate to strong correlation scores. The second cluster is influenced by the exponential growth of the number of bugs. The third cluster is grouped by the distribution of the number of days of the projects.

For example, F01 and F02 projects have the same distribution scales and a moderate cross-correlation score. Hence, they are grouped in the same cluster. On the other hand, the F12 project shows exponential growth for the number of bugs and a different data occurrence pattern. Building a model for the F01 project using F12 would overestimate the prediction result. Hence, DC-SRGM achieves better performance when applying it in the middle of the projects to build a model using a similar group of projects.

5.2. RQ1: Effectiveness of DC-SRGM

The experiments in RQ1 compared DC-SRGM to the Logistic and LSTM models. Tables 6 and 7 present the AE values of the three models for the industrial datasets and OSS datasets, respectively. Table 8 describes the results of the statistical test between DC-SRGM and the two other models. For the industrial datasets, DC-SRGM yielded the largest improvement. On average, it improved the AE by 24.6% and 50% compared to the LSTM and Logistic model, respectively.

Table 6. Comparison of DC-SRGM with the LSTM and Logistic models by the AE values. Bold denotes the best AE values. W/L is the number of datasets that each method is better and worse than. "# DS Threshold below 0.1" is the number of datasets for which each model's performance is lower than the threshold.

Project	DC-SRGM	LSTM	Logistic
F01	0.067	**0.040**	0.266
F02	**0.071**	0.080	0.146
F03	0.192	**0.130**	0.142
F04	**0.091**	0.260	0.377
F05	**0.075**	0.127	0.218
F06	**0.040**	0.090	0.211
F07	0.329	0.500	**0.146**
F08	**0.049**	0.104	0.187
F09	0.055	**0.048**	0.146
F10	**0.088**	0.121	0.214
F11	**0.068**	0.073	0.074
F12	**0.095**	0.161	0.359
F13	**0.211**	0.243	0.348
F14	0.107	**0.020**	0.183
F15	**0.126**	0.201	0.191
Average	**0.110**	0.146	0.220
Improved%	-	+24.6%	+50%
W/L	10/5	4/11	1/14
# DS Threshold below 0.1	10	6	1

Table 7. Prediction Accuracy of the models on OSS datasets by the AE values. Bold denotes the best AE Values. W/L is the number of datasets for which each method is better and worse than. "# DS Threshold below 0.1" is the number of datasets that each model's performance is lower than the threshold.

Project	DC-SRGM	LSTM	Logistic
Camel	**0.081**	0.099	0.440
Ignite	0.067	**0.063**	0.110
Jclouds	0.190	**0.029**	0.260
Karaf	**0.035**	0.105	0.830
Lucene	**0.270**	0.438	0.950
Maven	**0.120**	0.122	0.240
Shiro	**0.100**	0.139	0.110
Spark	0.201	**0.139**	0.190
Syncope	0.240	**0.128**	0.220
Tez	**0.037**	0.180	0.780
Zookeeper	**0.133**	0.190	0.140
Average	**0.134**	0.148	0.388
Improved%	-	+9.45%	+65.4%
W/L	7/4	4/7	0/11
# DS Threshold below 0.1	5	3	1

Table 8. Statistic results with the Nemenyi test for the effectiveness of DC-SRGM. * and ** denote that there were significant differences in the groups as the significance levels were 0.1 and 0.01, respectively.

	Models	*p*_Value
Industry	DC-SRGM and LSTM	0.0710 *
	DC-SRGM and Logistic	0.0045 **
OSS	DC-SRGM and LSTM	0.3657
	DC-SRGM and Logistic	0.0288 *

Table 6 compares the number of datasets where each model obtained better or worse (win or lose) scores across datasets. If a model achieved a score below the threshold (0.1), it was considered as an indicator of good accuracy. In most cases, DC-SRGM achieved better AE values. Figure 6a also expresses the median of AE values among the three models. The red line represents the threshold. The DC-SRGM model had lower AE values with a median below 0.1, implying a higher accuracy than the other two models. The LSTM model was close to the threshold, and the Logistic model showed the worst performance.

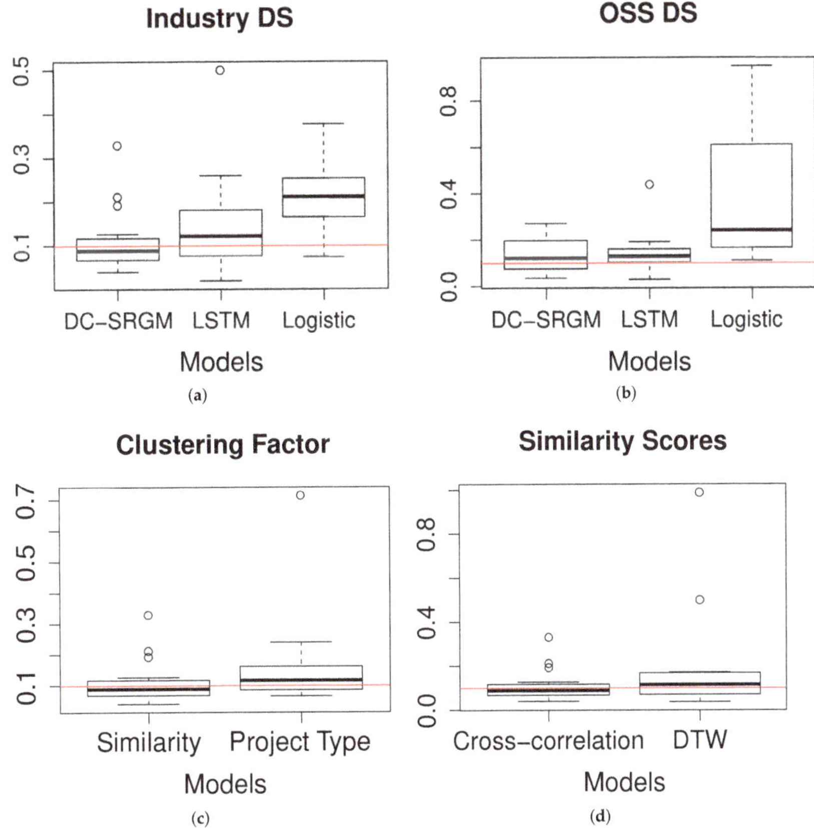

Figure 6. *Cont.*

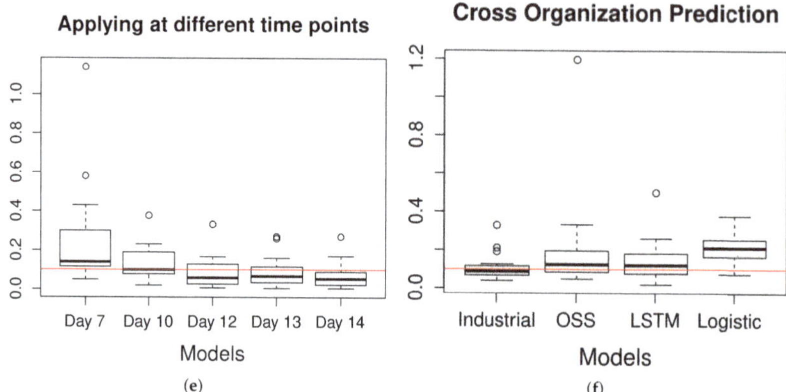

Figure 6. Comparison of the model prediction accuracy in terms of average error, AE. (**a**) Performance in industrial datasets (DS), (**b**) Performance in OSS datasets, (**c**) DC-SRGM based on project similarity and project domain type, (**d**) DC-SRGM based on cross-correlation and DTW, (**e**) DC-SRGM applied at the different number of days, and (**f**) DC-SRGM across organizations.

In the case of the OSS datasets (Figure 6b and Table 7), the results slightly differed, which is most likely due to the difference in the project nature between industrial and OSS projects. DC-SRGM achieved the best score. It showed 65.4% improvement compared to the Logistic model in terms of AE average and better scores in terms of W/L. However, the performance with the LSTM model did not pass the significant test, and its boxplot was bigger than the LSTM model. The LSTM model increased its accuracy in the OSS environment due to the larger amount of training data. OSS datasets have a different development environment and style; specifically, having a larger project size provides better accuracy for the LSTM model using the current project prediction method.

There are two exceptional cases where the proposed DC-SRGM was less accurate: F03 and F14 prediction. In the clustering result, the F03 project was grouped in the third cluster, which was grouped according to the number of days despite having a strong correlation with the projects in the first cluster. This impacted F03 modeling and is why DC-SRGM provided less accurate results than the LSTM and Logistic models. In terms of the F14 project, its domain differed from the other projects, and it had a long duration, according to the domain experts of these experimental projects.

Figure 7a–d plot the results when applying DC-SRGM, LSTM, and Logistic models to the F02, F03, F04, and F10 datasets at the middle of the projects, respectively. The predicted number of bugs by DC-SRGM described the potential number of bugs more correctly than the other two models. Hence, the industrial and OSS datasets results indicated that DC-SRGM outperformed LSTM and the Logistic model and improved the prediction accuracy when applied in an ongoing stage of industrial development. For OSS projects, DC-SRGM significantly outperformed the Logistic model, and, on average, DC-SRGM was better than LSTM. However, its performance slightly decreased in the industrial environment while the performances of the LSTM model increased.

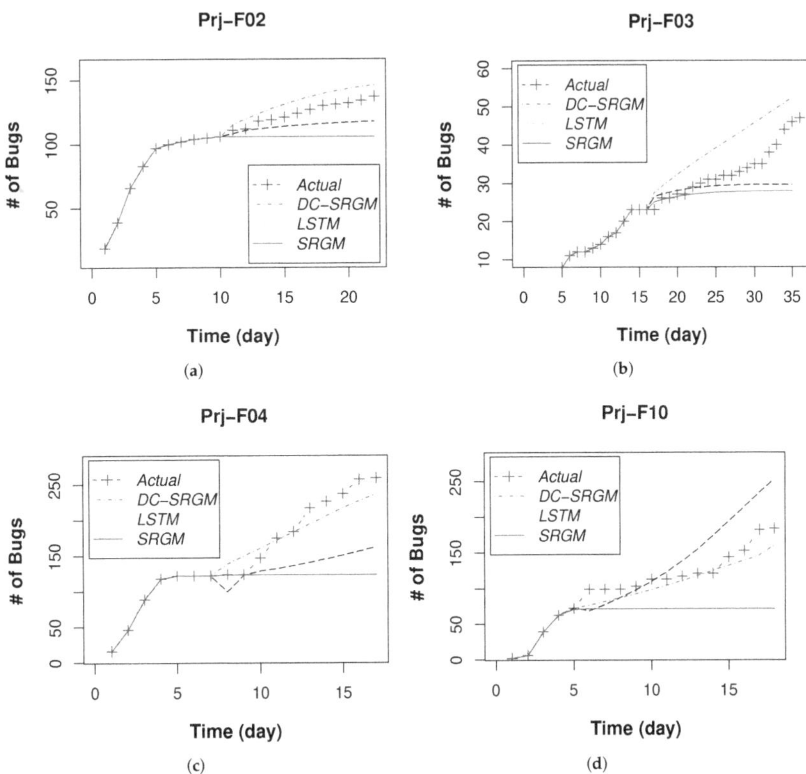

Figure 7. Predicted number of bugs at the middle of the projects. Actual, DC-SRGM, LSTM, SRGM represent the actual detected number of bugs, the prediction by DC-SRGM, the LSTM model, and the Logistic SRGM model, respectively. (**a**) Project F02, (**b**) F03, (**c**) F04, and (**d**) F10.

RQ1: Is DC-SRGM more effective in ongoing projects than other models?

The proposed DC SRGM outperforms the LSTM and Logistic models for most datasets as it has a lower mean AE value. The improvements are significant in industrial datasets. Hence, DC-SRGM is more effective in describing the future number of bugs correctly for ongoing software development projects.

5.3. RQ2: Impact of Clustering Factors on DC-SRGM

RQ2 examined the prediction accuracy of two different clustering factors on DC-SRGM. Two models were built. One used the project similarity score, a cross-correlation, and the other used the project domain type to identify important factors for modeling. Figure 6c shows boxplots for AE values from the predictions using the two different clustering factors. "Project Similarity" and "Project Domain Type" in Table 9 report the details of the AE values, where bold denotes the better result. Blank cells are projects which cannot be determined in the selected experiment datasets. The project similarity-based DC-SRGM obtained better scores in most cases, and the median was below the threshold.

Table 9. Comparison of the prediction accuracy of DC-SRGM using project similarity and project domain type as clustering factors. W/L is the number of datasets that each method is better and worse than. "# DS Threshold below 0.1" is the number of datasets for which each method's performance is lower than the threshold.

Project	Project Similarity	Project Domain Type
F01	**0.067**	0.074
F02	**0.071**	0.091
F03	0.192	**0.129**
F04	**0.091**	0.137
F05	0.075	–
F06	**0.040**	0.119
F07	**0.329**	0.714
F08	**0.049**	0.186
F09	**0.055**	0.113
F10	**0.088**	0.096
F11	0.068	**0.066**
F12	0.095	–
F13	**0.211**	0.239
F14	0.107	–
F15	0.126	**0.080**
Average	**0.110**	0.170
W/L	9/3	3/9
# DS Threshold below 0.1	10	7

On the other hand, the project domain type-based model was close to the threshold. Hence, project clustering by similarity scores affected the model's ability to obtain suitable project data to learn the number of bugs. Although the domain was the same, clustering by project domain type did not affect the model performance. There are irrelevant projects with very different growth patterns for bugs even though they are in the same domain. Therefore, DC-SRGM modeling should be performed using the project similarity scores as the priority rather than the project domain type.

RQ2: What factors influence the performance of DC-SRGM?

In most cases, DC-SRGM clustered by project similarity scores outperforms the model clustered by project domain type on AE values, indicating that project similarity is an important factor in the clustering process for good predictions results.

5.4. RQ3: Impact of Similarity Measurements on DC-SRGM

RQ3 compared the performances of DC-SRGM based on cross-correlation and DTW to assess the similarity measurement technique's impact and determine a better similarity measurement for DC-SRGM. Figure 6d shows boxplots for AE values of both methods. DC-SRGM based on the cross-correlation had lower AE values with a median below the threshold. On the other hand, the DTW-based model was close to the threshold, implying that cross-correlation shows a better performance. "Cross-correlation" and DTW in Table 10 represent details of the AE values, where bold denotes the better method. Across 15 datasets, although there is no obvious difference between the two methods in the number of datasets with the lower AE value, the cross-correlation-based model outperformed the DTW-based model on average and achieved a value lower than the threshold in more cases.

Table 10. Comparison of the prediction accuracy DC-SRGM using cross-correlation and DTW as similarity measures. W/L is the number of datasets that each method is better and worse than. "# DS Threshold below 0.1" is the number of datasets for which each method's performance is lower than the threshold.

Project	Cross-Correlation	DTW
F01	0.067	**0.037**
F02	0.071	**0.039**
F03	**0.192**	0.499
F04	0.091	**0.081**
F05	0.075	**0.048**
F06	**0.040**	0.170
F07	**0.329**	0.988
F08	**0.049**	0.166
F09	**0.055**	0.089
F10	**0.088**	0.115
F11	**0.068**	0.169
F12	**0.095**	0.115
F13	0.211	**0.165**
F14	0.107	**0.060**
F15	0.126	**0.089**
Average	**0.110**	0.188
W/L	8/7	7/8
# DS Threshold below 0.1	10	7

Clustering based on DTW could not always classify relevant datasets or eliminate the irrelevant datasets for the target project. One reason is that the DTW function returned the scores based on the shape of the dataset sequence, whereas cross-correlation returned the scores based on the value and pattern of the dataset. Another reason is that the cross-correlation scores can describe the correlation level, such as significant or non-significant. In DTW, it is difficult to identify the threshold in the variations of datasets. Therefore, changing the applied similarity measurement technique impacted the model performance. To identify similar project groups correctly, the cross-correlation technique is better suited for DC-SRGM.

RQ3: Do different similarity measurements affect the prediction quality of DC-SRGM?

Cross-correlation-based DC-SRGM achieves better accuracy than DTW. To enhance source project selection, cross-correlation is a better technique for DC-SRGM from the SRGM modeling viewpoint.

5.5. *RQ4: Impact of Applying DC-SRGM at Different Time Points*

To determine the impact of the amount of data from an ongoing project applied in DC-SRGM modeling, the experiment was conducted using the target datasets from industrial data on days 7, 10, 12, 13, and 14. The model's performances at different time points were compared to determine a suitable time frame to apply DC-SRGM in ongoing development stages. Table 11 shows the AE values of the models at each time point. Figure 6e compares the median of AE values at each prediction time point. Accurate results were not obtained when applying DC-SRGM on day 7 of ongoing projects, but a few projects had significant improvement upon using them on day 10. Applying the model on day 12 or later improved the AE values. Overall, the proposed method can identify the correct clusters and achieve stable results starting from day 12. Therefore, DC-SRGM can be applied to ongoing software development projects beginning on day 12.

Table 11. Comparison of DC-SRGM for different numbers of days. "# DS Threshold below 0.1" is the number of datasets for which each model's performance is lower than the threshold.

Project	Day 7	Day 10	Day 12	Day 13	Day 14
F01	0.070	0.078	0.072	0.060	0.060
F02	0.050	0.030	0.045	0.040	0.050
F03	0.580	0.377	0.167	0.160	0.170
F04	0.100	0.087	0.073	0.031	0.028
F05	0.130	0.070	0.029	0.024	0.020
F06	0.140	0.225	0.039	0.043	0.030
F07	1.140	0.780	0.333	0.270	0.140
F08	0.410	0.098	0.009	0.011	0.015
F09	0.160	0.111	0.007	0.005	0.005
F10	0.190	0.112	0.143	0.110	0.079
F11	0.140	0.020	0.006	0.007	0.007
F12	0.190	0.230	0.058	0.066	0.060
F13	0.430	0.190	0.025	0.260	0.270
F14	0.130	0.100	0.131	0.120	0.100
F15	0.080	0.190	0.125	0.087	0.050
Average	0.262	0.179	0.090	0.092	0.072
# DS Threshold below 0.1	4/15	7/15	10/15	10/15	12/15

RQ4: Can DC-SRGM precisely describe ongoing projects' status?

The model applied on day 12 of the ongoing projects provides a more stable and improved accuracy than the other models. Hence, managers can start using DC-SRGM on day 12 to describe the reliability of a project correctly.

5.6. RQ5: Predicting the Performance by Cross Organization Datasets

For RQ5, the experiment was designed to validate the effectiveness of the DC-SRGM model applied using cross-organization OSS datasets for predictions of industrial projects. DC-SRGM models trained by OSS datasets were used to predict the second half of the industrial datasets. The performance was compared with the results of models trained by industrial datasets.

Table 12 shows the AE values predicted utilizing industrial datasets and OSS datasets along with the performances of the LSTM model and Logistic model. Figure 6f shows the median of AE values. Among the models, DC-SRGM based on industrial datasets achieved the best performance on average. However, the industry-based model and OSS-based model produced the same number of best cases. Therefore, OSS datasets can be applied to predict industrial projects when source project data is unavailable.

RQ5: Can DC-SRGM trained with OSS datasets indicate the industrial project's situation?

DC-SRGM trained with OSS datasets obtains a better accuracy than LSTM and Logistic models. However, its accuracy is not better than the industrial projects-based model. Thus, OSS projects can be applied when previous source project data are unavailable.

Table 12. Accuracies of DC-SRGM built with industrial datasets and cross-organization datasets (OSS) are compared with the LSTM model and Logistic model. W/L is the number of datasets that each method is better and worse than. "Threshold below 0.1" is the number of datasets for which each method's performance is lower than the threshold.

Project	DC-SRGM		LSTM	Logistic
	Industry DS	Cross-org DS		
F01	0.067	**0.051**	0.040	0.266
F02	**0.071**	0.104	0.080	0.146
F03	0.192	**0.107**	0.130	0.142
F04	**0.091**	0.124	0.260	0.377
F05	0.075	**0.049**	0.127	0.218
F06	**0.040**	0.136	0.090	0.211
F07	0.329	0.196	0.500	**0.146**
F08	**0.049**	0.333	0.104	0.187
F09	0.055	0.196	**0.048**	0.146
F10	**0.088**	0.120	0.121	0.214
F11	0.068	**0.066**	0.073	0.074
F12	0.095	**0.066**	0.161	0.359
F13	0.211	**0.205**	0.243	0.348
F14	0.107	0.172	**0.020**	0.183
F15	**0.126**	0.196	0.201	0.191
Average	**0.110**	0.141	0.146	0.220
W/L	6/9	6/9	2/13	1/14
# DS Threshold below 0.1	10	4	6	1

5.7. Case Study

Practitioners from e-Seikatsu Co., Ltd. wanted to focus on the situation of the ongoing software development projects because it helps with effective test planning and resource arrangements.

Because the traditional reliability growth model could not describe the growth of the number of bugs for a project, we attempted to model with an advanced methodology, a deep learning-based LSTM model. However, due to the lack of training data of the same project, the model's performance required additional refinement.

Fortunately, the company had a lot of data from previously developed and released projects. Thus, by applying data from previous projects, we developed the DC-SRGM to use in the middle or earlier stages of development projects. By implementing DC-SRGM in the ongoing projects of e-Seikatsu, the proposed model provided a more accurate prediction than the other models considered. This case study confirmed that the proposed approach is applicable when the past data are unavailable in the initial stage of the current development projects.

6. Threats to Validity

In this study, we treated the number of bugs growing as a time-dependent variable for model construction. However, there may be other related factors. For example, the number of detected bugs may depend upon testing efforts. In addition, the experiment was conducted with one LSTM architecture, although the LSTM network architecture may impact its prediction performance. Moreover, when collecting data from open sources, data validity in reporting defect data [28] may be an issue. These are threats to internal validity.

We tested only DC-SRGM with two datasets from two organizations. This is insufficient to make generalizations. In the future, testing of more datasets from many organizations needs to be performed. Additionally, when comparing models, the Logistic model was used as a traditional method since it has been well adopted in SRGMs [11,13,29] and is the most suitable for fitness for the collected experimental datasets. However, the literature reports many other traditional SRGMs. These are threats to external validity.

The training process of our method would not take much time since it usually uses a set of time series sequences where each sequence would be around a few dozen days to several hundred days at most, depending on the length of each similar past project. In contrast, the project clustering process may take some time and manual efforts if various other factors are examined for clustering. This is another threat to external validity from the viewpoint of the practical usefulness of our method.

One threat to construct validity is that we supposed that identifying correct clusters means the group of projects with the same or similar attributes, such as the project scale and growth pattern of the number of bugs rather than the project domains. Therefore, the project domains may differ within the same cluster in actual cases.

7. Conclusions and Future Work

Herein we proposed a new software reliability growth modeling method DC-SRGM using a combination of an LSTM model and a cluster-based project selection method based on similar characteristics of projects via a similarity scoring process. This proposed method alleviates issues regarding insufficient previous data and is an improvement compared to traditional methods for reliability growth modeling.

We conducted experiments using both industrial and OSS data to evaluate DC-SRGM with a statistical significance test. The case studies showed that DC-SRGM is superior to all other evaluated models. It achieved the highest accuracy in industrial datasets, indicating that the project similarity is more important than the project domain type when clustering projects. Moreover, cross-correlation performed better than DTW in specifying project similarity from a defect prediction viewpoint. The experiment involving different time points indicated that DC-SRGM can be used for a project with 12 days of defect data to stably and accurately predict the number of bugs that might be encountered in subsequent days. Finally, DC-SRGM in ongoing projects can assist managers in decision-making for testing activities by understanding reliability growth.

As our future work, we will explore other process metrics (such as testing efforts) and product metrics [30,31] (such as code size) for project clustering and prediction model construction. We plan to extend experiments to confirm the usefulness and generalizability of our method by testing more datasets from many organizations and comparing with other prediction models, including other traditional machine learning-based approaches reported in the literature.

From the viewpoint of practical usage, our method is expected to be implemented within existing development tools and environments, especially continuous integration tools with quality dashboards [32,33] to monitor cumulative numbers of bugs and continuous future prediction on a daily basis. Such tool integration should also facilitate the adoption of measurements and records of necessary failure and related data of (un)distributed team development projects in target organizations.

Furthermore, to improve the quality and continuous monitoring, our method should be extended to provide more reliability metrics beyond predicting the number of bugs.

Author Contributions: Conceptualization and methodology, K.K.S.; literature review and analysis, all authors. All authors have read and agreed to the published version of the manuscript.

Funding: This research received no external funding.

Institutional Review Board Statement: Not applicable.

Informed Consent Statement: Not applicable.

Data Availability Statement: Not applicable.

Conflicts of Interest: The authors declare no conflict of interest.

References

1. Wang, J.; Zhang, C. Software reliability prediction using a deep learning model based on the RNN encoder-decoder. *Reliab. Eng. Syst. Saf.* **2018**, *170*, 73–82. [CrossRef]
2. Washizaki, H.; Honda, K.; Fukazawa, Y. Predicting Release Time for Open Source Software Based on the Generalized Software Reliability Model. In Proceedings of the 2015 Agile Conference, AGILE 2015, National Harbor, MD, USA, 3–7 August 2015; pp. 76–81. [CrossRef]
3. Xu, Z.; Pang, S.; Zhang, T.; Luo, X.; Liu, J.; Tang, Y.; Yu, X.; Xue, L. Cross Project Defect Prediction via Balanced Distribution Adaptation Based Transfer Learning. *J. Comput. Sci. Technol.* **2019**, *34*, 1039–1062. [CrossRef]
4. Okumoto, K.; Asthana, A.; Mijumbi, R. BRACE: Cloud-Based Software Reliability Assurance. In Proceedings of the 2017 IEEE International Symposium on Software Reliability Engineering Workshops, ISSRE Workshops, Toulouse, France, 23–26 October 2017; pp. 57–60. [CrossRef]
5. Honda, K.; Nakamura, N.; Washizaki, H.; Fukazawa, Y. Case Study: Project Management Using Cross Project Software Reliability Growth Model Considering System Scale. In Proceedings of the 2016 IEEE International Symposium on Software Reliability Engineering Workshops, ISSRE Workshops 2016, Ottawa, ON, Canada, 23–27 October 2016; IEEE Computer Society: Washington, DC, USA, 2016; pp. 41–44. [CrossRef]
6. Honda, K.; Washizaki, H.; Fukazawa, Y.; Taga, M.; Matsuzaki, A.; Suzuki, T. Empirical Study on Tendencies for Unstable Situations in Application Results of Software Reliability Growth Model. In Proceedings of the 2018 IEEE International Symposium on Software Reliability Engineering Workshops, ISSRE Workshops, Memphis, TN, USA, 15–18 October 2018; Ghosh, S., Natella, R., Cukic, B., Poston, R.S., Laranjeiro, N., Eds.; IEEE Computer Society: Washington, DC, USA, 2018; pp. 89–94. [CrossRef]
7. Bin, Y.; Zhou, K.; Lu, H.; Zhou, Y.; Xu, B. Training Data Selection for Cross-Project Defection Prediction: Which Approach Is Better? In Proceedings of the 2017 ACM/IEEE International Symposium on Empirical Software Engineering and Measurement, ESEM 2017, Toronto, ON, Canada, 9–10 November 2017; Bener, A., Turhan, B., Biffl, S., Eds.; IEEE Computer Society: Washington, DC, USA, 2017; pp. 354–363. [CrossRef]
8. Turhan, B.; Menzies, T.; Bener, A.B.; Stefano, J.S.D. On the relative value of cross-company and within-company data for defect prediction. *Empir. Softw. Eng.* **2009**, *14*, 540–578. [CrossRef]
9. San, K.K.; Washizaki, H.; Fukazawa, Y.; Honda, K.; Taga, M.; Matsuzaki, A. DC-SRGM: Deep Cross-Project Software Reliability Growth Model. In Proceedings of the IEEE International Symposium on Software Reliability Engineering Workshops, ISSRE Workshops 2019, Berlin, Germany, 27–30 October 2019; Wolter, K., Schieferdecker, I., Gallina, B., Cukier, M., Natella, R., Ivaki, N.R., Laranjeiro, N., Eds.; IEEE Computer Society: Washington, DC, USA, 2019; pp. 61–66. [CrossRef]
10. Goel, A.L. Software Reliability Models: Assumptions, Limitations, and Applicability. *IEEE Trans. Softw. Eng.* **1985**, *11*, 1411–1423. [CrossRef]
11. Ullah, N.; Morisio, M. An Empirical Study of Reliability Growth of Open versus Closed Source Software through Software Reliability Growth Models. In Proceedings of the 19th Asia-Pacific Software Engineering Conference, APSEC 2012, Hong Kong, China, 4–7 December 2012; Leung, K.R.P.H., Muenchaisri, P., Eds.; IEEE Computer Society: Washington, DC, USA, 2012; pp. 356–361. [CrossRef]
12. Rana, R.; Staron, M.; Berger, C.; Hansson, J.; Nilsson, M.; Törner, F. Evaluating long-term predictive power of standard reliability growth models on automotive systems. In Proceedings of the IEEE 24th International Symposium on Software Reliability Engineering, ISSRE 2013, Pasadena, CA, USA, 4–7 November 2013; IEEE Computer Society: Washington, DC, USA, 2013; pp. 228–237. [CrossRef]
13. Honda, K.; Washizaki, H.; Fukazawa, Y. Generalized Software Reliability Model Considering Uncertainty and Dynamics: Model and Applications. *Int. J. Softw. Eng. Knowl. Eng.* **2017**, *27*, 967. [CrossRef]
14. Salehinejad, H.; Baarbe, J.; Sankar, S.; Barfett, J.; Colak, E.; Valaee, S. Recent Advances in Recurrent Neural Networks. *arXiv* **2017**, arXiv:1801.01078.
15. Bengio, Y.; Simard, P.Y.; Frasconi, P. Learning long-term dependencies with gradient descent is difficult. *IEEE Trans. Neural Netw.* **1994**, *5*, 157–166. [CrossRef] [PubMed]
16. Mikolov, T.; Joulin, A.; Chopra, S.; Mathieu, M.; Ranzato, M. Learning Longer Memory in Recurrent Neural Networks. In Proceedings of the 3rd International Conference on Learning Representations, ICLR 2015 Workshop Track Proceedings, San Diego, CA, USA, 7–9 May 2015.
17. Zhang, X.; Ben, K.; Zeng, J. Cross-Entropy: A New Metric for Software Defect Prediction. In Proceedings of the 2018 IEEE International Conference on Software Quality, Reliability and Security, QRS 2018, Lisbon, Portugal, 16–20 July 2018; pp. 111–122. [CrossRef]
18. Hochreiter, S.; Schmidhuber, J. Long Short-Term Memory. *Neural Comput.* **1997**, *9*, 1735–1780. [CrossRef] [PubMed]
19. Zhu, W.; Lan, C.; Xing, J.; Zeng, W.; Li, Y.; Shen, L.; Xie, X. Co-Occurrence Feature Learning for Skeleton Based Action Recognition Using Regularized Deep LSTM Networks. In Proceedings of the Thirtieth AAAI Conference on Artificial Intelligence, Phoenix, AZ, USA, 12–17 February 2016; Schuurmans, D., Wellman, M.P., Eds.; AAAI Press: Palo Alto, CA, USA, 2016; pp. 3697–3704.
20. Porto, F.R.; Minku, L.L.; Mendes, E.; Simão, A. A Systematic Study of Cross-Project Defect Prediction with Meta-Learning. *arXiv* **2018**, arXiv:1802.06025.
21. Kitchenham, B.A.; Mendes, E.; Travassos, G.H. Cross versus within-Company Cost Estimation Studies: A Systematic Review. *IEEE Trans. Softw. Eng.* **2007**, *33*, 316–329. [CrossRef]

22. Lokan, C.; Mendes, E. Investigating the Use of Chronological Splitting to Compare Software Cross-company and Single-company Effort Predictions. In Proceedings of the 12th International Conference on Evaluation and Assessment in Software Engineering, EASE 2008, Workshops in Computing, Bari, Italy, 26–27 June 2008; Visaggio, G., Baldassarre, M.T., Linkman, S.G., Turner, M., Eds.; BCS: London, UK, 2008.
23. Liu, C.; Yang, D.; Xia, X.; Yan, M.; Zhang, X. Cross-Project Change-Proneness Prediction. In Proceedings of the 2018 IEEE 42nd Annual Computer Software and Applications Conference, COMPSAC 2018, Tokyo, Japan, 23–27 July 2018; Reisman, S., Ahamed, S.I., Demartini, C., Conte, T.M., Liu, L., Claycomb, W.R., Nakamura, M., Tovar, E., Cimato, S., Lung, C., et al., Eds.; IEEE Computer Society: Washington, DC, USA, 2018; Volume 1, pp. 64–73. [CrossRef]
24. Chidamber, S.R.; Kemerer, C.F. A Metrics Suite for Object Oriented Design. *IEEE Trans. Softw. Eng.* **1994**, *20*, 476–493. [CrossRef]
25. Jureczko, M.; Madeyski, L. Towards identifying software project clusters with regard to defect prediction. In Proceedings of the 6th International Conference on Predictive Models in Software Engineering, PROMISE 2010, Timisoara, Romania, 12–13 September 2010; Menzies, T., Koru, G., Eds.; p. 9. [CrossRef]
26. Egri, A.; Horváth, I.; Kovács, F.; Molontay, R.; Varga, K. Cross-correlation based clustering and dimension reduction of multivariate time series. In Proceedings of the 2017 IEEE 21st International Conference on Intelligent Engineering Systems (INES), Larnaca, Cyprus, 20–23 October 2017; pp. 000241–000246. [CrossRef]
27. Izakian, H.; Pedrycz, W.; Jamal, I. Fuzzy clustering of time series data using dynamic time warping distance. *Eng. Appl. Artif. Intell.* **2015**, *39*, 235–244. [CrossRef]
28. Herzig, K.; Just, S.; Zeller, A. It's not a bug, it's a feature: how misclassification impacts bug prediction. In Proceedings of the 35th International Conference on Software Engineering, ICSE '13, San Francisco, CA, USA, 18–26 May 2013; Notkin, D., Cheng, B.H.C., Pohl, K., Eds.; IEEE Computer Society: Washington, DC, USA, 2013; pp. 392–401. [CrossRef]
29. Huang, C.; Lyu, M.R.; Kuo, S. A Unified Scheme of Some Nonhomogenous Poisson Process Models for Software Reliability Estimation. *IEEE Trans. Softw. Eng.* **2003**, *29*, 261–269. [CrossRef]
30. Tsuda, N.; Washizaki, H.; Honda, K.; Nakai, H.; Fukazawa, Y.; Azuma, M.; Komiyama, T.; Nakano, T.; Suzuki, H.; Morita, S.; et al. WSQF: Comprehensive software quality evaluation framework and benchmark based on SQuaRE. In Proceedings of the 41st International Conference on Software Engineering: Software Engineering in Practice, ICSE (SEIP) 2019, Montreal, QC, Canada, 15–31 May 2019; Sharp, H., Whalen, M., Eds.; pp. 312–321. [CrossRef]
31. He, P.; Li, B.; Liu, X.; Chen, J.; Ma, Y. An empirical study on software defect prediction with a simplified metric set. *Inf. Softw. Technol.* **2015**, *59*, 170–190. [CrossRef]
32. Honda, K.; Nakai, H.; Washizaki, H.; Fukazawa, Y.; Asoh, K.; Takahashi, K.; Ogawa, K.; Mori, M.; Hino, T.; Hayakawa, Y.; et al. Predicting Time Range of Development Based on Generalized Software Reliability Model. In Proceedings of the 21st Asia-Pacific Software Engineering Conference, APSEC 2014, Jeju, Korea, 1–4 December 2014; Volume 1: Research Papers; Cha, S.S., Guéhéneuc, Y., Kwon, G., Eds.; IEEE Computer Society: Washington, DC, USA, 2014; pp. 351–358. [CrossRef]
33. Nakai, H.; Honda, K.; Washizaki, H.; Fukazawa, Y.; Asoh, K.; Takahashi, K.; Ogawa, K.; Mori, M.; Hino, T.; Hayakawa, Y.; et al. Initial Industrial Experience of GQM-Based Product-Focused Project Monitoring with Trend Patterns. In Proceedings of the 21st Asia-Pacific Software Engineering Conference, APSEC 2014, Jeju, Korea, 1–4 December 2014; Volume 2: Industry, Short, and QuASoQ Papers; Cha, S.S., Guéhéneuc, Y., Kwon, G., Eds.; pp. 43–46. [CrossRef]

 *mathematics* 

Article

Modeling and Verifying the CKB Blockchain Consensus Protocol †

Meng Sun [1], Yuteng Lu [1], Yichun Feng [1] and Qi Zhang [1] and Shaoying Liu [2,*]

[1] School of Mathematical Sciences, Peking University, Beijing 100871, China; sunm@pku.edu.cn (M.S.); luyuteng@pku.edu.cn (Y.L.); yichunfeng@pku.edu.cn (Y.F.); zhang.qi@pku.edu.cn (Q.Z.)
[2] Graduate School of Advanced Science and Engineering, Hiroshima University, Higashi Hiroshima City 739-8527, Japan
* Correspondence: sliu@hiroshima-u.ac.jp
† This paper is an extended version of our papers published in Proceedings of SEKE 2021, pp. 150–153, KSI Research Inc. and Knowledge Systems Institute, 2021, and Proceedings of BlockSys 2020, CCIS 1267, pp. 3–17, Springer, 2020.

Abstract: The Nervos CKB (Common Knowledge Base) is a public permissionless blockchain designed for the Nervos ecosystem. The CKB consensus protocol is the key protocol of the Nervos CKB, which improves the limit of the consensus's performance for Bitcoin. In this paper, we developed the formal model of the CKB consensus protocol using timed automata. Based on the model, we formally verified various important properties of the Nervos CKB to provide a sufficient trustworthiness assurance. Especially, the security of the Nervos CKB against the selfish mining attacks to the protocol was investigated.

Keywords: Nervos CKB; consensus protocol; model checking; UPPAAL

Citation: Sun, M.; Lu, Y.; Feng, Y.; Zhang, Q.; Liu, S. Modeling and Verifying the CKB Blockchain Consensus Protocol. *Mathematics* **2021**, *9*, 2954. https://doi.org/10.3390/math9222954

Academic Editor: Frank Werner

Received: 25 October 2021
Accepted: 12 November 2021
Published: 19 November 2021

1. Introduction

Blockchains are distributed digital ledgers for which there are numerous benefits such as decentralization, persistency, and anonymity. A continuously growing ledger of transactions being represented as a chain of blocks is provided in a blockchain, where the transactions are distributed and maintained over a peer-to-peer network [1]. Blockchain has become a popular technology since it was first proposed by Satoshi Nakamoto in 2008 to support Bitcoin [2] and has been successfully applied in many scenarios due to its power to create, transfer, and own assets in crypto-economy networks. Ethereum [3] extends the application range of blockchain and allows developers to write smart contracts and create different decentralized applications. Both Bitcoin and Ethereum have shown their exciting potential for building a powerful crypto-economy network and have attracted much attention from governments and industry.

Developing a secure and trustworthy blockchain is highly challenging because of the vulnerabilities and the complexity of the distributed execution environment. In addition to the security issues, the processing speed is also an important concern. However, both Bitcoin and Ethereum have a limit of processing large transactions per time unit. In other words, their processing capability is severely limited by the scalability. To alleviate this problem for long-term sustainability, the Nervos team proposed the Common Knowledge Base (CKB) [4], which uses a decentralized and secure layer and provides common knowledge for the peer-to-peer network.

Since the CKB has become the trust root of the decentralized secure crypto-economy system, guaranteeing the security and consistency of the CKB consensus protocol have become very important. In fact, there are some protocols in which vulnerabilities were discovered after they have been taken as correct and used for a long time [5]. In the literature, there are some existing works for formal modeling and verification of blockchain systems.

For example, the work in [6] proposed a novel approach for verifying the properties of Ethereum smart contracts using statistical model checking, and a formal model of the Bitcoin protocol was proposed and verified with the UPPAAL model checker [7,8] in [9].

The CKB *consensus protocol* [10] is the key protocol being used in the CKB to build the secure and optimal crypto-economy system [4]. The protocol aims to overcome the two drawbacks of Bitcoin consensus: the *low transaction processing throughput* and the *vulnerability to selfish mining attacks*. It limits the time of connecting the sender in the search of a lost transaction. Such a restriction improves the efficiency of transaction processing without compromising the security of the blockchain. Furthermore, the protocol adopts a novel *"two-step confirmation"*, which can be used for selfish mining attack resistance.

Since the CKB is becoming more and more popular and its applications are constantly increasing, the security properties of the CKB should have more attention paid to them. The security of the CKB calls for a detailed investigation, and its ability to resist selfish mining attacks has not been formally checked. In this paper, we propose the formal model of the CKB consensus protocol using timed automata. Based on the formal model, we verified the corresponding important properties with mathematical rigor with UPPAAL, which is a model checker that has been successfully used in various case studies [9,11,12]. Model checking [13] is a formal method of verification, which requires mathematical formalisms for both the desired properties and systems and assures system correctness w.r.t. the properties specified in given specifications automatically. Meanwhile, model checking is also helpful for finding and fixing bugs in the system implementation.

The work in this paper is an extension of our previous studies [14,15], where we initially discussed the formal models of the CKB block synchronization protocol and consensus protocol, respectively, and the verification of some important properties of these two protocols. In this paper, we further improved the formal models of the CKB consensus protocol and investigated its robustness against malicious attacks, especially selfish mining attacks. The main contributions of this paper are as follows:

- The formal models of the CKB consensus protocol in timed automata are proposed;
- A family of properties is formally defined and verified in UPPAAL;
- The ability of the CKB to resist malicious selfish mining attacks is proven.

The rest of this paper is organized as follows. The Nervos CKB and CKB consensus protocol are briefly described in Section 2. Section 3 presents the formal model of the CKB consensus protocol. Then, a family of properties of the protocol is formally defined and verified using UPPAAL in Section 4. Section 5 discusses the ability of the CKB to resist selfish mining attacks. Related work is provided in Section 6. Finally, Section 7 concludes the paper and discusses possible future studies.

2. Preliminaries

This section gives an introduction to the Nervos CKB, the CKB consensus protocol, and attacks.

2.1. The Nervos Network

The CKB is an open, public, and PoW-based blockchain, which was proposed in the Nervos Network [16]. It was inspired by Bitcoin, but provides higher scalability and lower transaction costs compared to Bitcoin. There are mainly two ways to improve the scalability of blockchain: increasing the block space to store more transactions and moving part of the operations off-chain for execution. The Nervos Network [16] uses the second approach and creates a two-layer environment. Figure 1 shows its layered architecture, which separates the state and computation and provides better scalability and more flexibility to each layer.

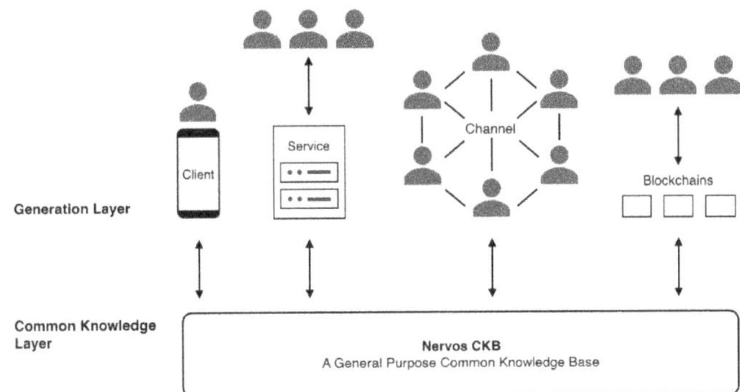

Figure 1. The CKB layered architecture.

The CKB layer, designed as a public permissionless blockchain for a layered crypto-economy network, is the first layer in the Nervos Network. It is responsible for providing the decentralized and secure infrastructure. In addition, it also includes the operation of state verification. In order to settle the assets that come in and out of the second layer, the CKB layer ensures the decentralization and sustainability of the entire blockchain. The second layer is the environment of generating transactions and calculating and is mainly responsible for generating states and protecting privacy. For different needs, it can be designed separately to match various scenarios. The encryption of the CKB layer protects the activities in the second layer. The second layer's operation can be expanded to a large extent under the security provided by the CKB layer.

Applications on the second layer can choose the proper generation methods based on their particular needs. The CKB layer provides common knowledge custody for the crypto-economy network, and its design target focuses on states. Common knowledge refers to states verified by global consensus, and crypto-assets are examples of common knowledge. CKB can generate trust and extend this trust to the second layer, making the whole network trustworthy.

The operations of the Nervos Network consist of three parts: state generation executed off-chain, the state-verification-based CKB virtual machine, and storing the states in the cell. Once a new state is generated by the second layer, it will be placed into the transaction. Then, the transaction will be broadcast to the whole network. To overcome the shortcomings of Bitcoin and Ethereum, as mentioned above, the CKB consensus protocol increases the output and enhances the security. The two-step confirmation is used for transaction verification where the two steps are defined as the proposal step and the commitment step. All transactions must go through the two-step confirmation. In the Nervos Network, users can participate in activities as three types of nodes: The *mining nodes*, which are responsible for collecting transactions and generating blocks, the *full nodes*, which responsible for the verification, and the *light nodes*, which only focus on the data they need and use the least resources. All nodes can freely enter or exit the blockchain.

2.2. CKB Consensus Protocol

The CKB consensus protocol is a variant of Nakamoto Consensus (NC) and complies with the PoW mechanism. While retaining the advantage of NC, the CKB consensus protocol improves the performance limit and resistance to selfish mining attacks by adopting a two-step confirmation, as shown in Figure 2. The block structures in the CKB include the proposal zone and the commitment zone [4,10]. When a blockchain user wants to record a transaction on the blockchain, a new transaction is generated. Based on the design of the CKB, miners put these new transactions into the proposal zone of a block. The proposal step starts once the proposal zone receives the transactions. This step will mainly go through two operations. The first one is to check *txpid*, which is defined as the first few bits of the transaction ID. In the second operation, full nodes confirm whether they have received the transaction and then verify it. When a transaction passes the above operations, it is considered to be "proposed". Next, the commitment step starts once the transaction is put into the commitment zone by the miner. In this step, full nodes confirm that the transaction is not a duplicate and it would not conflict with previous transactions. It is assumed that the transaction's *txpid* appears in the proposal zone of one block and the commitment zone of another block, then full nodes confirm that the distance between these two blocks on the chain is kept within a predefined range. The transaction is "committed" after the commitment step is complete.

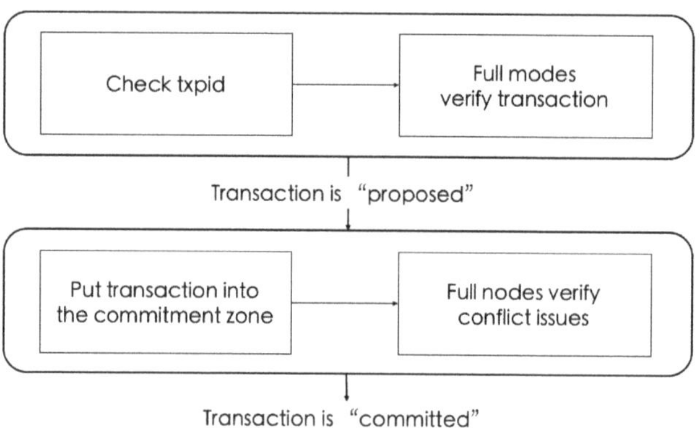

Figure 2. Two-step confirmation.

The block propagation mechanism adopted in the consensus protocol checks whether the transaction in a block is lost while avoiding extra round trips. In selfish mining attacks, some transactions are concealed by the malicious miners. If such missing transactions are continuously requested, an extra round trip will occur. The block propagation mechanism regulates the maximum number of steps for the round trip through the following two operations. In the first operation, when a committed transaction is previously unknown, its sending node will be requested. There exist some transactions that are indeed proposed, but they are not broadcast. The sending node must provide these transactions and put them in the prefilled transaction list. If the sending node and the receiving node have the same list that contains these nonbroadcast transactions, these transactions can be considered valid. In the second operation, if a transaction is still missing, the sending node will be queried again. When the sending node does not provide this transaction within the time limit, this node will be included in the blacklist. Just as Bitcoin consensus, the CKB cannot resist majority attack (51% attack) either.

2.3. Selfish Mining Attack

Some blockchain systems including Bitcoin have suffered from the selfish mining attack. In the worst case, the malicious miners can occupy the dominant position in the mining, and the decentralized characteristics may disappear. Then, the original advantages of the blockchain no longer exist.

The way to gain illegal benefits in a selfish mining attack is that the malicious miners create nonpublic blocks and use these blocks to replace the blocks created by the honest miners. When one malicious miner generates a new block and launches a selfish mining attack, he/she hides the block and waits for the opportunity to announce it. In general, multiple malicious miners join together to form a malicious group and share the computing power to improve the probability of success. The more blocks these malicious miners possess, the higher the profits they can obtain. Meanwhile, the other honest miners are competing with the malicious group for mining. When the computing power of honest miners far exceeds that of the malicious group, it will be difficult for the malicious group to gain an advantage to obtain benefits.

The key idea of the selfish mining attack is to create a secret branch from the chain, and the miners in the malicious group will work only on this private chain. When competing with the honest miners for mining, the malicious group waits until the private chain is longer than the public chain. By the time the private chain gains the upper hand, the malicious group announces it to the public, causing a fork. Since the newly announced chain is longer than the original chain, other miners choose the longer one to follow. Furthermore, the blocks added thereafter are successors of this private chain. Therefore, the private chain replaces the original one as the main blockchain. Since the original chain is abandoned, all the mining rewards of honest miners go down the drain. The malicious group is more profitable when the newly announced private chain becomes longer.

3. The Formal Model of the CKB Consensus Protocol

In this section, we propose the formal model of the consensus protocol using timed automata. To accurately simulate the operations of transaction verification in the CKB consensus protocol, this model formalizes both the verification process and the interactions among different nodes. Such a model consists of four automata: *two-step*, *miner*, *full node*, and *block-propagation*.

All the variables in the model are used to specify whether the operations are successful or not. The default initial values of the variables are all 0. Once the operations are complete, the corresponding values are assigned to the variables according to the results. The assigned value is 1 if the operation is successfully completed and greater than 1 if the operation is abnormal. The assigned variables are taken as parameters in the guard conditions on transitions.

3.1. *Two-Step Automaton*

The two steps in the two-step confirmation are "*proposal*" and "*commitment*", respectively. All the transactions that pass the two-step confirmation are taken as legal. After a node generates a new transaction, a miner collects the transaction and completes the PoW to generate a new block. The transaction is firstly written in the proposal zone of one block, and then, the proposal step begins.

The initial state of the two-step automaton in Figure 3 is *T*0, which represents the generation of a new transaction. The channel *collectP!* simulates the operation of mining and is used to synchronize with *collectP?*, which is a channel in the miner automaton. The variable *c* denotes the global time, which represents the time interval of each mining epoch. Variable *cp* is used to specify whether the transaction has been put into the proposal zone. According to the CKB consensus protocol, the difficulty of the PoW and the time interval will be adjusted to make full use of the hardware performance, maintaining high-efficiency production. Although the time interval in the protocol is not constant, setting time *c* to a fixed value in this model does not affect the simulation of verification and propagation.

The function of the proposal zone is to announce new transactions being processed by a miner to all nodes. Transactions that have not yet passed subsequent verification are not considered to be valid. Therefore, these transactions in the proposal zone do not affect the legality and spreadability of the blocks. The state *T*1 captures "start of proposal confirmation". There are 4 operations in the proposal confirmation: (1) to confirm that a transaction exists in the proposal zone; (2) to check the *txpid* of the transaction; (3) to confirm that the transaction has been received by the full node; (4) to verify the transaction content. The value of variable *checkT* is used to specify whether the transaction successfully passes the *txpid* check. The value of *checkT* is zero by default before the check, and a forced state transition will be made by the invariant.

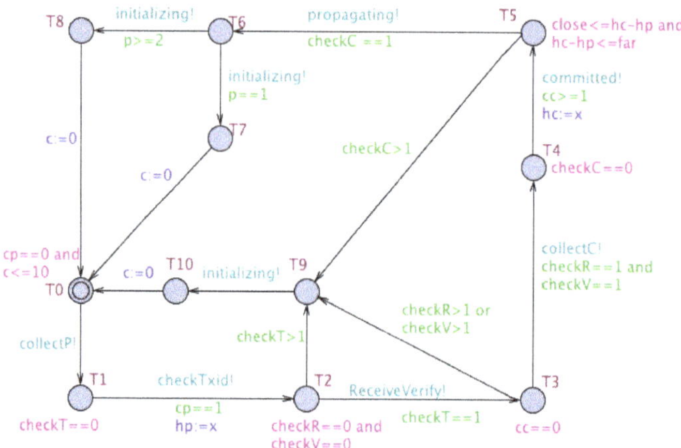

Figure 3. The two-step automaton.

In the CKB consensus protocol, *txid* checking is performed by the full nodes, so the channel *checkTxid!* is used to synchronize with the channel *checkTxid?* in the full node automaton. The full node automaton assigns the checking result to *checkT*. The value of variable x denotes the corresponding block height on the blockchain. Whenever a new block is added into the blockchain, the value of x increases by 1. The height of the block in which the transaction exists is recorded using the value of variable *hp*.

The process continues to move forward if all operations in the two-step confirmation process are successful. If any verification fails, the state transfers to *T*9, indicating that it is impossible to broadcast the transaction. *T*2 is the state for "verification of transaction". The channel *ReceiveVerify!* simulates the verification performed by a full node. Once the verification is finished, the full node automaton assigns value 1 to the variables *checkR* and *checkV*. The values of these two variables are used to indicate whether the transaction is successfully received and verified by the full node, respectively. *T*3 is the state in which the transaction is ready for "mining of the second step". Variable *cc* marks whether the transaction is put into the commitment zone. A transaction τ that has been verified in the first confirmation step is regarded as a "proposed transaction". If τ is in the proposal zone of a block with height *hp*, we say that τ is proposed at height *hp*.

Miners can collect all the transactions that have completed the first confirmation step and write them into a new block's commitment zone. The channel *collectC!* synchronizes with the channel *collectC?* in the miner automaton to simulate the mining behavior. The mining operations in the two steps are different in the locations in which the miners write the transaction. There are two blocks in which the transaction exists, and the height interval between these two blocks is limited in a previously defined range.

*T*4 is the state for "start of the commitment confirmation". It is reached once the transaction has been denoted as proposed and put in the commitment zone. The value of variable *checkC* shows whether this transaction conflicts with other transactions on the chain. Channel *committed!* synchronizes with channel *committed?* in the full node automaton and simulates the confirmation of the proposed transaction. A proposed transaction τ must meet the constraint $cc >= 1$ once it enters the confirmation. This constraint means that τ has been put into the commitment zone. The current height of this block on the chain is captured by the variable *hc*.

*T*5 is the state that conforms to the invariant $close <= hc\text{-}hp <= far$. The transaction appears in the proposal zone and the commitment zone of two different blocks. The time spent in the two-step confirmation process creates a difference in *hc* and *hp*. The values of the constants *close* and *far* are predefined according to the efficiency of the hardware equipment. The height interval between the two blocks can be regarded as the time required for the first step of confirmation. The setting of *close* is to ensure the time interval is long enough for the transaction to be propagated to the entire network. Each node has limited memory space in the local device, and the value of *far* is decided on the basis of the number of proposed transactions that can be stored in its device.

The state transfers to *T*6 once the constraint *checkC* == 1 is satisfied, while the channel *propagating!* is triggered simultaneously. All transactions that reach this state are regarded as "committed transactions". In the two-step confirmation process, if any of the variable values in *checkT*, *checkR*, *checkV*, and *checkC* is greater than 1, the verification is a failure. Transactions that fail in verification directly go to *T*9, which is defined as "transaction invalid". Invalid transactions do not undergo subsequent verification steps.

The channel *propagate!* synchronizes with *propagate?* in the block-propagation automaton. If there is a transaction τ in the commitment zone of a certain block that is either proposed or committed, then τ can be spread to the network. If a transaction is missing, the block-propagation automaton initiates contact and requests the missing part from the miner automaton. The miner should respond within a short time. Otherwise, he/she is disconnected and blacklisted.

*T*7 and *T*8 are the states that indicate "authorization of broadcast" and "prohibition of broadcast", respectively. The value of variable p denotes whether the transaction is propagable. If the transaction is legal and can be propagated, then $p == 1$. Otherwise, the value 2 indicates that the transaction cannot be propagated. When the state reaches *T*7, *T*8, or *T*10, it means the end of the transaction verification. When the next transaction is born, the automaton state returns to *T*0, and the global time and variables are reset.

3.2. Full Node Automaton

Once a new block is generated, the legitimacy and the PoW of blocks are checked by the full node before they are broadcast. Since the two-step confirmation is transaction-oriented rather than block-oriented for the verification process, the full node automaton is also transaction-oriented. In this automaton, all operations aim at a single transaction.

Figure 4 (The state marked **c** is committed. A state is committed if any of the locations in the state are committed. A committed state cannot delay, and the next transition must involve an outgoing edge of at least one of the committed locations.) depicts the full node automaton. In the first confirmation step, the full node performs the checking of transaction *txid* and the verification of contents, which are captured by the channels *checkTxid?* and *ReceiveVerify?*, respectively. State *F*1 is "checking of *txid*", and the variable *checkT* is the result. States *F*2 and *F*3 correspond to "confirmation of receiving" and "transaction verification", respectively. The variables *checkR* and *checkV* are used to note the results of these two operations.

When the full node reaches the second confirmation step, it becomes responsible for committing the transaction. Once the channel *committed?* synchronizes with the channel *committed!* in the two-step automaton, the state *F*4 is reached. The assignment of the variable *checkC* marks whether the operation is successful. The state moves to *F*5 once any operation fails. In this case, the transaction becomes invalid.

3.3. Miner Automaton

A miner's behavior is specified in Figure 5, where *M*1 captures the standby state. Once new transactions are generated, miners package these transactions and generate new blocks through the PoW. This automaton simulates the behavior of honest miners, so the mining results are all public. State *M*2 means "new block generation", which is reached after mining.

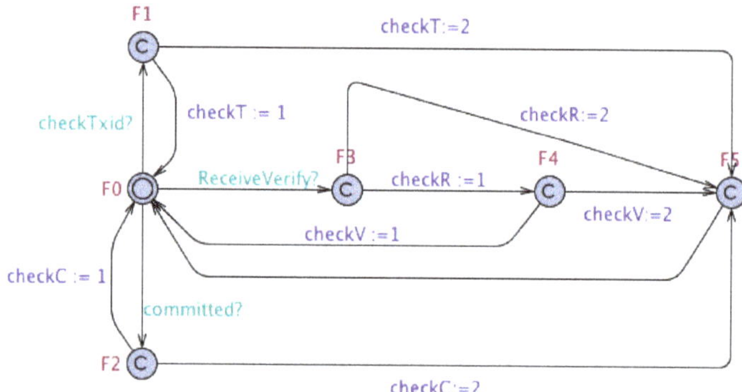

Figure 4. The full node automaton.

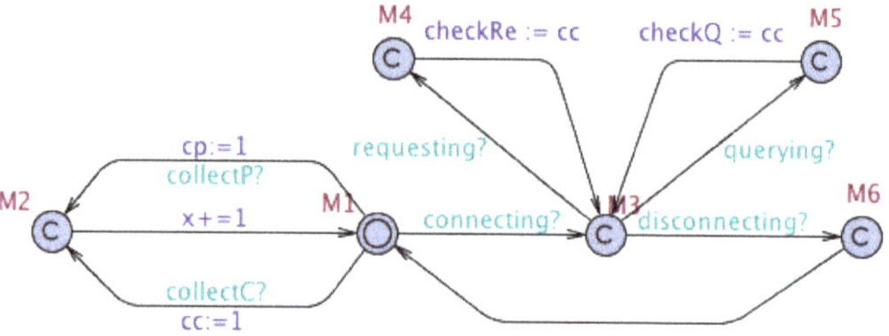

Figure 5. The miner automaton.

The automaton synchronizes with channel *connecting!* in the block-propagation automaton through the channel *connecting?* if a transaction is missing. Then, the state transfers to $M3$, which represents "the contact with the miner". Channel *requesting?* describes the process in which the miner is asked for the missing transaction. After that, the state transfers to $M4$, which stands for "response to the request", and the miner sends the requested transaction back. The assignment $checkRe := cc$ uses the operation result after the transaction is written in the commitment zone as the miner's reply.

When the transaction is still missing, the inquiry will be launched again. Channel *querying!* in the block-propagation automaton synchronizes with *querying?* in the miner automaton. State $M5$ means "reply to query". Variable *checkQ* represents the answer of the miner. Similarly, the value of *checkQ* is assigned to *cc*. The miner is taken as suspicious and blacklisted after two failed requests for the transaction. The channel *disconnecting?* is used to simulate this operation, which synchronizes with the channel *disconnecting!* in the block-propagation automaton and transfers to state $M6$ for "disconnection".

3.4. Block-Propagation Mechanism

The process of the block-propagation mechanism is described in Figure 6, which starts from the standby state $P0$ by synchronizing the channel *propagating?* with the channel *propagating!* in the two-step automaton. State $P1$ checks if the transaction is in the commitment zone or not. The value of variable p indicates whether the transaction can be propagated. The transaction can be broadcast if $p := 1$, and the broadcasting is forbidden if $p := 2$. If the value of cc is different from 1 ($cc \geq 2$), the transaction is not in the commitment zone of any public block, and the channel *connecting!* should be activated to synchronize with channel *connecting?* in the miner.

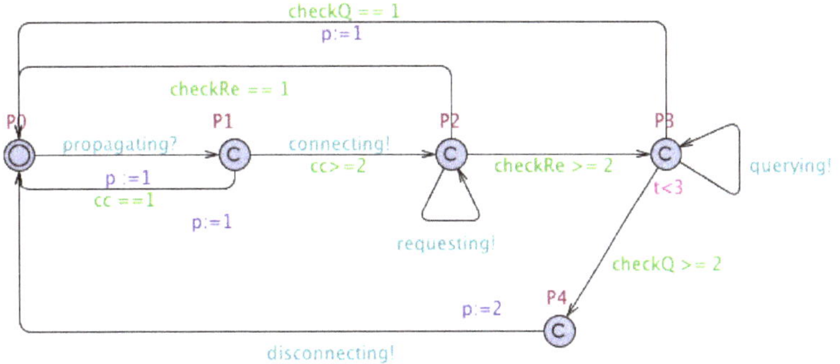

Figure 6. The block-propagation automaton.

State $P2$ means that the transaction is previously unknown. If *checkRe* $>= 2$, which means that the transaction is still not acquired, the state transfers to $P3$ for "failure in request". The channel *querying!* is used to synchronize with the channel *querying?* in the miner, which must reply in a short duration ($t < 3$ in Figure 6). State $P4$, which means "transaction invalidation", is reached if the missing part is still unknown. Then, the miner is blacklisted and disconnected. Such an operation is simulated by the transition labeled by the channel *disconnecting!* and an assignment $p := 2$, which tells the two-step automaton that this transaction should not be propagated. This transition leads the automaton back to the initial standby state.

4. Verification of the CKB Consensus Protocol in UPPAAL

We conducted a series of experiments to explore the credibility and consistency of the consensus protocol by formalizing and verifying its key properties. In this section, we did not consider properties related to the malicious attackers. The presence of malicious attackers will be discussed in the next section. In the following, we define a family of properties that should be satisfied in the CKB consensus protocol. Based on the proposed formal model, we conducted some experiments using the UPPAAL model checker to check the correctness and consistency of the protocol.

First of all, based on the design of the CKB consensus protocol, newly generated transactions must go through a process of being put into the proposal zone. We define this process as P1, in which T1 represents that the transaction is in the proposal zone. Subsequently, the information of such transactions will be received by other nodes, and the legality of the blocks and the propagation will not be affected by the validity of the transactions. The verification result in UPPAAL demonstrates that the protocol satisfies (1).

$$A <> TwoStep.T1 \tag{1}$$

Only after a transaction successfully passes the *txid* check, it can be considered as proposed. Therefore, transactions that have not completed the *txid* check are not "proposed transactions". This property is formalized as (2), and the verification result shows that the protocol satisfies (2) as well.

$$\begin{gathered} A\,[\,]\ TwoStep.T4\ imply\ (checkT == 1) \\ A\,[\,]\ not\ checkT == 1\ imply\ not\ TwoStep.T4 \end{gathered} \tag{2}$$

P3 formalizes the following property: the full node should receive and verify a transaction before it is proposed. On the other hand, the transaction cannot be considered proposed if the full node has not received the transaction or completed the verification of its content. In the proposal step, the transaction *txid* is processed first, and then, a notification is sent to the full nodes. As mentioned earlier, the transaction cannot be considered proposed until it passes the check *txid* ($checkT == 1$). Once the check fails, the transaction will never be considered as proposed. Furthermore, the transaction must have been received ($checkR == 1$) and verified ($checkV == 1$) by the full nodes. The state $T4$ in the two-step automaton indicates that the transaction is proposed. The verification in UPPAAL shows that (3) is satisfied.

$$A\,[\,]\ TwoStep.T4\ imply\ (checkR == 1\ and\ checkV == 1)$$

$$A\,[\,]\ (not\ checkR == 1)\ or\ (not\ checkV == 1)\ imply\ not\ TwoStep.T4 \tag{3}$$

Before the transaction is put into the commitment zone, it must have been received and verified by the full node. A transaction that has not been received or verified by the full node cannot appear in the commitment area. We formalize this property as (4). The state $T5$ means that the transaction is put in the commitment zone. In fact, the second step of the two-step confirmation will be activated if and only if the miner finishes placing the transaction in the commitment area. Only through the verification of the proposal step, the transaction will be put into the commitment zone. (4) is satisfied based on the verification result.

$$A\,[\,]\ TwoStep.T5\ imply\ checkR == 1\ and\ checkV == 1$$

$$A\,[\,]\ not\ (checkR == 1\ and\ checkV == 1)\ imply\ not\ TwoStep.T5 \tag{4}$$

A transaction must be located in the commitment zone with height *hc* and satisfy the condition: $close \leq hc - hp \leq far$ when it is committed. Such a property is formalized as (5), in which $T6$ means commitment of the transaction. The value of *checkC* is used to indicate whether the transaction is in the commitment area. This property is satisfied according to the verification in UPPAAL.

$$\begin{aligned} A\,[\,]\ TwoStep.T6\ imply\ checkC == 1 \\ and\ (close <= hc - hp\ and\ hc - hp <= far) \end{aligned} \tag{5}$$

If a transaction is missing and cannot be obtained by the miner after the requesting and querying operations, the miner will be blacklisted and disconnected. This property is formalized as (6).

$$\begin{aligned} A\,[\,]\ BlockPropagation.P3\ and\ BlockPropagation.P4 \\ imply\ MiningNode.M6 \end{aligned} \tag{6}$$

The model is repeatable, and there is no deadlock, formalized as (7).

$$A\,[\,]\ not\ deadlock \tag{7}$$

Both (6) and (7) are satisfied based on the verification.

5. Consistency and Robustness Analysis with Attacks

In reality, malicious attacks are always inevitable. In this section, we added attacks to our models and checked the security properties of the protocol.

The security of CKB consensus protocol against selfish mining attacks is discussed in this subsection. In the attack scenario, the other automaton models remain the same, but the miner's behavior is different. The malicious miner deliberately hides a block when generating it. We verified whether the protocol can defend against selfish mining attacks. The security properties are specified in the CTL formula and were proven in UPPAAL.

Figure 7 offers an automaton that simulates the behavior of a malicious miner. Compared to the honest miner in Figure 5, this automaton has an additional state $M7$, while the remaining states and transitions stay unchanged. When the malicious miners collect proposed transactions and put them into the commitment zone of the new block, the state transfers to $M7$, which is defined as "attack start". When $M7$ transfers to state $M2$, which is "new block generation", the automaton synchronizes channel *attack!* with *attack?* in the SelfishMining automaton to perform the attack. According to the result of the attack, the selfish mining automaton returns the parameter *cc*, and the two-step automaton decides whether the transition should be fired based on the value of *cc*. The condition $cc > 0$ means that the transaction has been collected by the miner and put into the block.

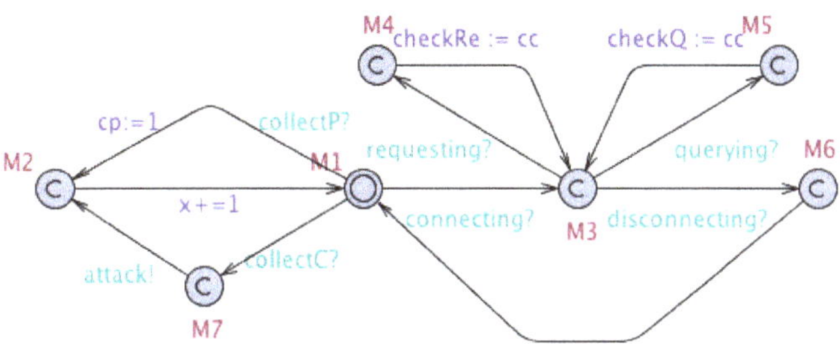

Figure 7. The malicious miner automaton.

The mining competition between the malicious group and the honest miners can be described as the following three scenarios. In the first scenario, the malicious group leads the honest miners and generates blocks more quickly. As a result, the private chain has an absolute advantage. If the length of the private chain is already longer than the public chain by two, the malicious group can choose to announce the private chain immediately. At the moment, the public chain is shorter, so it will be discarded. The malicious group can also choose not to publish the private chain and continue mining. When the length of the public chain is about to catch up with the length of the private chain, that is to say, the gap between the two chains is only one, the malicious group will announce the private chain. In the second scenario, honest miners take the lead to find the new block and put it in the public chain. Once length of the private chain lags behind the length of the public one, the malicious group will directly abandon the private chain. In the third scenario, the malicious group has the same computing power as the honest miners, namely, the honest miners and the malicious group would find blocks at the same time. There is no advantage in the private chain. At this time, the malicious group could announce the private chain, and then, the full nodes would choose the public chain or private chain to follow.

The malicious group could also continue to bet until the game is over. In the first scenario, the selfish mining attack succeeds, and the malicious group will receive rewards for all blocks on the private chain. In the second scenario, the attack fails, and the malicious group receives nothing. In the last scenario, if a subsequent block is added to the private chain, the malicious group can still obtain the reward corresponding to the blocks on the private chain. Conversely, if the public chain is chosen, all blocks in the private chain will be discarded, and the malicious group will not be able to profit.

Figure 8 is a rough attempt to illustrate the selfish mining algorithm. $S0$ is the initial state. When the channel *attack?* fires synchronously with *attack!* in the malicious miner automaton, the state transfers to $S1$, which is regarded as "start of attack". In state $S1$, there are two nondeterministic branches. The upper branch represents the first scenario of mining competition, while the lower branch moves toward the second and third scenarios of mining competition. The variable *private* represents the private chain's length held by the malicious group, and the variable *public* is the length of the public chain maintained by other honest miners. Note that *public* is not the length of the main blockchain; it only represents the length of the public branch when the private branch is generated. The default values of *private* and *public* are both initially zero.

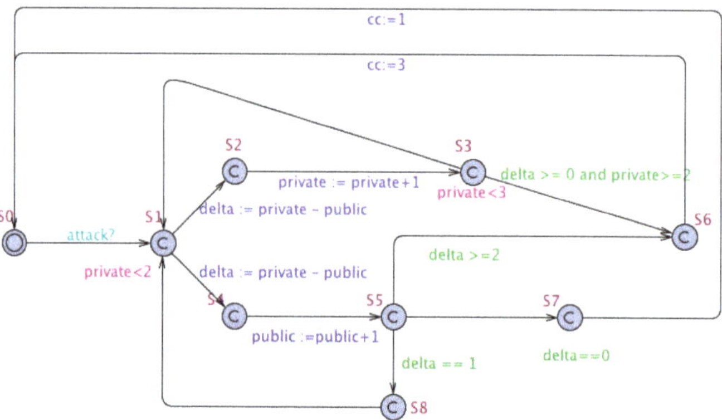

Figure 8. The selfish mining automaton.

It is indicated that the malicious group and honest miners are competing for mining at the same starting point. The variable *delta* is the difference in length between the private chain and the public chain and is used to distinguish the current competition between the two. When the automaton fires the transition from $S1$, the variable *delta* is updated first. At the beginning of the attack, since *private* and *public* are both defaulted to zero, the value of *delta* should be zero regardless of whether the state transfers to the upper or lower branch. If the automaton chooses the upper branch, the state transfers to $S2$. At this time, the malicious group successfully generates a block and adds it to the private chain. The assignment $private := private + 1$ implies that the length of the private chain increases by one. While the malicious group is mining, honest miners are also competing. At this time, if other malicious miners of the same group find the second block, the private chain is determined to be ahead of the public chain. Then, the private chain could be announced, and the selfish mining attack is successful. The state $S6$ after the transition indicates "a successful attack". In state $S3$, if the malicious group is unable to obtain a new block faster than honest miners twice in a row, the state transfers back to $S1$, and the malicious group continues to compete with honest miners.

For simplicity, we did not consider the case that the malicious group holds a favorable position in computing power and keeps the private chain longer than the public chain. Therefore, the state invariant was set to enforce the transition. When *private* is greater than two, the private chain should be announced. The automaton can select the lower branch to state $S4$, which implies "honest miners generate new block". When the new block is added to the public chain, the variable *public* increases by one. Then, the second or third scenario of competition may occur. In the second scenario, the guard condition $delta == 0$ indicates that the private chain has fallen behind the public chain of the honest miner, so the malicious group can only immediately discard the private chain. In this case, the state transfers to $S7$, indicating "attack failure". In the third scenario, the guard $delta == 1$ means that the private chain and the public chain have the same length at this time, and the state transfers to $S8$, which implies that the two sides are equal in strength. Hence, the malicious group would like to compete again until the outcome is clear.

Unlike the honest miners, a malicious miner hides the blocks it generates. Such a behavior of hiding a block can happen in both the proposal and commitment steps, so we discuss the possible selfish mining attacks in these two steps separately. First, we assert that the attack launched at the proposal step will prevent malicious miners from gaining benefits. According to property P3, if the transaction cannot be received by the full nodes, it will not be regarded as proposed. In other words, this transaction is not a "valid transaction"; the state of the two-step automaton directly transfers to $T9$, and this transaction is not adopted. Some transactions may pass the proposal verification, but they are not broadcast. In this case, these transactions are placed in the prefilled list and sent to the miners during the commitment step.

Next, we explored the scenario of starting a selfish mining attack during the commitment step and analyzed whether the CKB consensus protocol can effectively combat the attack. All the following properties were successfully verified in UPPAAL.

The property "all transactions must have appeared in the proposal zone before the commitment process" is formalized as (8). State *T5* is the first stage of the commitment process, and *cp* is a sign representing "the transaction already exists in the proposal zone". The commitment process must only be the second step of the two-step confirmation. In other words, no transaction can skip the first step. According to (1), all transactions are processed by being put into the proposal zone. Before a transaction performs the second step, the nodes were informed of the transaction in the proposal zone of a block.

$$A\,[\,]\ TwoStep.T5\ imply\ (cp == 1) \tag{8}$$

Before a transaction is committed, the full nodes should receive this transaction and verify its validity. This property is formalized as (9). State *T6* means "transaction is regarded as committed"; the condition $checkR == 1$ denotes that the full nodes have received the transaction; $checkV == 1$ is the sign that the full nodes have verified the transaction. The property (9) indicates that when the transaction is proposed, the full nodes have been informed of the transaction and the content of this transaction is confirmed. According to (8) and (9), a transaction cannot remain unknown after it is generated. Attributed to the role of the proposal step, the transaction must be announced in the first step.

$$A\,[\,]\ TwoStep.T6\ imply\ (checkR == 1\ and\ checkV == 1) \tag{9}$$

Assuming that the malicious miner wants to hide the block in the second step, we have the following property (10): as long as the selfish mining attack is successful and a block and its included transactions are hidden in the commitment process, the block will not be propagated. State *S6* represents a successful selfish mining attack, and state *T7* stands for "block-propagation". There is a case in which a transaction is regarded as proposed, but it does not appear in the commitment zone. This case only happens when the malicious miners launch a selfish mining attack. According to the protocol, the full nodes will request these missing transactions. If the malicious miner does not disclose the private blocks and transactions in time, the protocol prohibits the propagation of these blocks.

$$A\,[\,]\ SelfishMining.S6\ imply\ not\ TwoStep.T7 \tag{10}$$

The properties (8)–(10) together reveal then that CKB consensus protocol could prevent malicious miners from making unfair profits in selfish mining.

6. Related Work

There have been some results in the literature on the verification of blockchains and smart contracts. Based on these studies, we can see the practical meaning of applying formal verification techniques to blockchains.

Model checking approaches have been successfully applied in industry, especially for verification of hardware and communication systems, and also adopted recently in the verification of blockchain models. A formal model of the Bitcoin protocol using automata was developed in [9], in which the probability for double-spending attacks was also studied. The decentralized smart contract protocol (DSCP) was analyzed using game theory and the Markov decision process in [17], and the PRISM model checker was used to verify a family of DSCP properties. In [18], smart contracts were formally specified using Promela and verified in SPIN. The work in [19] adopted interface automata as the semantic model for smart contracts and used the NuSMV model checker to detect violations of the agreements. In [6], the Behavior Interaction Priorities framework (BIP) was used to specify the behavior of smart contract implementation, and the blockchain behavior was verified using the statistical model-checking tool SMC.

Timed automata were adopted in [12] to develop a modeling framework for the Bitcoin contracts, and some security properties were verified based on this model. The runtime verification approach was investigated in [20,21], in which the formal model of the smart contract was provided using some form of automata. In [22], the behavior of EVM was formally defined in Why3, and a framework combining proofs and testing for the analysis of EVM and smart contracts was developed.

Meanwhile, there also exist some works on blockchain consensus. In [23], a detailed study of some network consensus algorithms was proposed. It is significant to compare different consensus algorithms as they are the key components in blockchain protocols. Based on model checking techniques, Reference [24] presented an interesting semi-automatic approach for asynchronous consensus algorithms.

To guarantee the trustworthiness of the CKB blockchain, we need to formally verify the CKB protocols. In previous works [14,15,25], we discussed this topic, and this paper extends our previous results by further investigating the robustness of the CKB protocols against malicious attacks. This work is helpful for the trustworthiness of the CKB blockchain.

7. Conclusions and Future Work

In this paper, we proposed a formal model of the CKB consensus protocol using timed automata and verified a family of properties related to the correctness and consistency of the CKB blockchain for different cases with or without malicious attacks in the UPPAAL model checker. We simulated potential malicious attacks in the experiments and investigated the impacts of such attacks. The properties that were formally verified provided a reference for possible scenarios of CKB applications. We hope that users of the CKB may understand the behavior of the CKB consensus protocol more precisely with the help of the formal model. According to the verification results, we can reasonably conclude that the CKB protocols are able to counter malicious attacks.

The CKB framework is still under development, and some possible optimizations might be adopted for the protocol to make better use of the bandwidth and computation resource. In the future, we hope to further develop the formal model to incorporate the optimization and provide better enhanced assurance for the trustworthiness of the consensus protocol. We will also investigate the formal model of the consensus protocol further to check the result under other kinds of attacks, such as the Sybil attack, etc. Additional investigation of different concrete application scenarios and the impact of the transport layer protocol on the CKB are in our scope as well.

Author Contributions: Conceptualization, M.S. and S.L.; methodology, Y.L., Q.Z. and Y.F.; software, Y.F. and Q.Z.; validation, M.S., S.L. and Y.L.; formal analysis, Y.F. and Q.Z.; investigation, Y.L.; resources, M.S. and S.L.; data curation, Y.F. and Q.Z.; writing—original draft preparation, Y.L., Y.F. and Q.Z.; writing—review and editing, M.S. and S.L.; visualization, Y.F.; supervision, M.S. and Y.L.; project administration, M.S.; funding acquisition, S.L. and M.S. All authors have read and agreed to the published version of the manuscript.

Funding: The research was supported by the Guangdong Science and Technology Department (Grant No. 2018B010107004), the National Natural Science Foundation of China under Grant No. 62172019, 61772038, and 61532019, ROIS NII Open Collaborative Research 2021-(21FS02), and Hiroshima University.

Institutional Review Board Statement: Not applicable.

Informed Consent Statement: Not applicable.

Data Availability Statement: Not applicable.

Acknowledgments: The authors are grateful to the members of Cryptape, and the Nervos team for their helpful discussions.

Conflicts of Interest: The authors declare no conflict of interest.

References

1. Zheng, Z.; Xie, S.; Dai, H.; Chen, X.; Wang, H. Blockchain challenges and opportunities: A survey. *Int. J. Web Grid Serv.* **2018**, *14*, 352–375. [CrossRef]
2. Nakamoto, S. Bitcoin: A Peer-to-Peer Electronic Cash System. 2008. Available online: https://bitcoin.org/bitcoin.pdf (accessed on 2 October 2021).
3. Ethereum. Available online: https://github.com/ethereum (accessed on 2 May 2020).
4. Xie, J. Nervos CKB: A Common Knowledge Base for Crypto-Economy. 2018. Available online: https://github.com/nervosnetwork/rfcs/blob/master/rfcs/0002-ckb/0002-ckb.md (accessed on 1 October 2021).
5. Bhargavan, K.; Blanchet, B.; Kobeissi, N. Verified models and reference implementations for the TLS 1.3 standard candidate. In Proceedings of the 2017 IEEE Symposium on Security and Privacy, San Jose, CA, USA, 22–26 May 2017; pp. 483–502.
6. Abdellatif, T.; Brousmiche, K.-L. Formal verification of smart contracts based on users and blockchain behaviors models. In Proceedings of the 2018 9th IFIP International Conference on New Technologies, Mobility and Security (NTMS), Paris, France, 26–28 February 2018; pp. 1–5.
7. Behrmann, G.; David, A.; Larsen, K.G. A Tutorial on Uppaal. In *Formal Methods for the Design of Real-Time Systems*; Ser. LNCS; Springer: Berlin/Heidelberg, Germany, 2004; Volume 3185, pp. 200–236.
8. David, A.; Larsen, K.G.; Legay, A.; Mikučionis, M.; Poulsen, D.B. Uppaal SMC tutorial. *Int. J. Softw. Tools Technol. Transf.* **2015**, *17*, 397–415. [CrossRef]
9. Chaudhary, K.; Fehnker, A.; van Pol, J.; Stoelinga, M. Modeling and verification of the bitcoin protocol. In Proceedings of the Workshop on Models for Formal Analysis of Real Systems, MARS 2015, Ser. EPTCS. Suva, Fiji, 23 November 2015; pp. 46–60. Available online: https://arxiv.org/abs/1511.04173 (accessed on 13 September 2021).
10. Zhang, R. CKB Consensus Protocol. 2019. Available online: https://github.com/nervosnetwork/rfcs/blob/master/rfcs/0020-ckb-consensus-protocol/0020-ckb-consensus-protocol.md (accessed on 1 October 2021).
11. Lu, Y.; Sun, M. Modeling and verification of IEEE 802.11i security protocol in UPPAAL for internet of things. *Int. J. Softw. Eng. Knowl. Eng.* **2018**, *28*, 1619–1636. [CrossRef]
12. Andrychowicz, M.; Dziembowski, S.; Malinowski, D.; Mazurek, L. Modeling bitcoin contracts by timed automata. In *Formal Modeling and Analysis of Timed Systems, Proceedings of the 12th International Conference, FORMATS 2014, Florence, Italy, 8–10 September 2014*; Ser. LNCS; Springer: Cham, Switzerland, 2014; Volume 8711.
13. Clarke, E.M.; Henzinger, T.A.; Veith, H.; Bloem, R. (Eds.) *Handbook of Model Checking*; Springer: Cham, Switzerland, 2018.
14. Zhang, Q.; Lu, Y.; Sun, M. Modeling and Verification of the Nervos CKB Block Synchronization Protocol in UPPAAL. In *Blockchain and Trustworthy Systems, Proceedings of the Second International Conference, BlockSys 2020, Dali, China, 6–7 August 2020*; Springer: Singapore, 2020; pp. 3–17.
15. Feng, Y.-C.; Lu, Y.; Sun, M. Modeling and Verification of CKB Consensus Protocol in UPPAAL. In Proceedings of the 33rd International Conference on Software Engineering & Knowledge Engineering (SEKE 2021), Pittsburgh, PA, USA, 1–10 July 2021; KSI Research Inc. and Knowledge Systems Institute: Skokie, IL, USA, 2021; pp. 150–153.
16. Nervos Network Homepage. 2020. Available online: http://www.nervos.org (accessed on 1 October 2021).
17. Bigi, G.; Bracciali, A.; Meacci, G.; Tuosto, E. Validation of decentralised smart contracts through game theory and formal methods. In *Programming Languages with Applications to Biology and Security—Essays Dedicated to Pierpaolo Degano on the Occasion of His 65th Birthday*; Ser. LNCS; Springer: Cham, Switzerland, 2015; Volume 9465, pp. 142–161.
18. Bai, X.; Cheng, Z.; Duan, Z.; Hu, K. Formal modeling and verification of smart contracts. In Proceedings of the 2018 7th International Conference on Software and Computer Applications (ICSCA 2018), Kuantan, Malaysia, 8–10 February 2018; ACM: New York, NY, USA, 2018; pp. 322–326.
19. Madl, G.; Bathen, L.A.D.; Flores, G.H.; Jadav, D. Formal verification of smart contracts using interface automata. In Proceedings of the 2019 IEEE International Conference on Blockchain (Blockchain), Atlanta, GA, USA, 14–17 July 2019; pp. 556–563.
20. Ellul, J.; Pace, G.J. Runtime verification of ethereum smart contracts. In Proceedings of the 2018 14th European Dependable Computing Conference (EDCC), Iasi, Romania, 10–14 September 2018; IEEE Computer Society: Washington, DC, USA, 2018; pp. 158–163.
21. Azzopardi, S.; Colombo, C.; Pace, G.J. Model-Based Static and Runtime Verification for Ethereum Smart Contracts. In *Model-Driven Engineering and Software Development, Proceedings of the 8th International Conference, MODELSWARD 2020, Valletta, Malta, 25–27 February 2020*; Revised Selected Papers, Ser. CCIS; Springer: Cham, Switzerland, 2021; Volume 1361, pp. 323–348.
22. Zhang, X.; Li, Y.; Sun, M. Towards a Formally Verified EVM in Production Environment. In *Coordination Models and Languages, Proceedings of the 22nd IFIP WG 6.1 International Conference, COORDINATION 2020, Valletta, Malta, 15–19 June 2020*; Ser. LNCS; Springer: Cham, Switzerland, 2020; Volume 12134, pp. 341–349.
23. Duan, Z.; Mao, H.; Chen, Z.; Bai, X.; Hu, K.; Talpin, J.-P. Formal modeling and verification of blockchain system. In Proceedings of the 10th International Conference on Computer Modeling and Simulation (ICCMS 2018), Sydney, Australia, 8–10 January 2018; pp. 231–235.
24. Tsuchiya, T.; Schiper, A. Verification of consensus algorithms using satisfiability solving. *Distrib. Comput.* **2011**, *23*, 341–358. [CrossRef]
25. Bu, H.; Sun, M. Towards Modeling and Verification of the CKB Block Synchronization Protocol in Coq. In *Formal Methods and Software Engineering, Proceedings of the 22nd International Conference on Formal Engineering Methods, ICFEM 2020, Singapore, 1–3 March 2021*; Springer: Cham, Switzerland, 2020; pp. 287–296.

MDPI
St. Alban-Anlage 66
4052 Basel
Switzerland
Tel. +41 61 683 77 34
Fax +41 61 302 89 18
www.mdpi.com

Mathematics Editorial Office
E-mail: mathematics@mdpi.com
www.mdpi.com/journal/mathematics

www.ingramcontent.com/pod-product-compliance
Lightning Source LLC
LaVergne TN
LVHW071033170726
843515LV00006B/1275

* 9 7 8 3 0 3 6 5 3 7 9 9 3 *